Marty Kra

Elements of Modern X-ray Physics

Elements of Modern X-ray Physics

Jens Als-Nielsen

Ørsted Laboratory
Niels Bohr Institute
Copenhagen University
Denmark

Des McMorrow

Risø National Laboratory
Denmark

John Wiley & Sons, Ltd

New York • Chichester • Weinheim • Brisbane • Singapore • Toronto

Baffins Lane, Chichester,
West Sussex PO19 1UD, England

National 01243 779777
International (+44) 1243 779777
e-mail (for orders and customer service enquiries):
cs-books@wiley.co.uk
Visit our Home Page on http://www.wiley.co.uk
or http://www.wiley.com

Other Wiley Editorial Offices

John Wiley & Sons, Inc., 605 Third Avenue,
New York, NY 10158-0012, USA

WILEY-VCH Verlag GmbH, Pappelallee 3,
D-69469 Weinheim, Germany

Jacaranda Wiley Ltd, 33 Park Road, Milton,
Queensland 4064, Australia

John Wiley & Sons (Asia) Pte Ltd, 2 Clementi Loop #02-01,
Jin Xing Distripark, Singapore 129809

John Wiley & Sons (Canada) Ltd, 22 Worcester Road,
Rexdale, Ontario M9W 1L1, Canada

Library of Congress Cataloging-in-Publication Data

Als-Nielsen, Jens, 1937-
Elements of modern x-ray physics/ Jens Als-Nielsen, Des McMorrow.
p. cm.
Includes bibliographical references and index.
ISBN 0-471-49857-2 (hb: alk.paper) – ISBN 0-471-49858-0 (pb: alk.paper)
1. X-rays. I. McMorrow, Des. II. Title.

QC481 .A47 2000
539.7′222–dc21

00-063342

British Library Cataloguing in Publication Data

A catalogue record for this book is available from the British Library

ISBN 0 471 498572 (HB) 0 471 49858 0 (PB)

Typeset by the authors
Printed and bound in Italy.
This book is printed on acid-free paper responsibly manufactured from sustainable forestry, in which at least two trees are planted for each one used for paper production.

Contents

List of Symbols

α_c	critical angle
β	imaginary part of deviation of refractive index from unity
β_e	electron velocity in units of c
δ	real part of deviation of refractive index from unity
$\Delta\Omega$	element of solid angle
$\hat{\varepsilon}$	polarization unit vector of X-ray electric field
ϵ_0	permittivity of free space
γ	ratio of electron energy in storage ring to rest mass energy, $\dfrac{\mathcal{E}_e}{mc^2}$
γ^{-1}	opening angle of synchrotron radiation cone
$\hbar$	Planck's constant
λ	wavelength of X-ray
λ_1	fundamental wavelength of undulator radiation
λ_u	undulator spatial period
λ_{C}	Compton scattering length
Λ_{ext}	extinction depth
μ	linear absorption coefficient for intensity
μ_0	permeability of free space
ω	angular frequency
ω_{o}	orbital angular frequency of electron in a synchrotron
$\omega_u t'$	emitter phase
$\omega_1 t$	observer phase
Φ_0	incident photon flux (photons/s/unit area)
ρ	electron number density
ρ_a	atomic number density
ρ_m	atomic mass density
$\rho(\mathcal{E})$	density of states
σ_a	absorption cross-section
θ	Bragg angle
ζ_{D}	Darwin width, relative bandwidth

$a(a^\dagger)$	annihilation (creation) operator
a_0	Bohr radius
A	atomic mass number
$\mathbf{A}$	vector potential of photon field
b	scattering length
b	asymmetry parameter
B_T	Debye-Waller factor
c	speed of light
d	lattice plane spacing
$-e$	electronic charge
$\mathcal{E}$	photon energy
$\mathcal{E}_e$	electron energy
$\mathbf{E}$	electric field
E_0	magnitude of electric field
$f(\mathbf{Q})$	atomic form factor (scattering factor)
$f^0(\mathbf{Q})$	non-resonant atomic scattering factor
f'	real part of atomic dispersion correction
f''	imaginary part of atomic dispersion correction
$F(\mathbf{Q})$	unit cell structure factor
$F^{\mathrm{CTR}}(\mathbf{Q})$	crystal truncation rod structure factor
$F^{\mathrm{mol}}(\mathbf{Q})$	molecular structure factor
$\mathbf{G}$	reciprocal lattice vector
h, k, l	Miller indices
$\mathbf{H}$	magnetic field
$\mathcal{H}_I$	interaction Hamiltonian
$\mathcal{H}_{\mathrm{rad}}$	Hamiltonian of radiation field
I_0	incident photon intensity (photons/s)
I_{sc}	scattered photon intensity (photons/s)
$\mathbf{k}$	wavevector of X-ray
K	undulator parameter

m	mass of electron
m_n	mass of neutron
m_{hkl}	multiplicity of Bragg reflection
M_{if}	matrix element between initial i and final f states
n	refractive index
N_{A}	Avogadro's number
P	polarization factor
$\mathcal{P}$	principal value
$\mathbf{p}$	momentum of electron
$\mathbf{q}$	wavevector of photoelectron
$\mathbf{Q}$	wavevector transfer (scattering vector)
r_0	Thomson scattering length (classical electron radius)
$\mathbf{R_n}$	lattice vector
$\mathbf{v}$	velocity of electron
v_c	volume of unit cell
w_{D}	Darwin width, angular
Z	atomic number

Preface

The construction of the first dedicated X-ray beamlines at synchrotron sources in the late 1970's heralded the start of a new era in X-ray science. In the intervening years tremendous progress has been made, both with respect to improvements to the sources, and with our knowledge of how to exploit them. Today's third-generation sources deliver extremely bright beams of radiation over the entire X-ray band (c. 1–500 keV), and with properties such as polarization, energy resolution, etc., that can be tailored to meet almost any requirement. These improvements have driven a surge of activity in X-ray science, and phenomena over a diverse range of disciplines can now be studied with X-rays that were undreamt of before the advent of synchrotron sources.

In light of these developments we believed that it was timely to produce a textbook at an introductory level. Our intention is to offer a coherent overview, which covers the basic physical principles underlying the production of X-rays, their interaction with matter, and also to explain how these properties are used in a range of applications. The main target audience for this book are final year undergraduates, and first year research students. Although the book has been written from the perspective of two physicists, we hope that it will be useful to the wider community of biologists, chemists, material scientists, etc., who work at synchrotron radiation facilities around the world. The main challenge in writing for a wider audience has been to convey the physical concepts without obscuring them in too much mathematical rigour. Therefore, many of the more difficult mathematical manipulations and theorems are explained in shaded boxes that may be studied separately. In addition appendices covering some of the required introductory physics have been included.

It is also our hope that this book will have appeal to more experienced research workers. Synchrotron radiation facilities are large laboratories where many different groups work on disparate areas of science. Cross fertilization of ideas is often the driving force of scientific progress. In order that these different groups, often working on neighbouring beamlines, can communicate their ideas a common background is required. It is our intention that this book should provide at least some of this background knowledge. In addition, many X-ray techniques are becoming viewed as standard analytical tools, and it is no longer necessary to understand every aspect of the design of an instrument in order to be able to perform experiments. While this is undoubtedly a positive development, it can also be argued that a greater knowledge of the underlying principles not only adds to the overall feeling of satisfaction, but also allows better experiments to be designed.

This book has emerged from a lecture course that has been running for several years at the University of Copenhagen. The material covered in this book is taught in one semester, and is augmented by practical lessons both in an X-ray laboratory at the university, and also during a week long trip to the HASYLAB synchrotron facility. The list of subjects covered in this book inevitably reflects to some degree our own areas of specialization. There is, for example, very little on the vast and important subject of imaging. It was also decided at an early stage not to focus on subjects, such as classical crystallography, that we felt were well described in other texts. In spite of these shortcomings we hope that the reader, whatever his or her background, will learn something by studying this book, and be inspired to think of new ways to exploit the great opportunities that the development of synchrotron radiation offers.

Jens Als-Nielsen and Des McMorrow

Copenhagen, September 2000

Acknowledgements

This book has grown out of our experiences of performing experiments at various synchrotron sources around the world. Our main thanks goes to our colleagues from these laboratories and elsewhere. In particular we would like to express our thanks to Henrik Bruus, Roger Cowley, Robert Feidenhans'l, Joseph Feldthaus, Francois Grey, Peter Gürtler, Wayne Hendrickson, Per Hedegaard, John Hill, Mogens Lehmann, Les Leiserowitz, Gerd Materlik, David Moncton, Ian Robinson, Jochen Schneider, Horst Schulte-Schrepping, Sunil Sinha, and Larc Tröger for their detailed comments on different parts of the book. We are also indebted to the students at the Niels Bohr Institute who have attended the course on Experimental X-ray Physics. They have not only spotted untold numbers of typographical errors in early drafts, but also helped refine the material, and through their enthusiasm have ensured that teaching the course has been a rewarding, and even at times entertaining, experience. Birgitte Jacobsen deserves a special mention for her careful reading of the manuscript. Finally, we would like to thank Felix Beckmann, C.T. Chen, T.-C. Chiang, Trevor Forsyth, Watson Fuller, Malcolm McMahon, Benjamin Perman, and Michael Wulff for providing examples of their work, and Keld Theodor for help in preparing some of the figures.

This book has been typeset using LATEX, and we would like to express our thanks to everyone who has helped develop this system over the years, and in particular to Henrik Rønnow for helping with some of the trickier typesetting issues.

The image on the front cover was provided courtesy of Michael Wulff, ESRF, Grenoble, France.

Notes on the use of this book

The material in this book follows a more or less linear development. The scene is set in the first chapter, where the predominant mechanisms for the interaction of X-rays and matter are described. Many of the important concepts and results are introduced in this chapter, and forward references are made to the remaining chapters where these concepts are discussed more fully and the results derived. An attempt has been made to reduce to a minimum the level of mathematical skill required to follow the arguments. This has been done by placing most of the more taxing manipulations and theorems in shaded boxes, or in one of the appendices.

Computers are of course now an indispensable tool for helping to visualize mathematical and physical concepts. For this reason we have chosen to include a listing in the last appendix of some of the computer programmes that were used to generate the figures in this book. The hope is that this will ease the process of turning mathematical formulae into computer algorithms, and also aid the design of more complex programmes required for the analysis of data, etc. The programmes have been written using the MATLAB® programming environment, although the way that they are derived from the mathematics is transparent enough that they can easily be converted to other languages. Figures for which programme listings are given are indicated by a star, $\star$.

1. X-rays and their interaction with matter

X-rays were discovered by W.C. Röntgen in 1895. Since that time they have become established as an invaluable probe of the structure of matter. The range of problems where X-rays have proved to be decisive in unravelling the structure of a given material is truly staggering. These include at one limit of complexity simple compounds, through to more complex and celebrated examples, such as DNA, and in more recent times to the structure of proteins, and even up to the functional units of living organisms. Progress in both our theoretical understanding of the interaction of X-rays with matter, and in our knowledge of how to exploit them experimentally, was steady from the period covering their discovery through to the mid 1970s. The main limitation in this period was the source, which had remained essentially unchanged from about

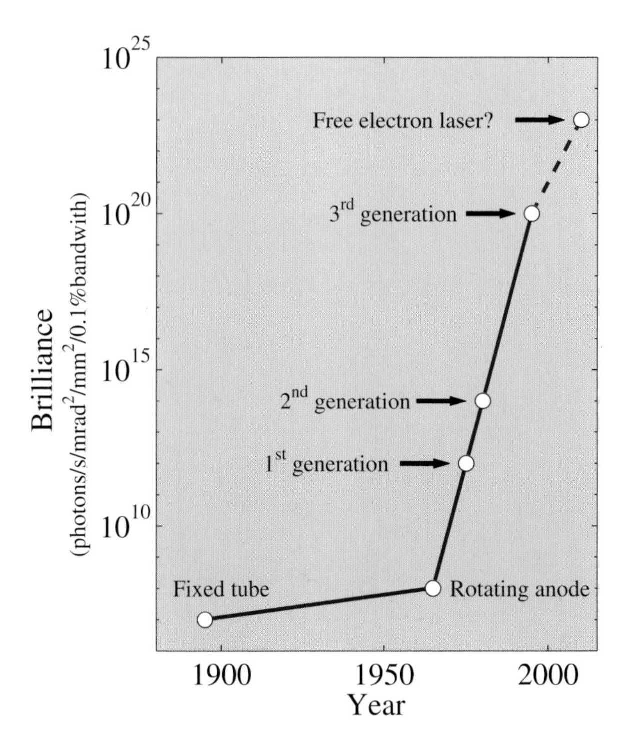

Figure 1.1: The brilliance of X-ray sources as a function of time. The brilliances of first, second, and third-generation synchrotron sources are indicated. Sources based on free-electron lasers will have an even higher brilliance. The brilliance of a source is defined and discussed in Chapter 2, along with the principles underlying the production of X-rays from synchrotrons and free-electron lasers.

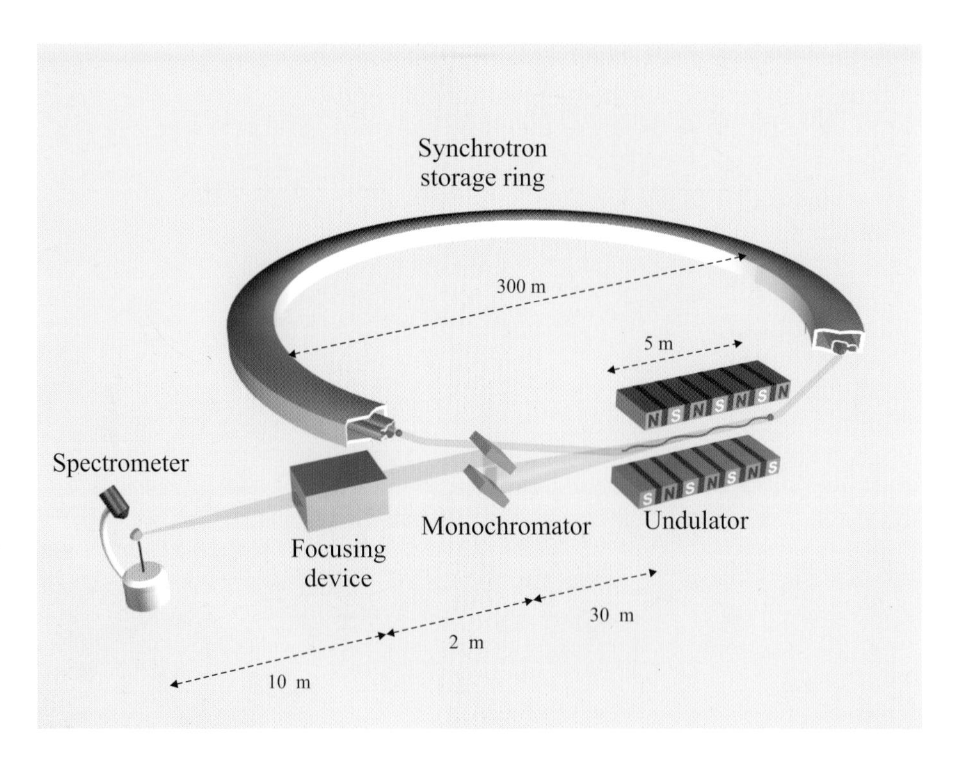

Figure 1.2: A schematic of a typical X-ray beamline at a third generation X-ray source. Bunches of charged particles (electrons or positrons) circulate in a storage ring (typical diameter around 300 m). The ring is designed with straight sections, where an insertion device, such as undulator, is placed. The lattice of magnets in an insertion device forces the particles to execute small oscillations which produce intense beams of radiation. This radiation then passes through a number of optical elements, such as a monochromator, focusing device, etc., so that a beam of radiation with the desired properties is delivered to the sample. Typical distances are indicated.

1912. In the 1970s it was realized that the synchrotron radiation emitted from charged particles circulating in storage rings constructed for high energy nuclear physics experiments was potentially a much more intense and versatile source of X-rays. Indeed synchrotrons have proven to be such vastly better sources that many storage rings have been constructed around the world dedicated solely to the production of X-rays. This has culminated to date in the so-called third-generation synchrotron sources, which are a factor of approximately 10^{12} times brighter than the early lab-based sources, as indicated in Fig. 1.1. With the advent of synchrotron sources the pace of innovation in X-ray science increased markedly (though perhaps not a trillion fold!), and today shows no signs of slowing. The next generation of sources based on the free-electron laser is already in the planning stage, and when they become operational further important breakthroughs will undoubtedly follow.

In Fig. 1.2 we show a schematic of the key components of a typical experimental beamline at a third-generation source. The details will of course vary considerably depending on the particular requirements, but many of the components shown will be found in one form or another on most beamlines. First there is the source itself. In this case the electrons do not follow a purely

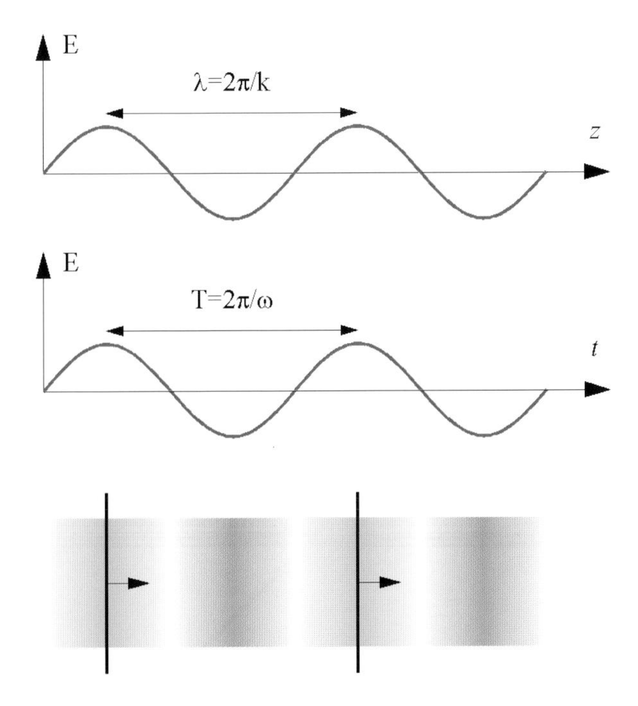

Figure 1.3: Three representations of an electromagnetic plane wave. Only the electric field **E** is shown. Top: spatial variation, described by the wavelength λ or the wavenumber k, at a given instant in time. Middle: temporal variation, described by the period T or the cyclic frequency ω, at a given point in space. Bottom: Top view of a plane wave with the wave crests indicated by the heavy lines, and the direction of propagation by the arrows. The shading indicates the spatial variation of the amplitude of the field.

circular orbit in the storage ring, but traverse through straight sections where lattices of magnets, so-called undulator insertion devices, force them to execute small-amplitude oscillations. At each oscillation X-rays are emitted and, if the amplitude of the oscillations is small, then the different contributions from the passage of a single electron add coherently, and a very intense beam of X-rays results. The second key component is the monochromator, as in many applications it is required to work at a particular average wavelength. It may also be desirable to choose the wavelength bandwidth, and monochromators made from perfect crystals through to multilayers allow for a considerable variation in this parameter. Thirdly, if working with small samples it may be desirable to focus the monochromatic beam down to as small a size as achievable. This is accomplished by devices such as X-ray mirrors and refractive Fresnel lenses. Finally, X-rays are delivered to the sample itself on which the experiment is performed.

One of the main goals of this book is to explain the physical principles underlying the operation of the key components shown in Fig. 1.2. As a first step it is necessary to understand some of the basic aspects of the interaction of X-rays with matter.

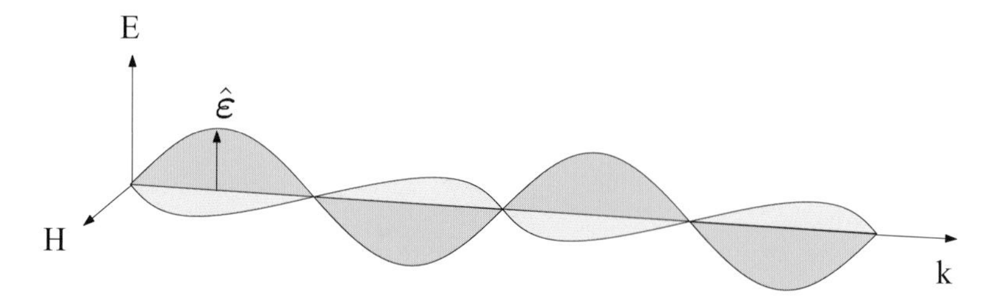

Figure 1.4: An X-ray is a transverse electromagnetic wave, where the electric and magnetic fields, **E** and **H**, are perpendicular to each other and to the direction of propagation **k**.

1.1 X-rays – waves and photons

X-rays are electromagnetic waves with wavelengths in the region of an Ångström (10^{-10}m). In many cases one is interested in a monochromatic beam of X-rays as depicted in Fig. 1.3. The direction of the beam is taken to be along the z-axis, perpendicular to the the electric, **E**, and magnetic, **H**, fields. For simplicity, we shall start by considering the electric field only and neglect the magnetic field. The top part of Fig. 1.3 shows the spatial dependence of the electromagnetic field at a given instance of time. It is characterized by the wavelength λ, or equivalently the wavenumber $\mathrm{k} = 2\pi/\lambda$. Mathematically the electric field amplitude is expressed as a sine wave, either in its real form, $\mathrm{E}_0 \sin(\mathrm{k}z)$, or in its more compact complex form, $\mathrm{E}_0 \mathrm{e}^{i\,\mathrm{k}z}$.

The lower part of Fig. 1.3 is an alternative illustration of the monochromatic plane wave. Only the wave crests are shown (full lines perpendicular to the z-axis), emphasizing that it is a plane wave with an electric field that is constant anywhere in a plane perpendicular to the z-axis. Although a beam is never ideally collimated, the approximation of a plane wave is often valid. The spatial and temporal variation of a plane wave propagating along the z-axis can be encompassed in one simple expression, $\mathrm{E}_0\, \mathrm{e}^{i\,(\mathrm{k}z-\omega t)}$. More generally in three dimensions the polarization of the electric field is written as a unit vector $\hat{\varepsilon}$, and the wavevector along the direction of propagation as **k**, so that

$$\mathbf{E}(\mathbf{r}, t) = \hat{\varepsilon}\, \mathrm{E}_0\, \mathrm{e}^{i(\mathbf{k}\cdot\mathbf{r}-\omega t)}.$$

Since electromagnetic waves are transverse we have $\hat{\varepsilon} \cdot \mathbf{k}$=0, and $\mathbf{k} \cdot \mathbf{E}$= $\mathbf{k} \cdot \mathbf{H}$=0 as shown in Fig. 1.4.

This is the classical description of a linearly polarized, electromagnetic plane wave. From a quantum mechanical perspective, a monochromatic beam is viewed as being quantized into photons, each having an energy $\hbar\omega$ and momentum $\hbar\mathbf{k}$. The intensity of a beam is then given by the number of photons passing through a given area per unit time. As the intensity is also proportional to the square of the electric field, it follows that the magnitude of the field is quantized. Instead of quantizing the **E** and **H** fields separately, it turns out to be more convenient to work with the vector potential **A**, since both **E** and **H** can be derived from **A**. In Appendix C it is explained how the vector potential is quantized, and the explicit form of the quantum mechanical Hamiltonian of the electromagnetic field is given. In this book we shall move freely between the classical and quantum descriptions, choosing whichever one leads us to the quickest and clearest understanding of the problem at hand.

The numerical relation[1] between wavelength λ in Å and photon energy $\mathcal{E}$ in keV is

$$\lambda\,[\text{Å}] = \frac{hc}{\mathcal{E}} = \frac{12.398}{\mathcal{E}\,[\text{keV}]}$$

An X-ray photon interacts with an atom in one of two ways: it can be scattered or it can be absorbed, and we shall discuss these processes in turn. When X-rays interact with a dense medium consisting of a very large number of atoms or molecules it is sometimes more convenient to treat the material as a continuum, with an interface to the surrounding vacuum (or air). At the interface the X-ray beam is refracted and reflected, and this is an alternative way in which the interaction may be discussed. The scattering and refraction descriptions are of course equivalent. In Chapter 3 we derive the X-ray reflectivity equations, and exploit this equivalence to relate the reflectivity to the microscopic properties of the medium of interest.

1.2 Scattering

To start with we shall consider the scattering of an X-ray by a single electron. In the classical description of the scattering event the electric field of the incident X-ray exerts a force on the electronic charge, which then accelerates and radiates the scattered wave. Classically, the wavelength of the scattered wave is the same as that of the incident one, and the scattering is necessarily *elastic*. This is not true in general in a quantum mechanical description, where the incident X-ray photon has a momentum of $\hbar\mathbf{k}$ and an energy of $\hbar\omega$. Energy may be transferred to the electron with the result that the scattered photon has a lower frequency relative to that of the incident one. This *inelastic* scattering process is known as the Compton effect, and is discussed at the end of this section. However, the elastic scattering of X-rays is the main process that is exploited in investigations of the structure of materials, and in this case it mostly suffices to adopt a classical approach. Of course momentum may be transferred even in an elastic scattering event, and this leads to the definition of the vector $\mathbf{Q}$ as

$$\boxed{\hbar\mathbf{Q} = \hbar\mathbf{k} - \hbar\mathbf{k}'}$$

where $\hbar\mathbf{k}$ and $\hbar\mathbf{k}'$ are the initial and final momenta of the photon respectively. The vector $\mathbf{Q}$ is known as the *wavevector transfer* or *scattering vector*, and as we shall see it is the natural variable to describe elastic scattering processes. $\mathbf{Q}$ is usually expressed in units of Å^{-1}.

[1] In this book we shall limit the wavelength band to 0.1 – 2 Å corresponding to the energy band 120–6 keV. The first limit, 0.1 Å or 120 keV, ensures that relativistic effects are negligible since the X-ray energy is considerably lower than the rest mass of the electron, mc^2= 511 keV. The second limit, 2 Å or 6 keV, is a practical limit ensuring that the X-rays have a high penetration power through light materials, such as beryllium. In many X-ray tubes, and in synchrotron radiation beam lines, the X-rays must be transmitted through a Be window, and above 6 keV the transmission of a 0.5 mm Be window exceeds 90%. Lower energy X-rays are called soft X-rays and will not be dealt with in this book.

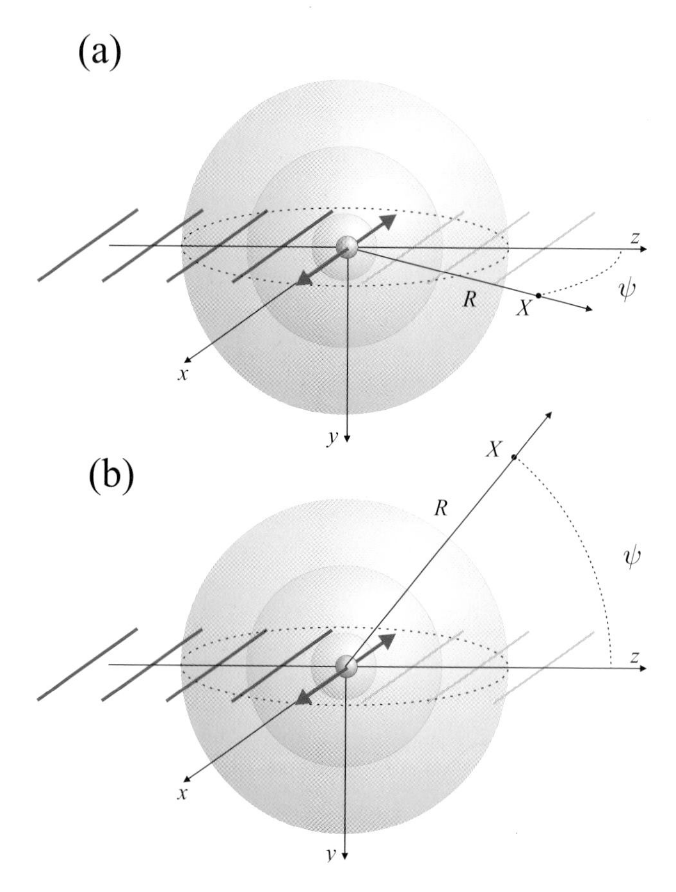

Figure 1.5: The classical description of the scattering of an X-ray by an electron. The electric field of an incident plane wave sets an electron in oscillation which then radiates a spherical wave. The incident wave propagates along the z axis and has its electric field polarized along x. The wave crests of the incident wave lie in between those of the scattered spherical wave because of the 180° phase shift in Thomson scattering. In the text the radiated field at an observation point X is calculated, and there are two distinct cases: (a) point X lies in the same plane as the polarization of the incident wave, and the observed acceleration has to be multiplied by a factor of $\cos\psi$; (b) the observation point X is in the plane normal to the incident polarization and the full acceleration of the electron is seen at all scattering angles ψ.

One electron

The elementary scattering unit of an X-ray in an atom is the electron. The ability of an electron to scatter an X-ray is expressed in terms of a *scattering length*, which we shall now derive. In the classical description of the scattering process an electron will be forced to vibrate when placed in the electric field of an incident X-ray beam, as illustrated in Fig. 1.5 for the case of an incident plane wave. A vibrating electron acts as a source, and radiates like a small dipole antenna. The problem then is to evaluate the radiated field at an observation point X, which is a distance R from the source, and at an angle ψ with respect to the direction of the incident beam. The radiated field at X is derived from Maxwell's equations in Appendix B. Here a heuristic argument is presented for the case that the observation point X lies in the same plane as the polarization of the incident wave (Fig. 1.5(a)).

The radiated energy density is proportional to the square of the radiated electric field, $\mathbf{E}_{\rm rad}$. This means that in the far-field limit, $\rm E_{rad}$ must decrease as $1/R$, since then the energy density decreases as $1/R^2$, and the *total* irradiated energy is independent of R, as required. Furthermore, the field is proportional to the charge of the electron, $-e$, and to the acceleration seen by the observer, $a_X(t')$, evaluated at a time t' earlier than the observation time t, since the radiation propagates at a finite velocity, c. In order to have the correct units (SI), the field must be proportional to $1/(4\pi\epsilon_0 c^2)$. Assembling these results together the radiated field is

$$\mathrm{E_{rad}}(R,t) = -\frac{-e}{4\pi\epsilon_0 c^2 R}\, a_X(t') \tag{1.1}$$

where $t' = t - R/c$, and the leading minus sign emerges from the rigorous derivation outlined in Appendix B.

As far as the acceleration seen from the observer is concerned, we note that for $\psi = \pi/2$ the observer does not see any acceleration at all, and that for $\psi = 0$ the full acceleration is observed. In general the true acceleration has to be multiplied by a factor of $\cos\psi$ for a scattering angle of ψ to obtain the observed acceleration. The acceleration, evaluated as the force divided by the mass, is

$$a_X(t') = \frac{-e}{m}\,\mathrm{E}_{x0}\,\mathrm{e}^{-i\omega t'}\cos\psi = \frac{-e}{m}\,\mathrm{E_{in}}\,\mathrm{e}^{i\omega\left(\frac{R}{c}\right)}\cos\psi \tag{1.2}$$

where $\mathrm{E_{in}} = \mathrm{E}_{x0}\mathrm{e}^{-i\omega t}$ describes the electric field of the incident wave. Eq. (1.1) can thus be rewritten as

$$\mathrm{E_{rad}}(R,t) = -\frac{-e}{4\pi\epsilon_0 c^2 R}\,\frac{-e}{m}\,\mathrm{E_{in}}\,\mathrm{e}^{i\omega\left(\frac{R}{c}\right)}\cos\psi \tag{1.3}$$

and hence the ratio of the magnitude of the radiated to incident field is

$$\frac{\mathrm{E_{rad}}(R,t)}{\mathrm{E_{in}}} = -\left(\frac{e^2}{4\pi\epsilon_0 mc^2}\right)\frac{\mathrm{e}^{i\mathrm{k}R}}{R}\cos\psi \tag{1.4}$$

By convention the prefactor of the spherical wave $\mathrm{e}^{i\,\mathrm{k}R}/R$ is denoted by the symbol r_0 and is known as the Thomson scattering length, or alternatively as the classical electron radius. In practical units the Thomson scattering length is

$$\boxed{r_0 = \left(\frac{e^2}{4\pi\epsilon_0 mc^2}\right) = 2.82\times10^{-5}\ \text{Å}}$$

For neutrons it is conventional to use the symbol b to represent the scattering length, and we shall sometimes also use the same symbol for the X-ray scattering length, particularly when the two techniques are being compared (see Appendix F). It is important to note that the leading minus sign in Eq. (1.4) means that the radiated field is 180° out of phase with the incident field, or in other words that the scattering process involves a phase shift of π.

X-ray detectors usually count single photons, and the measured intensity, I_{sc}, is then the number of photons per second recorded by the detector. This can be expressed as the energy per second, i.e. the power, flowing through the area of the detector divided by the energy of each photon. Let the incident beam intensity be I_0 with a cross-sectional area of A_0, and the cross-sectional area of the scattered beam be $R^2\Delta\Omega$, where $\Delta\Omega$ is the solid angle subtended by the detector. The energy per unit area of the beam is also proportional to the modulus squared of the electric field, independent of whether we consider the incident or radiated fields, so that

$$\frac{I_{sc}}{I_0} = \frac{|\mathbf{E}_{\mathrm{rad}}|^2 R^2 \Delta\Omega}{|\mathbf{E}_{\mathrm{in}}|^2 A_0} \tag{1.5}$$

It is usual to normalize the scattered intensity I_{sc}, by both the incident flux (I_0/A_0) and the solid angle of the detector, $\Delta\Omega$. This leads to the definition of the *differential cross-section*, $(d\sigma/d\Omega)$, as

$$\left(\frac{d\sigma}{d\Omega}\right) = \frac{(\text{Number of X-rays scattered per second into } \Delta\Omega)}{(\text{Incident flux})(\Delta\Omega)}$$

From Eq. (1.5) and (1.4) the differential cross-section for Thomson scattering is

$$\left(\frac{d\sigma}{d\Omega}\right) = \frac{I_{sc}}{(I_0/A_0)\Delta\Omega} = \frac{|\mathbf{E}_{\mathrm{rad}}|^2 R^2}{|\mathbf{E}_{\mathrm{in}}|^2} = r_0^2 \cos^2\psi$$

The *total* cross-section for Thomson scattering is found by integrating the differential cross-section over all possible scattering angles, and is equal to

$$\sigma_{\mathrm{T}} = \left(\frac{8\pi}{3}\right) r_0^2 = 0.665 \times 10^{-24}\mathrm{cm}^2 = 0.665 \text{ barn}$$

The classical cross-section, both the differential and total, for the scattering of an electromagnetic wave by a free electron is a constant, independent of energy.[2]

The factor of $\cos\psi$ in Eq. (1.2) was introduced to allow for the fact that the acceleration of the electron observed at X decreases as ψ is increased, and is zero when $\psi = \pi/2$. This assumes that the observer and the electric field of the incident wave are in the same plane (Fig. 1.5(a)). If, on the other hand, the incident wave is polarized perpendicular to the plane of observation (E_{y0} in the notation of Fig. 1.5(b)) then the full acceleration of the electron would be observed at all scattering angles ψ, and the factor of $\cos\psi$ should be replaced by 1. In general the electric vector of the incident ray points along a specific direction in the $x-y$ plane, with a magnitude E_0, and from Eq. (1.3) we have

$$|\mathrm{E}_{\mathrm{rad}}|^2 = \left(\frac{r_0^2}{R^2}\right)\left(\mathrm{E}_{y0}^2 + \mathrm{E}_{x0}^2 \cos^2\psi\right).$$

For an unpolarized source E_0 can point in any direction in the $x-y$ plane with equal probability, so that its average value is $\langle \mathrm{E}_0^2\rangle = \langle \mathrm{E}_{x0}^2\rangle + \langle \mathrm{E}_{y0}^2\rangle$. Given that the x and y directions are equivalent, $\frac{1}{2}\langle \mathrm{E}_0^2\rangle = \langle \mathrm{E}_{x0}^2\rangle = \langle \mathrm{E}_{y0}^2\rangle$. The modulus

[2]This is discussed further in Appendices A and B.

squared of the field radiated by an electron illuminated by an unpolarized source is therefore

$$|\mathrm{E}_{\mathrm{rad}}|^2 = \left(\frac{r_0^2}{R^2}\right)\langle E_0^2\rangle\,\frac{1}{2}\left(1+\cos^2\psi\right).$$

Collecting these results together, the differential cross-section for Thomson scattering is

$$\boxed{\left(\frac{d\sigma}{d\Omega}\right) = r_0^2\,P} \tag{1.6}$$

where P is the polarization factor which depends on the X-ray source:

$$P = \begin{cases} 1 & \text{synchrotron: vertical scattering plane} \\ \cos^2\psi & \text{synchrotron: horizontal scattering plane} \\ \frac{1}{2}\left(1+\cos^2\psi\right) & \text{unpolarized source} \end{cases} \tag{1.7}$$

The polarization factor for a synchrotron source arises from the fact that the electrons in a synchrotron orbit in the horizontal plane. It follows that their acceleration is also in this plane, and hence the emitted X-rays are linearly polarized in the horizontal plane[3]. A scattering experiment performed in the horizontal plane therefore corresponds to the case shown in Fig. 1.5(a), and the appropriate polarization factor is $P = \cos^2\psi$. For a vertical scattering geometry, Fig. 1.5(b), the full acceleration is observed at all scattering angles and $P = 1$.

One atom

Let us now proceed from the scattering by a single electron to consider the scattering by an atom with Z electrons.

To start with a purely classical description will be used, so that the electron distribution is specified by a number density, $\rho(\mathbf{r})$. The scattered radiation field is a superposition of contributions from different volume elements of this charge distribution. In order to evaluate this superposition one must keep track of the phase of the incident wave as it interacts with the volume element at the origin and the one at position $\mathbf{r}$, as shown in Fig. 1.6(a). The phase difference between two successive crests is 2π. The phase difference between the two volume elements is 2π multiplied by the ratio of $\mathbf{r}$, projected onto the incident direction, and the wavelength. This is nothing other than the scalar product of the two vectors $\mathbf{k}$ and $\mathbf{r}$. The simplicity of this expression is one of the reasons why it is so convenient to use the wavevector $\mathbf{k}$ to describe the incident wave. In the vicinity of the observation point X in Fig. 1.5, the scattered wave is locally like a plane wave with wavevector $\mathbf{k}'$. The phase difference, between the scattered wave from a volume element around the origin and one around $\mathbf{r}$ is $-\mathbf{k}'\cdot\mathbf{r}$. The resulting phase difference is thus

$$\Delta\phi(\mathbf{r}) = (\mathbf{k}-\mathbf{k}')\cdot\mathbf{r} = \mathbf{Q}\cdot\mathbf{r}$$

[3]X-rays from a synchrotron source are linearly polarized only if viewed in the orbit plane; out of this plane the polarization is elliptical.

(a) One atom

k k′
r
k · r k′ · r

(b) One molecule

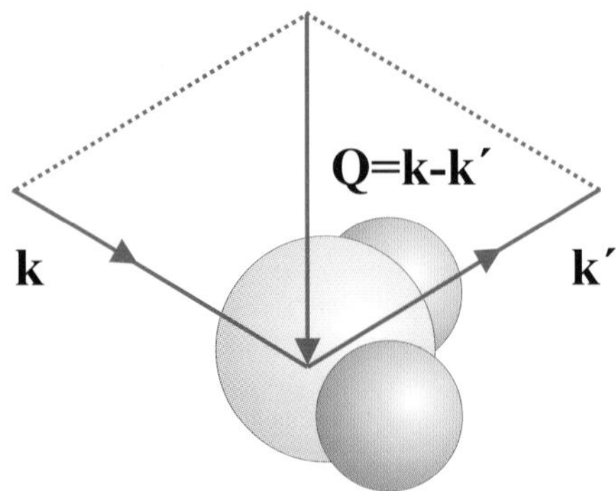

(c) A crystal

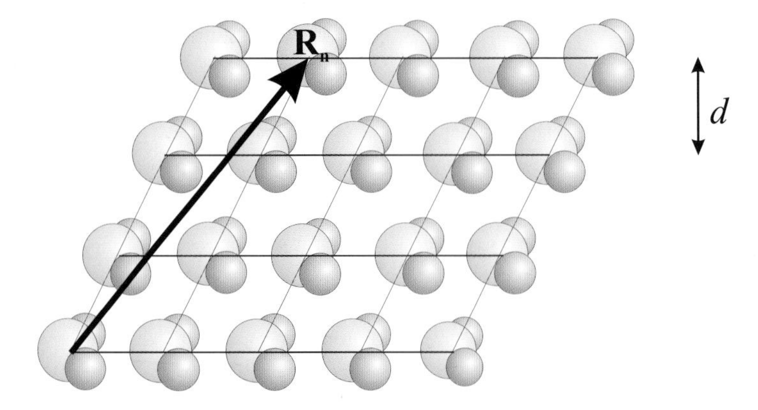

Figure 1.6: (a) Scattering from an atom. An X-ray with a wavevector $\mathbf{k}$ scatters from an atom to the direction specified by $\mathbf{k}'$. The scattering is assumed to be elastic, i.e. $|\mathbf{k}| = |\mathbf{k}'| = 2\pi/\lambda$. The difference in phase between a wave scattered at the origin and one at a position $\mathbf{r}$ is $(\mathbf{k} - \mathbf{k}') \cdot \mathbf{r} = \mathbf{Q} \cdot \mathbf{r}$. This defines the wavevector transfer $\mathbf{Q}$. (b) The scattering from a molecule. Here the scattering triangle is shown which relates $\mathbf{k}$, $\mathbf{k}'$ and $\mathbf{Q}$. (c) Scattering from a molecular crystal. The molecules are organized on a lattice with position vectors $\mathbf{R_n}$, and a lattice spacing of d.

where $\mathbf{Q}$ is the wavevector transfer. The scattering events depicted in Fig. 1.6 are elastic, with $|\mathbf{k}| = |\mathbf{k}'|$, so that from the scattering triangle we have $|\mathbf{Q}| = 2|\mathbf{k}| \sin\theta = (4\pi/\lambda)\sin\theta$.

Thus a volume element $d\mathbf{r}$ at $\mathbf{r}$ will contribute an amount $-r_0\rho(\mathbf{r})d\mathbf{r}$ to the scattered field with a phase factor of $e^{i\mathbf{Q}\cdot\mathbf{r}}$. The total scattering length of the atom is

$$-r_0 f^0(\mathbf{Q}) = -r_0 \int \rho(\mathbf{r})\, e^{i\mathbf{Q}\cdot\mathbf{r}}\, d\mathbf{r} \tag{1.8}$$

where $f^0(\mathbf{Q})$ is known as the *atomic form factor*. In the limit that $Q \to 0$ all of the different volume elements scatter in phase so that $f^0(\mathbf{Q} = 0) = Z$, the number of electrons in the atom. As Q increases from zero the different volume elements start to scatter out of phase and consequently $f^0(Q \to \infty) = 0$. The right hand side of Eq. (1.8) is recognizable as a Fourier transform. Indeed one of the recurrent themes of this book is that the scattering length may be calculated from the Fourier transform of the distribution of electrons in the sample. It should be clear that to calculate the scattered intensity we have to evaluate Eq. (1.8) and multiply by its complex conjugate.

Atomic electrons are of course governed by quantum mechanics, and have discrete energy levels. The most tightly bound electrons are those in the K shell, which have energies comparable to those of a typical X-ray photon. If the X-ray photon has an energy much less than the binding energy of the K shell, the response of these electrons to an external driving field is reduced by virtue of the fact that they are bound. Electrons in shells that are less tightly bound (L, M, etc) will be able to respond to the driving field more closely, but overall we expect that the scattering length of an atom to be reduced by some amount, which is by convention denoted f'. At energies much greater than the binding energy the electrons can be treated as if they are free and f' is zero. For energies in between these limits f' displays resonant behaviour at energies corresponding to atomic absorption edges, which are discussed in Section 1.3. In addition to altering the real part of the scattering length, we also expect that, by analogy with a forced harmonic oscillator, the response of the electron to have a phase lag with respect to the driving field. This is allowed for by including a term if'', which represents the dissipation in the system, and, as we shall see in Chapter 3 and 7, it is related to the absorption. Collecting these results together means that the atomic form factor is

$$f(\mathbf{Q}, \hbar\omega) = f^0(\mathbf{Q}) + f'(\hbar\omega) + if''(\hbar\omega) \tag{1.9}$$

where f' and f'' are known as the dispersion corrections[4] to f^0. We have written f' and f'' as functions of the X-ray energy $\hbar\omega$ to emphasize that their behaviour is dominated by tightly bound inner-shell electrons, and as a consequence cannot have any appreciable dependence on $\mathbf{Q}$. As might be expected from these introductory remarks, f' and f'' assume their extremal values when the X-ray energy is equal to one of the absorption edge energies of the atom. This resonant behaviour is manifestly element specific, and in Chapter 7 it is explained how it may be exploited to solve the structure of complex materials.

[4]These are also sometimes referred to as the anomalous dispersion corrections, but it is generally agreed that there is in fact nothing anomalous about them. It should be noted that with our sign convention f'' is negative.

One molecule

So far we have introduced the scattering length for an electron and subsequently for an atom composed of electrons. The next step in complexity is naturally molecules composed of atoms (Fig. 1.6(b)). It is obvious that just as the scattering length of an atom has a form factor, so will the scattering length of a molecule. Labelling the different atoms in the molecule by index j we may write

$$F^{\mathrm{mol}}(\mathbf{Q}) = \sum_{\mathbf{r}_j} f_j(\mathbf{Q})\, \mathrm{e}^{i\mathbf{Q}\cdot\mathbf{r}_j}$$

where as before $f_j(\mathbf{Q})$ is the atomic form factor of the j'th atom in the molecule, and it must be remembered to include the multiplicative factor of $-r_0$ if the intensity is required in absolute units. If one can determine $\left|F^{\mathrm{mol}}(\mathbf{Q})\right|^2$ experimentally for sufficiently many values of scattering vector $\mathbf{Q}$ then one can (at least by trial and error) determine the positions $\mathbf{r}_j$ of the atoms in the molecule. However, the scattering length of a single molecule is not sufficient to produce a measurable signal, even in the very intense X-ray beams produced by today's synchrotron sources. For that the molecules need to be assembled into a crystal.

A crystal

The defining property of a crystalline material is that it is periodic in space[5], as shown for a molecular crystal in Fig. 1.6(c). In elementary treatments of the scattering of X-rays from a crystal lattice, Bragg's law

$$m\lambda = 2d\sin\theta$$

is derived, where m is an integer. This is the condition for the constructive interference of waves which have an angle of incidence θ to a set of lattice planes a distance d apart. While this is a useful construction, it does have its limitations, principal among which is that it does not enable us to calculate the intensity of the scattering for which constructive interference occurs.

For that we need to build on what we have already developed and write down the scattering amplitude of the crystal. To do so we note that a crystal structure may be specified in the following way. First, a lattice of points is defined in space, which must reflect the symmetry of the crystal, and then a choice of unit cell is made, in other words a choice is made over which atoms to associate with each lattice site. If $\mathbf{R_n}$ are the lattice vectors that define the lattice, and $\mathbf{r}_j$ the position of the atoms with respect to any one particular lattice site, then the position of any atom in the crystal is given by $\mathbf{R_n}+\mathbf{r}_j$. It follows that the scattering amplitude for the crystal factorizes into the product of two terms, which we write as

$$F^{\mathrm{crystal}}(\mathbf{Q}) = \overbrace{\sum_{\mathbf{r}_j} f_j(\mathbf{Q})\mathrm{e}^{i\,\mathbf{Q}\cdot\mathbf{r}_j}}^{\text{Unit cell structure factor}} \;\overbrace{\sum_{\mathbf{R_n}} \mathrm{e}^{i\,\mathbf{Q}\cdot\mathbf{R_n}}}^{\text{Lattice sum}} \tag{1.10}$$

where the first term is the *unit cell structure factor*, the second term is a sum over lattice sites, and where again we have neglected a leading factor of $-r_0$.

[5] See, however, Section 4.4.5 on quasicrystals.

In applications, such as solid state physics it is the structure of the material that is of interest in its own right. For many other applications, such as in molecular and protein crystallography, the lattice is of no interest whatsoever, and assembling the molecules on a lattice merely serves to amplify the signal.

All the terms in the lattice sum given in Eq. (1.10) are phase factors located on the unit circle in the complex plane. The sum will therefore be of order unity unless the scattering vector happens to fulfill

$$\mathbf{Q} \cdot \mathbf{R_n} = 2\pi \times \text{integer} \tag{1.11}$$

in which case it becomes of order N, the number of unit cells. The lattice vectors $\mathbf{R_n}$ are of the form

$$\mathbf{R_n} = n_1\,\mathbf{a}_1 \;+\; n_2\,\mathbf{a}_2 \;+\; n_3\,\mathbf{a}_3$$

where $(\mathbf{a}_1, \mathbf{a}_2, \mathbf{a}_3)$ are the basis vectors of the lattice and (n_1, n_2, n_3) are integers. A unique solution to Eq. (1.11) can be found by introducing the important concept of the reciprocal lattice. This new lattice is spanned by the *reciprocal lattice basis* vectors which are defined by

$$\mathbf{a}_1^* = 2\pi\frac{\mathbf{a}_2 \times \mathbf{a}_3}{\mathbf{a}_1 \cdot (\mathbf{a}_2 \times \mathbf{a}_3)}, \quad \mathbf{a}_2^* = 2\pi\frac{\mathbf{a}_3 \times \mathbf{a}_1}{\mathbf{a}_1 \cdot (\mathbf{a}_2 \times \mathbf{a}_3)}, \quad \mathbf{a}_3^* = 2\pi\frac{\mathbf{a}_1 \times \mathbf{a}_2}{\mathbf{a}_1 \cdot (\mathbf{a}_2 \times \mathbf{a}_3)}$$

so that any lattice site in the reciprocal lattice is given by

$$\mathbf{G} = h\,\mathbf{a}_1^* \;+ k\,\mathbf{a}_2^* \;+\; l\,\mathbf{a}_3^*$$

where (h, k, l) are all integers. We can see that the product of a lattice vector in the reciprocal ($\mathbf{G}$) and direct ($\mathbf{R_n}$) spaces is

$$\mathbf{G} \cdot \mathbf{R_n} = 2\pi(hn_1 + kn_2 + ln_3) = 2\pi \times \text{ integer}$$

and hence the solution to Eq. (1.11) that we are seeking is to require that

$$\boxed{\mathbf{Q} = \mathbf{G}}$$

This proves that $F^{\text{crystal}}(\mathbf{Q})$ is non-vanishing if and only if $\mathbf{Q}$ coincides with a reciprocal lattice vector. This is the Laue condition for the observation of diffraction from a crystalline lattice which may be shown to be completely equivalent to Bragg's law (Chapter 4, page 122).

Scattering from a crystal is therefore confined to distinct points in reciprocal space. The intensity in each point is modulated by the absolute square of the unit cell structure factor. From a (large) set of intensities from a given crystal it is possible to deduce the positions of the atoms in the unit cell. These considerations may of course be generalized to crystals containing molecules. Indeed these methods have had an enormous impact on our knowledge of molecules. More than 95% of all molecular structures come from X-ray diffraction studies. Data sets from crystals of large molecules such as proteins or even viruses encompass tens of thousands of reflections and sophisticated methods have been developed to get from the measured intensities to the atomic positions in the molecule. In

Chapter 4 these concepts will be further developed, and the principles behind these methods will be explained.

In this section it has been tacitly assumed that the interaction between the X-ray and crystal is weak, since we have not allowed for the possibility that the scattered beam may be scattered a second or third time before leaving the crystal. This assumption leads to considerable simplicity and is known as the *kinematical approximation*. In Chapter 5 it is explained how this assumption breaks down when dealing with macroscopic perfect crystals, where multiple scattering effects become important, and we are then in what is known as the *dynamical* scattering limit.

Compton scattering by a free electron

The alternative to the classical description used so far in this section, is to view the incident X-ray as a beam of photons. For simplicity assume that the electron is initially at rest and is free. In a collision energy will be transferred from the photon to the electron, with the result that the scattered photon has a lower energy than that of the incident one. This is the Compton effect. Historically this was of considerable importance as it could not be explained using classical concepts, and thus gave further support to the then emerging quantum theory. The energy loss of the photon is readily calculated by considering the conservation of energy and momentum during the collision. The collision process is sketched in Fig. 1.7, while the kinematics of the collision are worked through in the box on page 16.

The result of the calculation is that the change in wavelength is proportional to the Compton scattering length defined by

$$\boxed{\lambda_C = \frac{\hbar}{mc} = 3.86 \times 10^{-3} \text{ Å}}$$

There are thus two fundamental scattering lengths for the X-ray, the Thomson scattering length, r_0, and the Compton scattering length, λ_C. The ratio of these two is the fine structure constant

$$\alpha = \frac{r_0}{\lambda_C} \approx \frac{1}{137}$$

The ratio of the final to initial energy of the photon is given in Eq. (1.12) and is plotted in Fig. 1.8. For a given scattering angle, the scattering becomes progressively more inelastic as the energy $\mathcal{E}$ of the incident X-ray is increased. The energy scale is set by the rest mass energy of the electron, $mc^2 = 511$ keV.

One important difference between Thomson and Compton scattering is that the latter is *incoherent*. It has already been shown how X-rays that are elastically scattered from a crystal add up coherently when Bragg's law (or equivalently the Laue condition) is fulfilled. The scattering is then restricted to lie at points on the reciprocal lattice. The same is not true for Compton scattering, as it is the interaction between a single photon and electron, and the variation of the Compton cross-section[6] varies only slowly with scattering angle. As far

[6]The calculation of the Compton cross-section is beyond the scope of this book. It is discussed by Lovesey and Collins [1996].

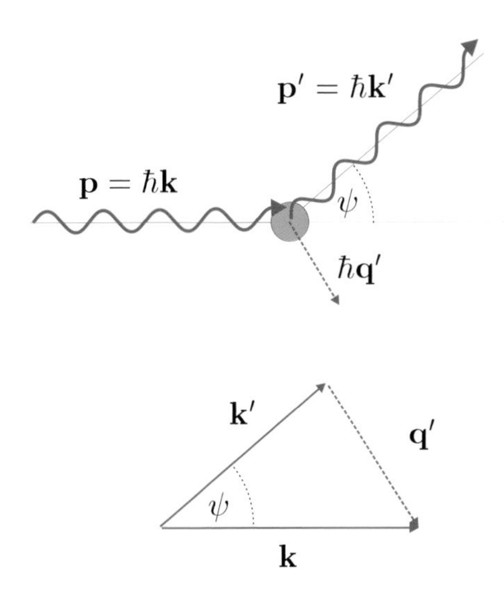

Figure 1.7: Compton scattering. A photon with energy $\mathcal{E} = \hbar c k$ and momentum $\hbar\mathbf{k}$ scatters from an electron at rest with energy mc^2. The electron recoils with a momentum $\hbar\mathbf{q}' = \hbar(\mathbf{k} - \mathbf{k}')$ as indicated in the scattering triangle in the bottom half of the figure.

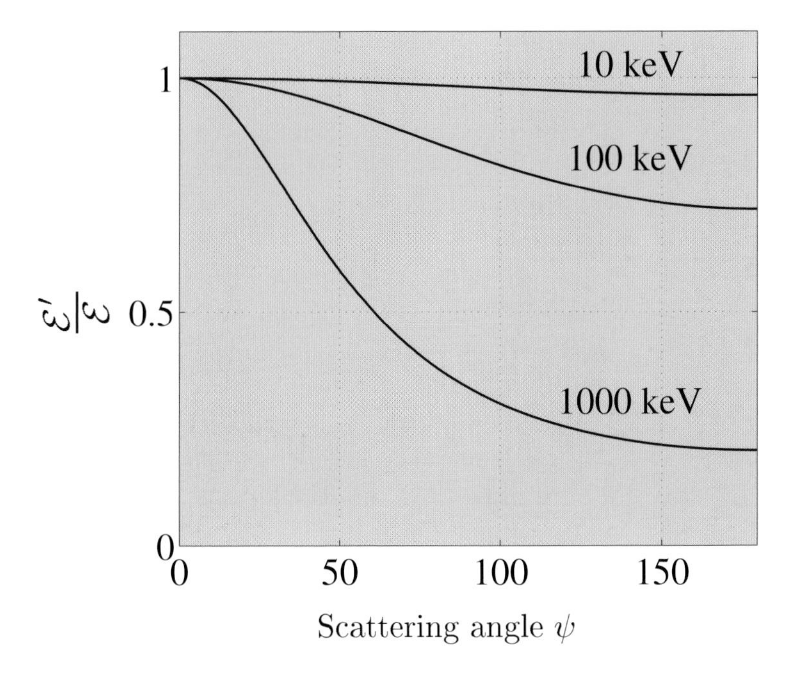

Figure 1.8: The ratio of the energy $\mathcal{E}'$ of the scattered photon to the energy $\mathcal{E}$ of the incident one as function of scattering angle. The curves have been calculated from Eq. (1.12) with $\lambda_C k = \mathcal{E}/mc^2 = \mathcal{E}[\text{keV}]/511$.

Kinematics of Compton scattering

Conservation of energy for the scattering of a photon by an electron shown in Fig. 1.7 leads to

$$mc^2 + \hbar c\mathrm{k} = \sqrt{(mc^2)^2 + (\hbar c\mathrm{q}')^2} + \hbar c\mathrm{k}'$$

Dividing both sides by mc^2, and using the definition of the Compton wavelength, $\lambda_C = \hbar c/(mc^2)$, leads to

$$1 + \lambda_\mathrm{C}(\mathrm{k} - \mathrm{k}') = \sqrt{1^2 + (\lambda_\mathrm{C}\mathrm{q}')^2}$$

This can be rewritten to obtain an expression for q'^2 by squaring both sides and collecting terms to give

$$\mathrm{q}'^2 = (\mathrm{k} - \mathrm{k}')^2 + 2\frac{(\mathrm{k} - \mathrm{k}')}{\lambda_\mathrm{C}}$$

Conservation of momentum (or equivalently wavevector) reads

$$\mathbf{q}' = \mathbf{k} - \mathbf{k}'$$

Taking the scalar product of $\mathbf{q}'$ with itself gives

$$\begin{aligned}\mathbf{q}' \cdot \mathbf{q}' = \mathrm{q}'^2 &= (\mathbf{k} - \mathbf{k}') \cdot (\mathbf{k} - \mathbf{k}') \\ &= \mathrm{k}^2 + \mathrm{k}'^2 - 2\mathrm{k}\mathrm{k}'\cos\psi\end{aligned}$$

Equating this with the expression for q'^2 derived from energy conservation yields

$$\mathrm{k}^2 + \mathrm{k}'^2 - 2\mathrm{k}\mathrm{k}'\cos\psi = \mathrm{k}^2 + \mathrm{k}'^2 - 2\mathrm{k}\mathrm{k}' + 2\frac{(\mathrm{k} - \mathrm{k}')}{\lambda_\mathrm{C}}$$

or

$$\mathrm{k}\mathrm{k}'(1 - \cos\psi) = \frac{(\mathrm{k} - \mathrm{k}')}{\lambda_\mathrm{C}}$$

which may be recast in the form

$$\frac{\mathrm{k}}{\mathrm{k}'} = 1 + \lambda_\mathrm{C}\mathrm{k}(1 - \cos\psi) = \frac{\mathcal{E}}{\mathcal{E}'} = \frac{\lambda'}{\lambda} \tag{1.12}$$

as diffraction experiments are concerned, Compton scattering gives rise to a smoothly varying background which sometimes needs to be subtracted from the data.

Compton scattering may be used to obtain unique information on the electronic structure of materials. So far we have assumed that the electron in the Compton scattering process is initially at rest. This assumption breaks down for electrons in a solid, which instead have a finite momentum. When the kinematics are worked through for this case, it turns out that the Compton cross-section gives a measure of the electronic momentum distribution.

1.3 Absorption

Now let us turn to the absorption process. It is depicted in Fig. 1.9(a). An X-ray photon is absorbed by the atom, and the excess energy is transferred to an electron, which is expelled from the atom, leaving the atom ionized.

The process is called *photoelectric absorption*. Quantitatively, the absorption is given by the linear absorption coefficient μ. By definition μdz is the attenuation of the beam through an infinitesimal sheet of thickness dz at a depth z from the surface (Fig. 1.10). The intensity $I(z)$ through the sample must therefore fulfill the condition

$$-dI = I(z)\,\mu\,dz \tag{1.13}$$

which leads to the differential equation

$$\frac{dI}{I(z)} = -\mu\,dz$$

The solution is found by requiring that $I(z = 0) = I_0$, the incident beam intensity at $z = 0$, and we have

$$I(z) = I_0\,\mathrm{e}^{-\mu z}$$

One can therefore easily determine μ experimentally as the ratio of beam intensities with and without the sample. The number of absorption events, W, in the thin sheet is proportional to I, and to the number of atoms per unit area, $\rho_a\,dz$, where ρ_a is the atomic number density. The proportionality factor is by definition the absorption cross-section, σ_a, so that

$$W = I(z)\,\rho_a\,dz\,\sigma_a = I(z)\,\mu\,dz$$

where in the last step we have used Eq. (1.13). The absorption coefficient is therefore related to σ_a by

$$\boxed{\mu = \rho_a\sigma_a = \left(\frac{\rho_m N_{\mathrm{A}}}{A}\right)\sigma_a} \tag{1.14}$$

where N_{A}, ρ_m and A are Avogadro's number, the mass density, and the atomic mass number, respectively.

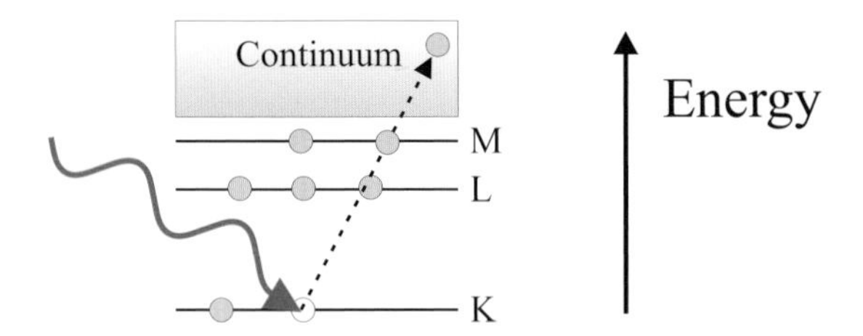

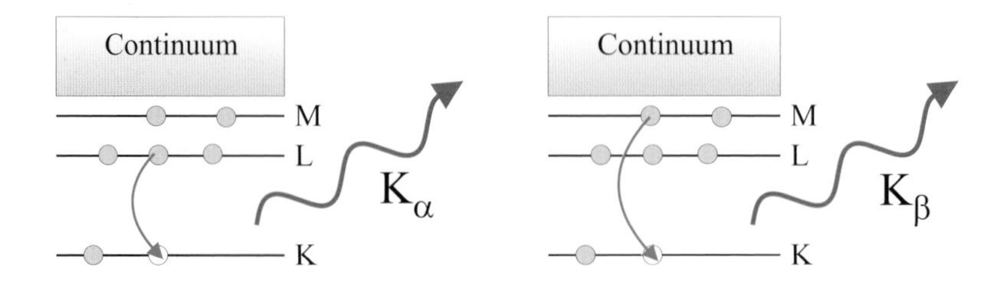

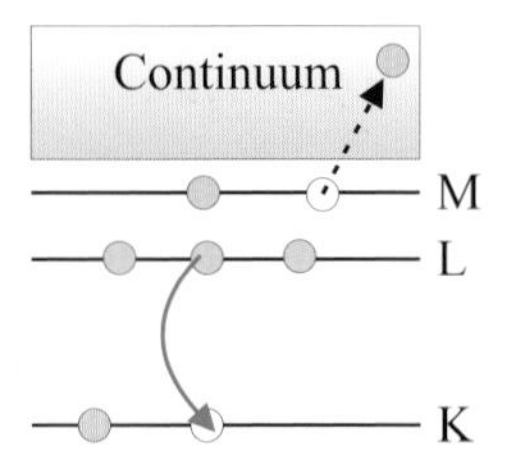

Figure 1.9: Schematic energy level diagram of an atom. For clarity we have indicated only the energy of the three lowest shells; the rest are merged into the continuum. (a) The photoelectric absorption process. An X-ray photon is absorbed and an electron ejected from the atom. The hole created in the inner shell can be filled by one of two distinct processes: (b) Fluorescent X-ray emission. One of the electrons in an outer shell fills the hole, creating a photon. In this example the outer electron comes either from the L or M shell. In the former case the fluorescent radiation is referred to as the K_α line, and in the latter as K_β. (c) Auger electron emission. The atom may also relax to its ground state energy by liberating an electron.

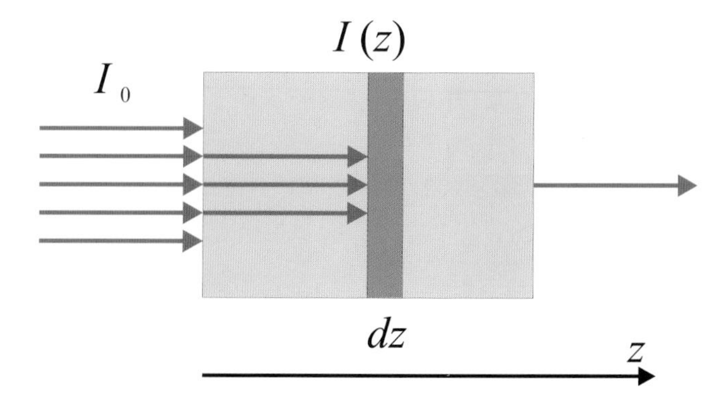

Figure 1.10: The attenuation of an X-ray beam through a sample due to absorption. The attenuation follows an exponential decay with a characteristic linear attenuation length $1/\mu$, where μ is the absorption coefficient.

When an X-ray photon expels an electron from an inner atomic shell it creates a hole in that shell. In Fig. 1.9(a) we illustrate this for the case of an electron excited from a K shell. The hole is subsequently filled by an electron from an outer shell, an L shell say, with the simultaneous emission of a photon with an energy equal to the difference in the binding energies of the K and L electrons (Fig. 1.9(b)). The emitted radiation is known as fluorescence. Alternatively, the energy released by an electron hopping from the L shell to the hole in the K shell can be used to expel yet another electron from one of the outer shells, as sketched in Fig. 1.9(c). This secondary emitted electron is called an Auger electron, named after the French physicist who first discovered the process. The monochromatic nature of fluorescent X-rays is a unique fingerprint of the kind of atom that produces the fluorescence. This method can be utilized for non-destructive chemical analysis of samples, and has the advantage that it is very sensitive. The radiation that creates the hole in the first place does not have to be an X-ray – it could also be from a beam of particles, such as of protons or electrons. The latter is a standard option in electron microscopes, enabling the chemical composition of samples to be determined with a very fine spatial resolution.

The absorption cross-section has a distinct dependence on photon energy. An example is shown in the top panel of Fig. 1.11 for the rare gas krypton. Below a photon energy of 14.32 keV the X-ray photon can only expel electrons from the L and M shells. The cross-section is approximately proportional to $1/\mathcal{E}^3$. At a characteristic energy, the so-called K-edge energy, the X-ray photon has enough energy to also expel a K electron, with a concomitant discontinuous rise in the cross-section of about one decade. From then on the cross-section continues to fall off as $1/\mathcal{E}^3$.

If we examine the fine structure of the absorption just around the edge it is apparent that it depends on the structure of the material. This is again illustrated for Krypton in Fig. 1.11 [taken from Stern and Heald, 1900]. The wiggles in the spectrum from two-dimensional crystalline Krypton on graphite demonstrate the phenomenon of Extended X-Ray Absorption Fine Structure (EXAFS) in condensed matter systems. We shall return to the interpretation of EXAFS data in Chapter 6.

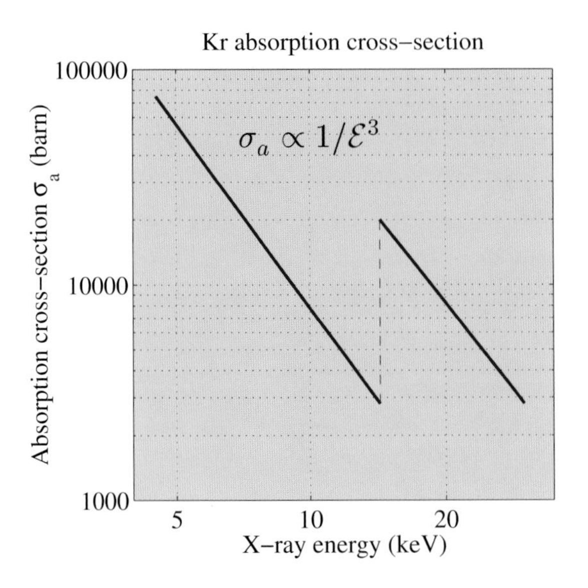

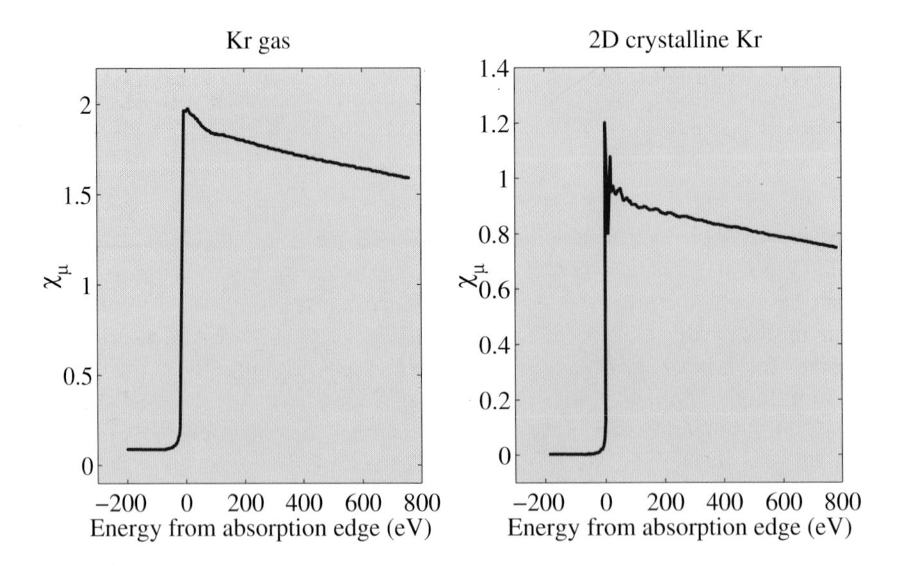

Figure 1.11: Top: The absorption cross-section of gaseous krypton. Above a photon energy of 14.325 keV a K shell electron can be expelled from the atom and a new "channel" for the absorption is then opened up. The double logarithmic plot illustrates that the cross-section varies as $1/\mathcal{E}^3$. Bottom: A comparison of the absorption spectra of gaseous krypton with krypton physiosorbed on graphite where the krypton atoms form a two-dimensional lattice. In the latter case fine structure, or wiggles, are evident which are known as EXAFS. The quantity χ_μ is proportional to the absorption cross-section σ_a.

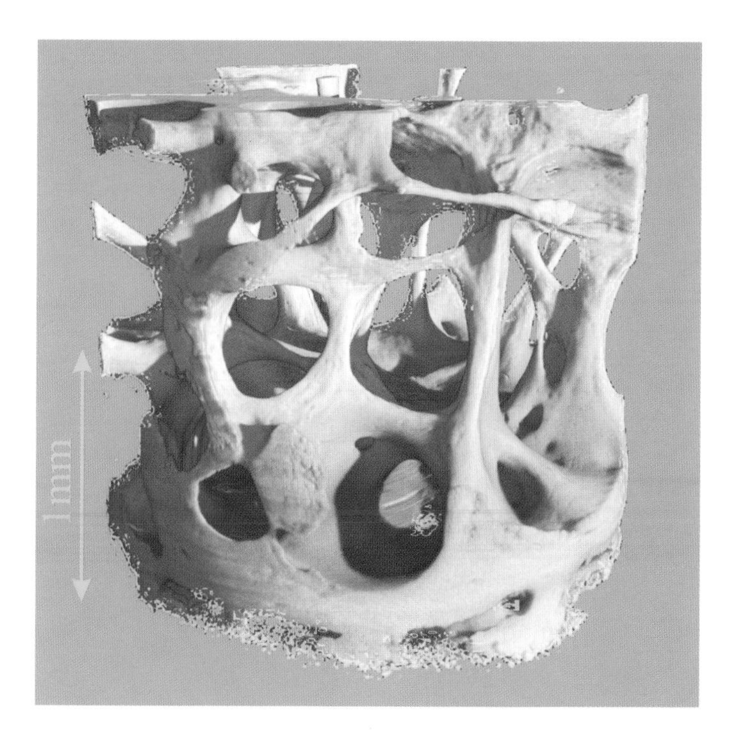

Figure 1.12: Three-dimensional micro-CAT reconstruction of a cylindrical human vertebral bone specimen scanned with 3.6 μm spatial resolution. Note the difference between the cortical end-plate and the underlying trabecular bone. (Image courtesy of a collaboration between Aarhus University, Denmark, and HASYLAB at DESY, Germany.)

The photoelectric absorption cross-section varies with the atomic number Z of the absorber, approximately as Z^4. It is this variation, and thus the contrast, between different elements that make X-rays so useful for imaging. Tissue is mainly water and hydrocarbons and thus has a $1/e$ thickness of many centimetres for hard X-rays, whereas bones in the body have a lot of Ca and a correspondingly smaller X-ray transmission. It was this, by now well-known, ability to look through the body that produced a sensation, when Wilhelm Conrad Röntgen discovered X-rays over a 100 years ago. When coupled with the computer power available today one can obtain the internal structure of parts of the body with remarkable precision. The technique is called CAT scanning, an acronym for Computer Axial Tomography (or Computer Aided Tomography). The idea is to take two-dimensional "shadow" pictures from many angles, and then reconstruct the three-dimensional object using a computer program. An example of the type of exquisite images that can be obtained with modern CAT scanning is given in Fig. 1.12. Another tomography application where computer power is essential, utilizes subtraction of pictures taken above and below the K edge of the element one is particularly interested in. In this way the element-sensitivity is enhanced dramatically.

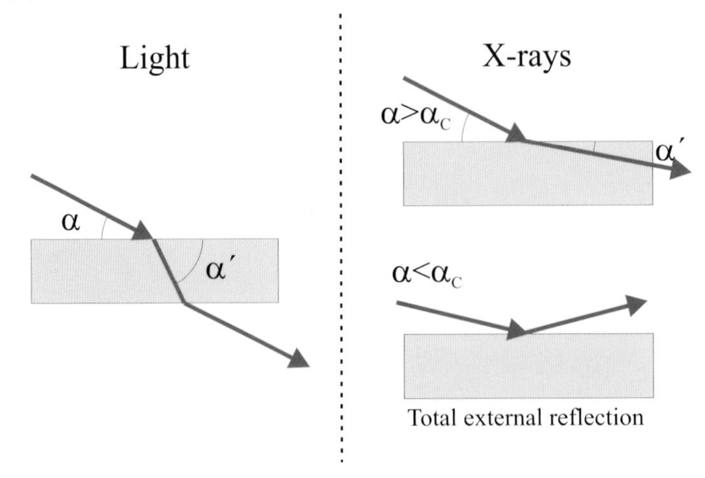

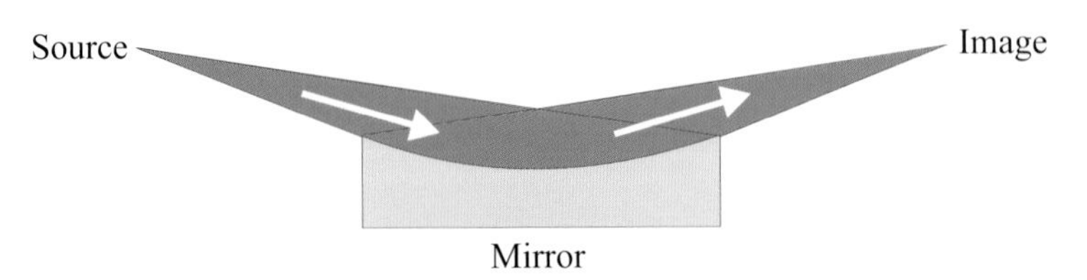

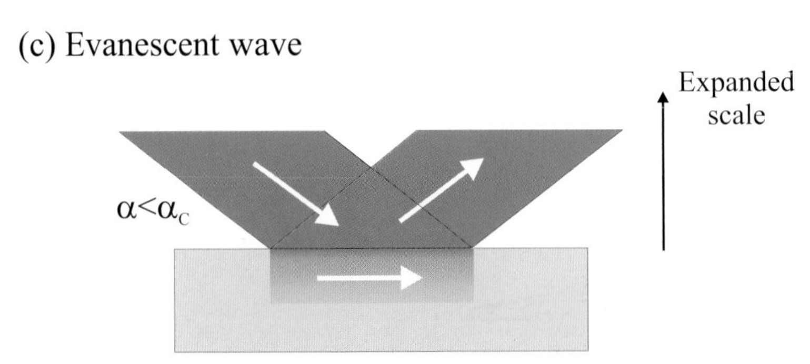

Figure 1.13: (a) The refraction of light shows that in the visible part of the spectrum the refractive index of glass is considerably greater than one. In contrast, the index of refraction for X-rays is slightly less than one, implying total external reflection at glancing angles below the critical angle α_c. (b) A focusing X-ray mirror can be constructed by arranging that the incident angle is below the critical angle for total external reflection. (c) At glancing angles below the critical angle the reflectivity is almost 100%, and the X-ray only penetrates into the material as an evanescent wave with a typical penetration depth of $\approx$10 Å. In this way X-rays can be a surface sensitive probe.

1.4 Refraction and reflection

The interaction of an X-ray photon with matter has so far been discussed at the atomic level. However, since X-rays are electromagnetic waves, one should also expect some kind of refraction phenomena at interfaces between different media. To describe such refractive phenomena, matter is taken to be homogeneous with sharp boundaries between the media, each having its own refractive index n. By definition the refractive index of vacuum is one. It is well known that for visible light in glass n is large and can vary considerably, ranging from 1.5 to 1.8 depending on the type of glass. This of course enables lenses to be designed for focusing light and thereby obtaining magnified images. For X-rays the difference from unity of n is very small, and as we shall see in Chapter 3 is of order 10^{-5} or so. In general for X-rays, the refractive index can be expressed as

$$\boxed{n = 1 - \delta + i\beta} \tag{1.15}$$

where δ is of order 10^{-5} in solid materials and only around 10^{-8} in air. The imaginary part β is usually much smaller than δ. Note that for X-rays, n is smaller than unity. The phase velocity inside the material is c/n and thus larger than the velocity of light, c. This does not, however, violate the law of relativity, which requires that only signals carrying "information" do not travel faster than the speed of light. Such signals move with the group velocity, not the phase velocity, and the group velocity is in fact less than c.

Snell's law relates the incident grazing angle α to the refracted grazing angle α' (see Fig. 1.13(a))

$$\cos\alpha = n \, \cos\alpha' \tag{1.16}$$

An index of refraction less than unity, implies that below a certain incident grazing angle called the critical angle, α_c, X-rays undergo total external reflection. Expansion of the cosine in Eq. (1.16) with $\alpha = \alpha_c$, $\alpha' = 0$ and using Eq. (1.15) allows us to relate δ to the critical angle α_c:

$$\alpha_c = \sqrt{2\delta}$$

where for simplicity we have taken $\beta = 0$. With δ being typically around 10^{-5}, α_c is of the order of a milli-radian. We shall see in Chapter 3, that the refractive constants δ and β can be derived from the scattering and absorption properties of the medium, respectively.

Total external reflection has several important implications for X-ray physics. First, total reflection from a curved surface enables focusing optics to be constructed as shown in the Fig. 1.13(b). A small source size is thus desirable, since from geometrical optics, a small source will be focused to a small image. A second consequence of total external reflection is that for $\alpha < \alpha_c$ there is a so-called evanescent wave within the refracting medium, see Fig. 1.13(c). It propagates parallel to the flat interface, and its amplitude decays rapidly in the material – typically with a penetration depth of only a few nano-meters. This should be compared with a penetration depth of several micro-meters at a glancing angle of several times α_c.

The much-reduced penetration of X-rays for angles less than α_c increases their surface sensitivity. This allows the scattering from the surface and near surface region to be studied, often in great detail, and indeed X-rays have become a valuable tool for the investigation of surfaces and interfaces.

1.5 Coherence

Throughout this introductory survey we have assumed that we are dealing with an X-ray beam in a perfect plane-wave state. This is obviously an idealization, and in this section we shall briefly discuss its limitation by recalling the concept of a *coherence length* of a real beam, and its relation to the source and monochromator. A real beam deviates from an ideal plane wave in two ways: it is not perfectly monochromatic, and it does not propagate in a perfectly well defined direction. Let us discuss these limitations in turn.

The top part of Fig. 1.14 shows two plane waves A and B with slightly different wavelengths, λ and $\lambda - \Delta\lambda$ say, but both propagating in exactly the same direction. The two waves are exactly in phase at the wavefront in P. The question is how far do we have to go away from P before the two waves are out of phase – this defines the *longitudinal* coherence length L_L. If the two waves are out of phase after travelling L_L, then they will be in phase again after travelling $2L_L$. Let that distance be N wavelengths λ, or equivalently $(N+1)(\lambda - \Delta\lambda)$, i.e.

$$2L_L = N\lambda = (N+1)(\lambda - \Delta\lambda)$$

The second equation implies that $(N+1)\Delta\lambda = \lambda$, or $N \approx \lambda/\Delta\lambda$, and using this result the first equation can be rearranged to read

$$\boxed{L_L = \frac{1}{2}\frac{\lambda^2}{\Delta\lambda}} \tag{1.17}$$

In the bottom part of Fig. 1.14 is shown the other case: two waves A and B of the same wavelength, but with slightly different directions of propagation, say by an angle of $\Delta\theta$. Their wavefronts coincide at point P, and the question is how far do we have to go from P along the wavefront of wave A before it is out of phase with wave B – by definition that distance is the *transverse* coherence length L_T. Clearly, if proceeding to a distance of $2L_T$, the two waves will be in phase again, and it is obvious from the figure that $2L_T\Delta\theta = \lambda$, i.e. $L_T=\lambda/(2\Delta\theta)$. Suppose that the different directions of propagation arise because the two waves originate from two different points on the source, let us say a distance D apart. If the distance from the observation point P to the source is R, then $\Delta\theta = D/R$ and we have

$$\boxed{L_T = \frac{1}{2}\frac{\lambda}{(D/R)} = \frac{\lambda}{2}\left(\frac{R}{D}\right)} \tag{1.18}$$

It is instructive to consider typical values for the coherence lengths, but to do so we need to make some assumptions about the source. At a third generation

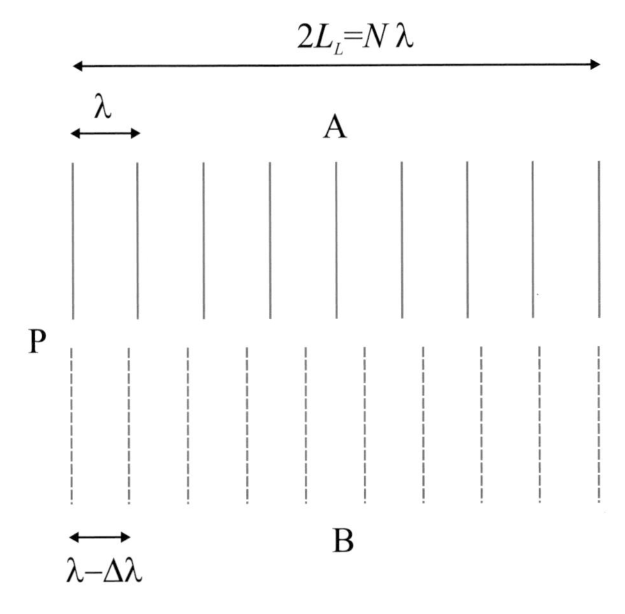

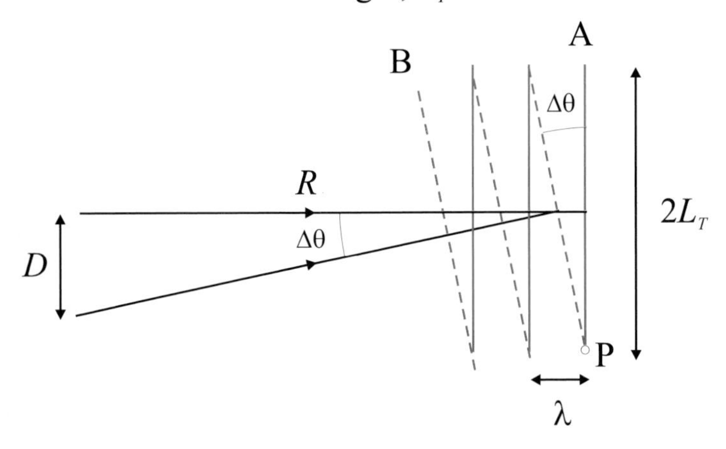

Figure 1.14: Longitudinal and transverse coherence lengths. (a) Two plane waves with different wavelengths are emitted in the same direction. For clarity we have shown the waves displaced from each other in the vertical direction. After a distance L_L, the longitudinal coherence length, the two are out of phase by a factor of π. (b) Two waves with the same wavelength are emitted from the ends of a finite sized source of height D.

synchrotron the vertical source size is around 100μm, and the experiment may be performed some 20 meters away, so that for 1 Å X-rays L_T is approximately $10 \mu m$ in the vertical plane. To calculate the longitudinal coherence length we need to make some additional assumption about the device used to monochromate the beam. If a perfect crystal is used, $\Delta\lambda/\lambda \approx 10^{-5}$ (see Chapter 5), and then, according to Eq. (1.17), L_L is $\approx 5\mu$m for 1 Å X-rays, similar in order of magnitude to L_T. The consequence of a finite coherence length is that it places an upper limit on the separation of two objects if they are to give rise to interference effects. To take a simple example, if two electrons are separated by much more than the coherence length then the total scattered intensity is the sum of scattered intensities from the individual electrons, and we do not first add the amplitudes and then take the square as has been described thus far.

1.6 Magnetic interactions

The discussion so far has centred on the interaction between the electric field of the X-ray and the charge of the electron. What has been neglected is the magnetic field of the X-ray and the spin of the electron. When these are included in a full treatment of the interaction, terms emerge in the scattering cross-section that are sensitive to the spin and orbital magnetic moments of the electron. In this way it is possible to use X-rays to investigate magnetic structures. The history of X-ray magnetic scattering is much more recent than that of classical X-ray diffraction. In fact the first observation of X-ray magnetic scattering had to wait until 1972 and the pioneering experiments of de Bergevin and Brunel on antiferromagnetic NiO [de Bergevin and Brunel, 1972].

The reason for this is simply that magnetic scattering is much weaker than charge scattering. The amplitude ratio of magnetic to charge scattering for a single electron is

$$\frac{A_{\text{magnetic}}}{A_{\text{charge}}} = \left(\frac{\hbar\omega}{mc^2}\right)$$

[Blume, 1985]. For 10 keV x-rays this ratio is $\approx$0.01, so the intensity of magnetic Bragg peaks that are purely magnetic in origin are weaker than the charge peaks by a factor of approximately 10^{-4}. Progress in the field of X-ray magnetic scattering was at first slow, but the routine availability of synchrotron radiation has given a tremendous boost to this subject, to the extent that it has now flourished into a field in its own right.

Sensitivity to magnetism is not restricted to scattering experiments, however, but also occurs in absorption processes. For example, the difference in absorption of left- and right-hand circularly polarized light by a solid (known as circular dichroism) can be directly related to the ferromagnetic magnetization density. It has also been found that magnetic scattering itself is a much richer phenomena than early expectations, with the discovery that resonant magnetic scattering processes occur when the energy of the incident X-ray is tuned close to certain atomic absorption edges [Namikawa et al., 1985, Gibbs et al., 1988]. These subjects take us beyond the scope of this volume, but it is important to realize that the study of the interaction of X-rays with matter is still an active field of research some 100 years or so after the discovery of the X-ray [see, for example, Lovesey and Collins, 1996].

Further reading

Röntgen Centennial - X-rays in Natural and Life Sciences, Eds. A Haase, G. Landwehr, and E. Umbach (World Scientific, Singapore, 1997)

Fifty Years of X-ray Diffraction, P. P. Ewald, (International Union of Crystallographers, N. V. A. Oosthoek's uitgeversmaatchappj, Utrecht, 1962)

X-rays 100 years later, Physics Today (special issue) **48**, (1995)

2. Sources of X-rays

2.1 Early history and the X-ray tube

X-rays were discovered by Wilhelm Conrad Röntgen in November 1895 in Würzburg. He was examining the light and other radiation associated with the discharge from electrodes in an evacuated glass tube. He had covered the tube, a so-called Geisler discharge tube, so that no visible light could escape. The laboratory was also darkened. All that could be seen was a faint yellow-green light from a fluorescent screen placed close to the tube. The fluorescent light was flickering, since the high voltage was supplied by the *ac* output of an induction coil, and could be seen even when the screen was several meters away from the tube. To his amazement the radiation from the tube passed through paper and wood, whereas metal pieces of equipment cast a shadow on the screen. The most stunning phenomenon occurred when he placed his hand into the space between the tube and the screen and saw the bones inside. Röntgen was a keen amateur photographer and he quickly had the idea to photograph the X-ray beam instead of using the fluorescent screen. The photographs were convenient scientific documentation of his discovery, which was first published in the annals of the local Würzburg Scientific Society in late December of 1895. The paper is entitled "Uber eine neue Art von Strahlen – vorläufige Mitteilung"[1]. The fact that one could now "see" inside the human body was a sensation that spread worldwide within a few weeks, and the implications for medical science during the following century can hardly be overstated.

It became clear from Röntgen's subsequent investigations that the imaging of bones in the body is based on the fact that X-ray absorption is strongly dependent on the atomic number of the elements; it varies approximately as Z^4. The other important application of X-rays, based on diffraction phenomena, showing how crystalline matter is built up by atoms forming a periodic lattice, had to wait until 1912 when von Laue and his coworkers obtained the first diffraction pattern from a crystal of copper sulfate. In the following year W.H. Bragg and W.L. Bragg (father and son) examined the diffraction of X-rays from a number of crystals and laid the foundations of the field of crystallography, which subsequently allowed one to determine the structure of molecules.

The young Bragg also found a particular simple way to interpret the diffraction patterns which proved unambiguously that X-rays are nothing other than electromagnetic radiation of short wavelength: Röntgen had also played with the same idea, and tried to prove it experimentally, but without success. His influence in German physics at the time was so great that even von Laue and coworkers were not tempted to reach the same conclusion from their diffraction experiments as Bragg.

[1] "On a new kind of radiation – preliminary communication"

The standard X-ray tube and the rotating anode

The X-ray tube Röntgen used was a tricky business to run reliably. It was therefore a tremendous practical step forward when in 1912 W. D. Coolidge from General Electric Research Laboratories in New York developed a new tube, where electrons were produced by a glowing filament and subsequently accelerated towards a water-cooled metal anode, see Fig. 2.1. Now one could vary the high voltage and the current independently, and the limitation of intensity was set only by the cooling efficiency. It turns out that the maximal power for such a device is around 1 kW. The Coolidge tube served as the standard X-ray tube for many decades with only marginal technical improvements.

Although it was appreciated early on that by spinning the anode the heat could be dissipated over a much larger volume than in a standard tube, allowing the total power to be correspondingly increased, it was not until the 1960's that so-called rotating anode generators became available on a commercial basis. One of the technical difficulties to overcome had been the problem of how to make a high-vacuum seal on the rotating shaft, inside which the cooling water must flow in and out.

The spectrum of X-rays generated from electrons impinging on a metal anode has two distinct components. There is a continuous part due to the electrons being decelerated, and eventually stopped in the metal. This is consequently known as *bremsstrahlung* radiation (after the German *bremsen* for brake), and has a maximum energy that corresponds to the high voltage applied to the tube. Superimposed on this broad spectrum is a sharper line spectrum. In a collision with an atom the incident electron may also cause an atomic electron to be removed from one of the inner shells, creating a vacancy. The subsequent relaxation of an electron from an outer shell into the vacancy produces an X-ray with a characteristic energy equal to the difference in energy between the two shells. This is the fluorescent radiation. For experiments requiring a monochromatic beam one often utilizes the K_α line which is several orders of magnitude more intense than the bremsstrahlung spectrum. However, only a very small fraction of the photons emitted into the solid angle of 2π can be utilized in a beam requiring an angular divergence of a few squared milli-radian. In addition, the line source is not continuously tunable so the optimal wavelength for the experiment cannot be chosen, or scanned, at will. As we shall see in the next section, X-rays generated from synchrotron sources do not have these drawbacks, and have a brilliance which is enormously higher than that of standard laboratory sources.

2.2 Introduction to synchrotron radiation

Synchrotron radiation takes its name from a specific type of particle accelerator. However, synchrotron radiation has become a generic term to describe radiation from charged particles travelling at relativistic speeds in applied magnetic fields which force them to travel along curved paths. Besides synchrotrons themselves, synchrotron radiation is produced in storage rings where electrons or positrons are kept circulating at constant energy. In a storage ring the synchrotron radiation is produced either in the bending magnets needed to keep the electrons in a closed orbit, or in insertion devices such as wigglers or undulators situated in

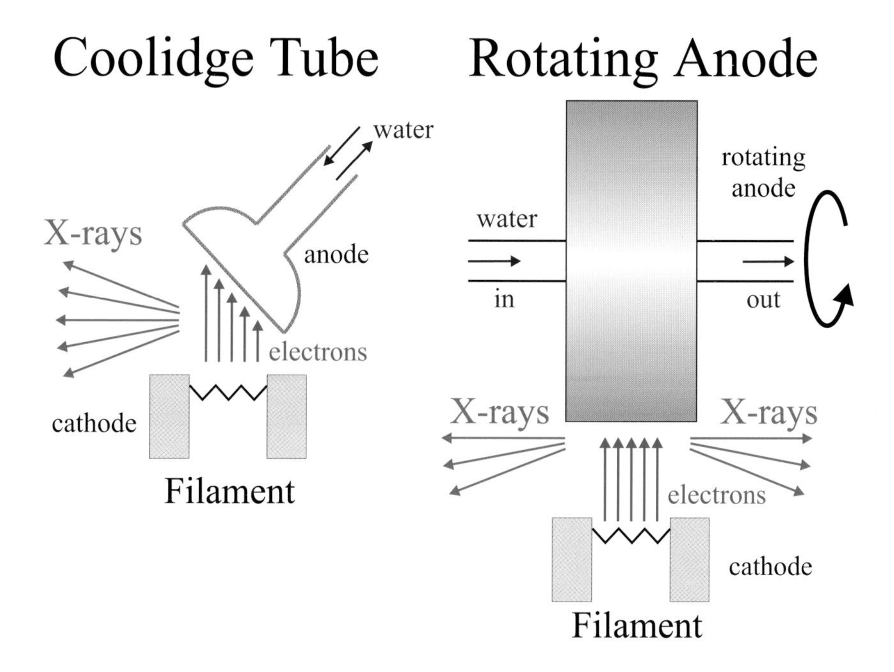

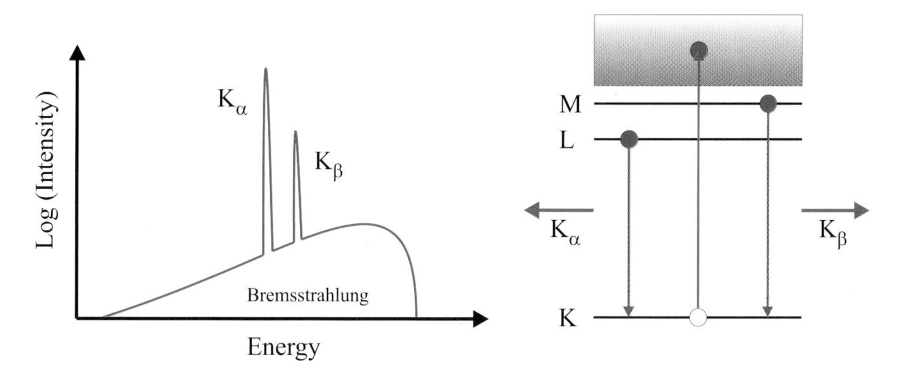

Figure 2.1: The standard X-ray tube (upper, left) was developed by Coolidge around 1912. The intensity limitation is set by the maximum power a cooled metal anode can withstand. The power can be increased by dissipating it over a larger volume which is achieved by rotating the anode (upper, right). The spectrum from an X-ray tube has discrete fluorescent lines superimposed on the continuous bremsstrahlung radiation (bottom, left). Schematic atomic energy level diagram (bottom, right): the K_α line results from transitions between an L and K shell, whereas the K_β comes from an M to K transition.

the straight sections of the storage ring. In these devices an alternating magnetic field forces the electrons to follow oscillating paths rather than moving in a straight line. In a wiggler the amplitude of the oscillations is rather large, and the radiation from different wiggles add incoherently, whereas in undulators, as we shall see, the small-amplitude oscillations from the passage of a single electron produce a coherent addition of the radiation from each oscillation. It is also interesting to note that synchrotron radiation in fact occurs naturally, and has been observed from plasmas around stellar nebula.

For X-ray research, however, practically all modern sources of synchrotron radiation are storage rings. The researcher new to the field of synchrotron X-ray science will encounter the usual set of abbreviations, such as S.R. (for synchrotron radiation), B.M. (from bending magnets), I.D. (from insertion devices) etc. We shall not use them in this book, but the reader is warned.

2.2.1 Characterising the beam: brilliance

Several aspects of an X-ray source determine the quality of the X-ray beam it produces. These aspects can be combined into a single quantity, called the brilliance, which allows one to compare the quality of X-ray beam from different sources. First of all, there is the number of photons emitted per second. Next, there is the collimation of the beam. This describes how much the beam diverges, or spreads out, as it propagates. Usually the collimation of the beam is given in milli-radian, both for the horizontal and for the vertical direction. Third, it may be of importance how large the source area is – if it is small, one may be able to focus the X-ray beam to a correspondingly small image size. The source area is usually given in mm^2. Finally, there is the issue of the spectral distribution. Some X-ray sources produce very smooth spectra, others have peaks at certain photon energies. So it matters, when making comparisons, what range of photon energies contribute to the measured intensity. The convention is therefore to define the photon energy range as a fixed relative energy bandwidth, which has been chosen to be 0.1%. There are several reasons why the relative rather than the absolute bandwidth is chosen. One reason is that monochromator crystals are often perfect crystals, and as we shall see in Chapter 5 the relative bandwidth for a perfect crystal in symmetric reflection geometry is independent of the photon energy, and depends only on the Miller indices of the reflection. Altogether then, one defines the figure-of-merit for the source as:

$$\boxed{\text{Brilliance} = \frac{\text{Photons/second}}{(\text{mrad})^2\ (\text{mm}^2\ \text{source area})\ (0.1\%\ \text{bandwidth})}} \tag{2.1}$$

The intensity in photons per second after the monochromator crystal is the product of the brilliance, angular divergences set by the horizontal and vertical apertures (in milli-radian), the source area (in mm^2), and the relative bandwidth of the monochromator crystal relative to 0.1%.

The brilliance is a function of the photon energy. The maximum brilliance from third generation undulators (see Fig. 1.1) is about 10 orders of magnitude higher than that from a rotating anode at the K_α line! This dramatic improvement has in many ways led to a paradigm shift in experimental X-ray science.

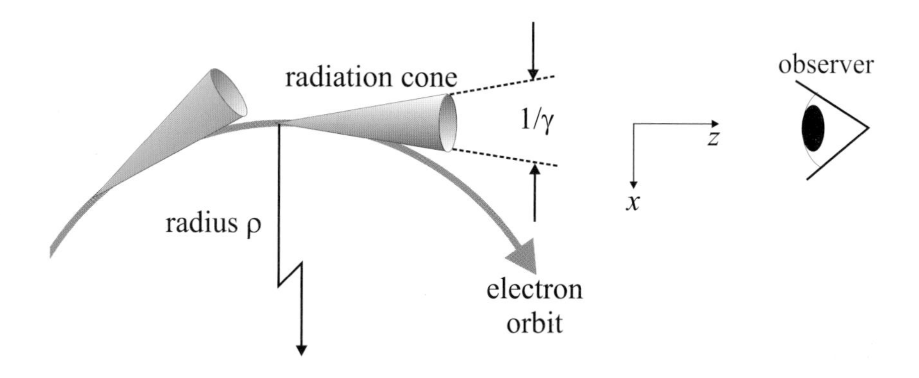

Figure 2.2: The trajectory of an electron is approximated by a circle of radius ρ. The radiation is confined to a narrow cone with an opening angle of $1/\gamma$ around the instantaneous velocity.

Experiments inconceivable only 10 to 20 years ago are now performed on almost a routine basis.

2.3 Synchrotron radiation from a circular arc

The radiation from an electron orbiting at relativistic speeds in a circle can be likened to a sweeping search light, as shown in the top part of Fig. 2.2. The characteristic features of the radiation depend on two key parameters: the cyclic frequency ω_o of the orbiting electron and γ, which is the electron energy in units of the rest mass energy, in other words $\gamma = \mathcal{E}_e/mc^2$.

The instantaneous direction of the radiation cone is that of the instantaneous velocity of the electron, and the opening angle of the cone is $\gamma^{-1} = mc^2/\mathcal{E}_e$. This is typically around 10^{-4}, or 0.1 milli-radian. The emitted spectrum is very broad, ranging from the far infrared to the hard X-ray region. However, the spectrum falls off quickly for photon frequencies higher than $\gamma^3\omega_o$. The angular frequency of an electron in the storage ring ω_o is typically of order 10^6 cycles per second, so the hard X-ray frequency cut-off is around 10^{18} cycles per second. We shall now show how one can understand these basic features from simple physical arguments. A few facts from the theory of relativity have to be recollected first, and these are given in the box on the next page. The other important ingredient of basic physics we need is the Doppler effect.

Before discussing an electron orbiting on a circular path, the simpler case is first considered of an electron travelling on a path comprised of short straight segments, with abrupt bends at points A, B, C, etc, as shown in Fig. 2.3. Subsequently the limit will be taken where the straight sections become infinitesimally small and the path a circular arc. While in linear, uniform motion the electron does not radiate, but at each bend it changes its velocity, and therefore it has a short period of acceleration during which it radiates. The observer is along the direction BC, and the time the electron spends in getting from one bend to the next is denoted $\Delta t'$. Consider in the top part of Fig. 2.3 the propagation of a wavefront emitted by the electron when passing the bend at B. During the time the electron spends in getting from B to C the wavefront has moved the

Relativistic formulae

The energy $\mathcal{E}_e$ of an electron at speed v is

$$\mathcal{E}_e = \frac{mc^2}{\sqrt{1-\left(\frac{\mathrm{v}}{c}\right)^2}}$$

It is convenient to use the electron energy γ, measured in units of its rest mass energy, $\gamma \equiv \mathcal{E}_e/mc^2$, and the speed β_e, measured in units of the velocity of light, $\beta_e \equiv \mathrm{v}/c$. The formula above then reads

$$\gamma \equiv \frac{1}{\sqrt{1-\beta_e^2}} \tag{2.2}$$

The electron energy in a typical X-ray synchrotron storage ring is 5 GeV. The rest mass of an electron is 0.511 MeV, so γ is of order 10^4. We can therefore expand Eq. (2.2) to obtain:

$$\beta_e = \left[1-\frac{1}{\gamma^2}\right]^{1/2} \cong 1-\frac{1}{2\gamma^2} \tag{2.3}$$

distance $c\Delta t'$ towards the observer, at which point a new wavefront is emitted from C, which is $\mathrm{v}\Delta t'$ closer to the observer than B. These two wavefronts will thus be $(c-\mathrm{v})\Delta t'$ apart, and the observer experiences that they arrive within a time interval $\Delta t = (c-\mathrm{v})\Delta t'/c = (1-\beta_e)\Delta t'$. The same kind of arguments can be made for the pair of wavefronts emitted when the electron was at A and B; the only difference is that the distance travelled by the electron towards the observer is $\mathrm{v}\Delta t' \cos\alpha$, α being the angle between the velocity and the direction to the observer. The wavefront from A is therefore not $(c-\mathrm{v})\Delta t'$ ahead of the wavefront emitted from B, but a distance $(c-\mathrm{v}\cos\alpha)\Delta t'$. In other words, the time compression of wavefronts – the Doppler effect – appears less pronounced to the observer.

The time interval, Δt, between wavefronts measured by the observer is related to the time interval $\Delta t'$ by

$$\Delta t = (1-\beta_e \cos\alpha)\Delta t'$$

Since β_e and $\cos\alpha$ are very close to unity we can expand them to obtain

$$\Delta t \approxeq \left[1-\left(1-\frac{1}{2\gamma^2}\right)\left(1-\frac{\alpha^2}{2}\right)\right]\Delta t' \approxeq \left[\frac{1+(\alpha\gamma)^2}{2\gamma^2}\right]\Delta t'$$

With $\alpha \approx 0$ and $\gamma \approx 10^4$ it is clear that the time compression of the wavefronts experienced by the observer is enormous. The Doppler effect is in fact maximal when α=0, and has decreased by a factor of two when $\alpha = 1/\gamma$. This then explains why the natural opening angle of synchrotron radiation is of order γ^{-1}. Note that this is the opening angle in all directions. In the vertical plane the angular divergence is γ^{-1}, whereas in the horizontal plane the angular

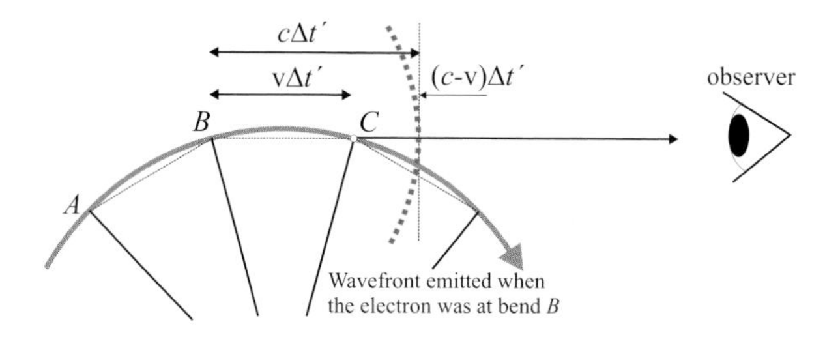

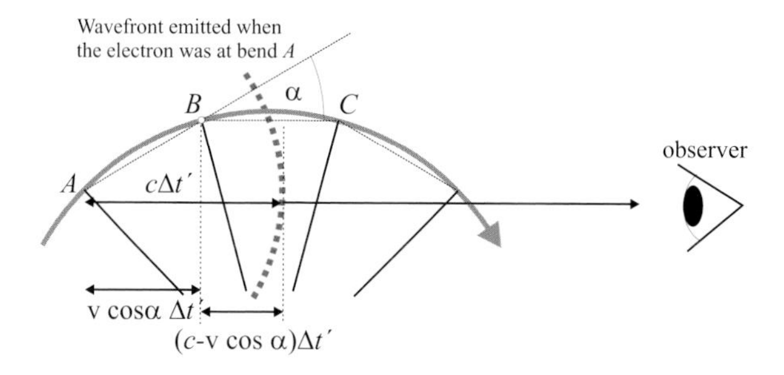

Figure 2.3: A circular arc is approximated by straight segments connected by bends at A, B, C, etc. When the electron passes a bend, a wavefront (thick dotted line) is emitted and propagates with velocity c. The wavefront in the top (bottom) was emitted from bend B (A). The electron velocity is $\mathbf{v}$, and the time for the electron to travel from one bend to the next is $\Delta t'$. The observer experiences a time interval of $\Delta t = (c - \mathrm{v}\cos\alpha)\Delta t'/c$ between wavefronts, where α is the angle between the electron velocity and the direction towards the observer.

divergence of the fan of radiation depends on how long a segment of the circular arc is viewed by the observer.

Taking the limit $\Delta t' \to 0$, the general relation between the retarded time t and the laboratory time t' is therefore given by the differential equation

$$\frac{dt}{dt'} = (1 - \beta_e \cos\alpha) \tag{2.4}$$

where β_e is the electron velocity in units of c, and α is the angle between the instantaneous velocity and the direction to the observer. We shall return to the solution of this differential equation in Section 2.4.1.

In Fig. 2.4 the entire circular orbit is shown. The radius ρ is determined by the magnetic field $\mathbf{B}$ in the following way. The magnitude of the Lorentz force is $e\mathrm{vB}$. Had the electron been non-relativistic this force would be equal to the centripetal acceleration v^2/ρ times the mass m. Noting that $m\mathrm{v} = \mathrm{p}$, the momentum, one obtains $\mathrm{p} = \rho e\mathrm{B}$. This relation is also valid for relativistic particles, in which case p is equal to $\gamma m\mathrm{v}$. For super-relativistic particles $\mathrm{v} \approxeq c$, so we find

$$\gamma\, m\, c = \rho\, e\, \mathrm{B} \tag{2.5}$$

or in practical units the radius is

$$\boxed{\rho[\mathrm{m}] = 3.3\,\frac{\mathcal{E}_e\,[\mathrm{GeV}]}{\mathrm{B}\,[\mathrm{T}]}} \tag{2.6}$$

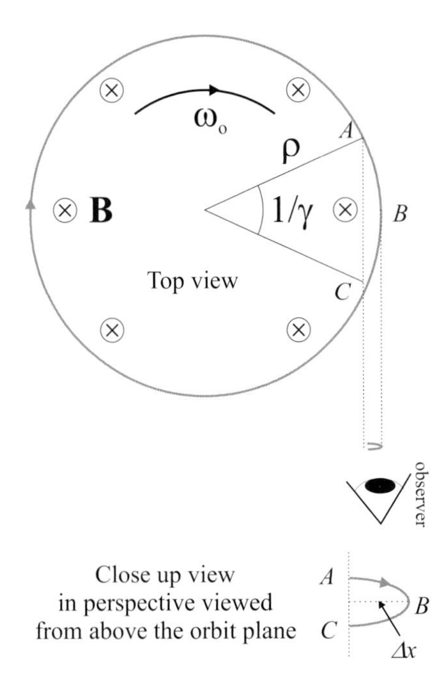

Figure 2.4: An electron is kept in a circular orbit by the magnetic field **B**. An observer in the direction of the tangent at point B will, due to the Doppler effect, see the electron as having a large acceleration when it is between points A and C. The close-up view indicates that the observer will experience something that looks like half a period of an oscillation.

The electron radiates during its entire cycle around its orbit, but the observer located in the direction of the tangent to point B only sees a significant amount of radiation while the electron passes from A to C, since the observer experiences that part of the travel as extremely fast, with an overwhelmingly large acceleration. The far-field radiation is proportional to this acceleration, and that is why we only have to consider this part of the circular path in our discussion.

It is important to note that the on-axis direction of the electric field of the electromagnetic wave is parallel to the acceleration of the emitting electron. The on-axis polarization of the X-ray is therefore linear in the horizontal plane. This is no longer true when the motion of the electron is viewed out of the orbit plane. As seen in the lower part of Fig. 2.4, when the arc ABC is viewed from above the orbit plane, it appears as part of an ellipse with the electron turning clockwise. If looked upon from below the orbit plane, the same figure can illustrate the electron path except that now A and C are interchanged. As the arrow must run from A to C via B, it is apparent that that the electron appears to move in a counter clockwise direction. Viewed out of the orbit plane the electron then has non-zero angular momentum, which it imparts to the emitted X-rays. From this it may be concluded that the radiation viewed above the orbit plane has a right-handed circular component, whereas below the orbit plane the circular polarization is in the opposite sense of rotation. This subject is returned to in Section 6.3 on X-ray magnetic dichroism.

The lower part of Fig. 2.4 shows a close-up of how the observer experiences

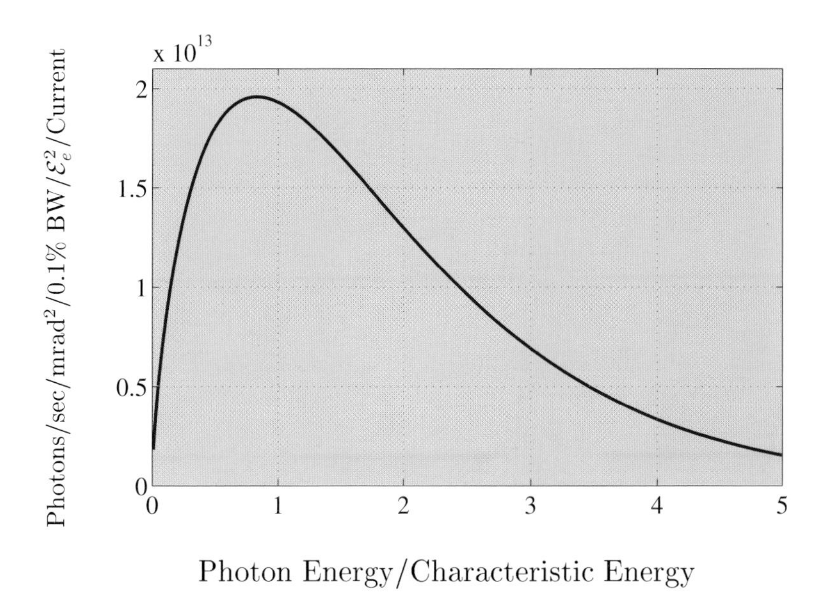

Figure 2.5: The brilliance spectrum from a bending magnet, normalized by the square of the electron energy and the beam current. The abscissa is $x = \hbar\omega/(\hbar\omega_c)$, i.e. the photon energy normalized by the characteristic energy $\hbar\omega_c = \left(\frac{3}{2}\right)\gamma^3\hbar\omega_o$. The numerical formula is $1.327 \times 10^{13} x^2 K^2_{2/3}(x/2)$, where $K_{2/3}(x/2)$ is a modified Bessel function. The electron energy $\mathcal{E}_e$ is in GeV, and the beam current in Amperes.

the large-acceleration part of the electron motion. It resembles one half of the period T of an entire oscillation. The time the electron spends in getting from A to C is $[\gamma^{-1}/(2\pi)]\,T = 1/(\gamma\omega_o)$, but the observer experiences the time to be γ^2 shorter, i.e. $1/(\gamma^3\omega_o)$. The highest frequency in Fourier transforming a pulse shape is of order the inverse pulse width, so the high cut-off frequency in the observed spectrum is of order $\gamma^3\omega_o$. In the more rigorous treatment it turns out to be convenient to define a characteristic frequency $\omega_c \equiv \left(\frac{3}{2}\right)\gamma^3\omega_o$. Since $\omega_o{=}2\pi/T = 2\pi/(2\pi\rho/c){=}c/\rho$, which by Eq. (2.5) is proportional to $\mathrm{B}/\mathcal{E}_e$, the corresponding characteristic photon energy is given in practical units by

$$\boxed{\hbar\omega_c[\mathrm{keV}] = 0.665\,\mathcal{E}_e^2[\mathrm{GeV}]\,\mathrm{B}[\mathrm{T}]} \tag{2.7}$$

The brilliance spectrum from a bending magnet is a universal function of (ω/ω_c) and is given in Fig. 2.5. It scales with the square of the electron energy, $\mathcal{E}_e$, and is proportional to the current in the storage ring.

So far we have considered only the radiation from a single electron. In a storage ring the electrons are stored in bunches. For certain applications one may choose to have a single bunch, but in general the storage ring is filled with a number of equi-spaced bunches. The duty cycle for one bunch in a 300 m long storage ring is 1 μs, and as the bunch length is of order a centimetre, the pulse duration is of order of a hundred pico-seconds. The resulting synchrotron radiation is consequently pulsed with a sub-nano second pulse width and a duty cycle in the μs range.

We can also estimate the radiated power from a section L along the electron path. The result, which is stated prior to its derivation, is

$$\boxed{\mathcal{P}[\mathrm{kW}] = 1.266\, \mathcal{E}_e^2[\mathrm{GeV}]\, \mathrm{B}^2[\mathrm{T}]\, L[\mathrm{m}]\, I[\mathrm{A}]} \tag{2.8}$$

It is beyond the scope of this book to derive the exact numerical prefactor in the above equation. Here a heuristic argument is used to obtain the dependence on the square of the electron energy and magnetic field; the proportionality to L, and to the ring current I, should be self evident.

The energy density of the radiated field is given by the Poynting vector, which has a magnitude $\mathrm{S} = \mathrm{B}_{\mathrm{rad}}\mathrm{E}_{\mathrm{rad}}/\mu_0 = c\epsilon_0 \mathrm{E}_{\mathrm{rad}}^2$. The electric field at a distance R from the source, in this case a single electron, is $\mathrm{E}_{\mathrm{rad}} = \mathcal{A}e/(4\pi\epsilon_0 c^2 R)$, where $\mathcal{A}$ is the apparent acceleration (see Eq. (1.1)). The acceleration is estimated from $\Delta x/(\Delta t)^2$, where Δx is the distance from B to the line AC in Fig. 2.4, and Δt is the time interval it takes for the electron to cover the distance as seen by the observer. As can be seen from Fig. 2.4, the distance $\Delta x = \rho(1 - \cos(\gamma^{-1}))$, from which it can be deduced that $\Delta x \sim \rho/\gamma^2$. Since the time interval is given by $\Delta t \sim \rho/(c\gamma^3)$, the acceleration is $\sim c^2\gamma^4/\rho$. The instantaneous energy radiated by the electron through an area element $R^2\Delta\Omega$ is the magnitude of the Poynting vector S times this area. The total energy is obtained by integrating the instantaneous energy over the time interval Δt. The result is

$$\begin{aligned}
\mathrm{S}\,(R^2\Delta\Omega)\Delta t &= c\epsilon_0\, \mathcal{A}^2 \left(\frac{e}{4\pi\epsilon_0 c^2 R}\right)^2 (R^2\Delta\Omega)\Delta t \\
&= c\epsilon_0 \left(\frac{c^4\gamma^8}{\rho^2}\right)\left(\frac{e^2}{(4\pi\epsilon_0)^2 c^4 R^2}\right)(R^2\gamma^{-2})\frac{\rho}{c\gamma^3} \\
&= \frac{1}{16\pi^2}\frac{e^2\gamma^3}{\epsilon_0\rho}
\end{aligned}$$

This is the emitted energy for an electron path length of ρ/γ, so per unit length the energy is $\sim \gamma^4/\rho^2$. From Eq. (2.6) we have $\rho \propto \mathcal{E}_e/\mathrm{B}$, and as $\gamma \propto \mathcal{E}_e$, we obtain the $\mathcal{E}_e^2\mathrm{B}^2$ dependence for the emitted power stated in Eq. (2.8).

2.3.1 Example: Bending magnet radiation at the ESRF

The European Synchrotron Radiation Facility (ESRF) in Grenoble, France, was the world's first third-generation X-ray source, and started regular operation for users in 1994. The storage ring consists of a number of straight sections where insertion devices may be placed, and in between these the electron beam passes through bending magnets where the electrons describe circular arcs. Similar third-generation sources have also been constructed in the United States (Advanced Photon Source, Argonne National Laboratory), and Japan (SPRING8).

The energy of the electrons in the storage ring at the ESRF is $\mathcal{E}_e = 6$ GeV, the ring electron current is typically around 200 mA, and the bending magnets produce a field of 0.8 T. The opening angle of the synchrotron beam from an

ESRF bending magnet is $1/\gamma = 5.11 \times 10^5/6 \times 10^9 = 0.08$ mrad. Here it is assumed that the bending magnet is viewed through a 1×1 mm^2 aperture at 20 m from the tangent point of the arc. The angular acceptance of the aperture is 1/20=0.05 mrad, somewhat smaller than the natural divergence of the radiation.

The radius of the electron orbit through the bending magnet can be found from Eq. (2.6). The result is

$$\rho = 3.3 \times \frac{6}{0.8} = 24.8\,\text{m}$$

The characteristic energy is given in Eq. (2.7) and is equal to

$$\hbar\omega_c = 0.665 \times 6^2 \times 0.8 = 19.2\,\text{keV}$$

In Fig. 2.5 the generic brilliance spectrum of a bending magnet is shown. To calculate the peak flux at the characteristic energy it is necessary to multiply by the solid angle of the aperture, the square of the electron energy and the current. The peak flux is then

$$\begin{aligned}\text{Flux} &= 1.95 \times 10^{13} \times \left(\frac{1}{20}\right)^2 \times 6^2 \times 0.2 \\ &= 3.5 \times 10^{11} \text{ photons/sec/0.1\% BW}\end{aligned}$$

in a bandwidth of 0.1%. According to Eq. (2.8) the observed radiated power from the bending magnet is determined by the length, L, of the electron orbit viewed through the aperture. As the radiation is viewed from the tangent point, L is equal to the radius of the electron orbit, ρ, multiplied by the acceptance angle in the horizontal plane of the aperture, i.e. L=24.8 m $\times$ 0.05 mrad= 1.24 mm. The power radiated is then

$$\mathcal{P} = 1.266 \times 6^2 \times 0.8^2 \times 1.24 \times 10^{-3} \times 0.2 = 7.3\,\text{W}$$

The observed power is smaller than this value for a number of reasons. First, the value given above is the value integrated over the vertical direction, and it is therefore necessary to correct for the finite angular acceptance of the slit. Second, there may be Be vacuum windows, and possibly other components such as filters, in a beamline at a synchrotron which reduce the measured value.

Beamlines which use a bending magnet as a radiation source usual make use of focusing optics to collect a fan of the emitted radiation in the horizontal plane. Typically the optics are designed to collect and focus a fan of 1 mrad. The values of the flux and power given above are then increased by a factor of 20.

Summary: radiation from a circular arc

We summarize the salient properties of the radiation from a circular arc as follows:

1. The radiation power is particularly intense at the moment when the instantaneous electron velocity points directly towards the observer, since at that instant the Doppler effect is maximal.

2. This glimpse of radiation dies away when the angle between the direction to the observer and the electron velocity is of order γ^{-1}.

3. A typical frequency in the spectrum is γ^3 times the cyclic frequency of the orbiting electron in the storage ring.

4. The on-axis radiation is linearly polarized in the horizontal plane, whereas a circularly component is obtained out of the orbit plane, with opposite helicities above and below the plane.

5. The radiation is pulsed, the pulse duration being the electron bunch length divided by c.

2.4 Insertion devices

There is a much more efficient way to produce X-ray beams from a synchrotron than by having the electrons orbiting in a purely circular arc. In a typical storage ring there are straight sections followed by circular arc segments. In any one of these straight sections a device can be inserted that forces the electron to execute oscillations in the horizontal plane as it traverses through the section. This is achieved by an array of magnets which produces a field that alternates from up to down along the path. These so-called insertion devices can be divided into two types.

Wigglers

As far as the trajectory of an electron is concerned a wiggler can be viewed as a series of circular arcs, turning successively to the left and to the right, as shown in Fig. 2.7. This leads to an enhancement in the intensity of the observed radiation by a factor of $2N$, where N is the number of periods. The spectrum from a wiggler is the same as that from a bending magnet of the same field strength. The formula for the emitted power is similar to that given by Eq. (2.8), except for one important difference. In a bending magnet the field $\mathbf{B}$ is constant along the length L, whereas in a wiggler the square of the average field is $\langle \mathrm{B}^2 \rangle = \mathrm{B}_0^2/2$, where B_0 is the maximum field. As a consequence Eq. (2.8) is altered to read

$$\mathcal{P}[\mathrm{kW}] = 0.633\, \mathcal{E}_e^2[\mathrm{GeV}]\, \mathrm{B}_0^2[\mathrm{T}]\, L[\mathrm{m}]\, I[\mathrm{A}]$$

The observed path length L of the electron is approximately equal to the length of the wiggler, which is typically of order 1 m. The radiated power is then of order 1 kW or more. Such a high heat load would distort, if not destroy, the optical performance of the perfect crystals that are used to monochromate the X-ray beam, and various methods have had to be devised to retain the optical quality. A dramatic illustration of the power of the X-ray beam from a wiggler at a third-generation synchrotron source is shown in Fig. 2.6. Here the intensity of the beam is so high that it ionizes the air rendering the path of the beam visible.

Figure 2.6: The white X-ray beam from the wiggler at the ID11 beamline, ESRF. The X-ray beam emerges from an evacuated beam tube with an intensity that is high enough to ionize air. (Image courtesy of Åke Kvick, ESRF.)

Undulators

In a wiggler the intensities of the radiation from each wiggle are added up to give the resulting intensity. However, it is possible to construct an insertion device such that the radiation emitted by a given electron at one oscillation is in phase with the radiation from the following oscillations. This implies that the amplitudes of the radiated waves are first added, and then the sum is squared to obtain the resulting intensity. A necessary condition for this to be true is that the electrons in the insertion device execute small amplitude oscillations on a scale set by γ^{-1}. The coherent addition of amplitudes is only valid at one particular wavelength (and its harmonics), and this has the further advantage that undulator radiation is quasi-monochromatic.

A comparison of the performance of a wiggler and undulator is given schematically in Fig. 2.7, and below we derive the essential equations describing undulator radiation.

2.4.1 Undulator radiation

For bending magnet radiation the basic parameters were γ and the cyclic frequency ω_0. For undulator radiation, the basic parameters are γ and the undulator spatial period λ_u. In addition we need something to characterize the amplitude of the oscillations. It could be the amplitude itself, but it turns out to be more convenient to use the maximum angular deviation from the undulator axis, as indicated in the drawing in the box on page 43. This maximum angle is some dimensionless number, of order unity and denoted K, times the natural opening angle for synchrotron radiation, γ^{-1}. Thus, in addition to γ and λ_u,

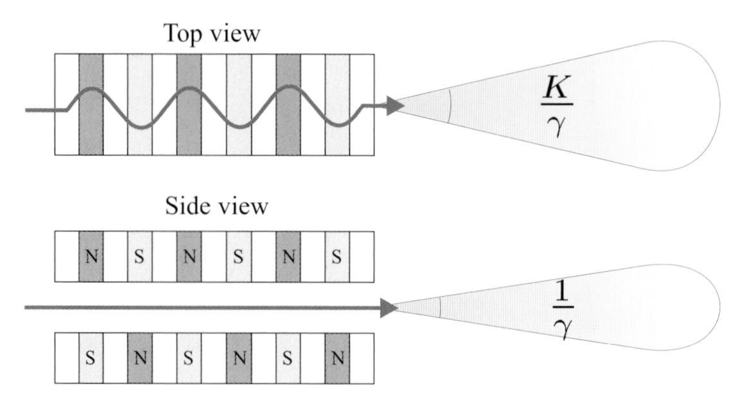

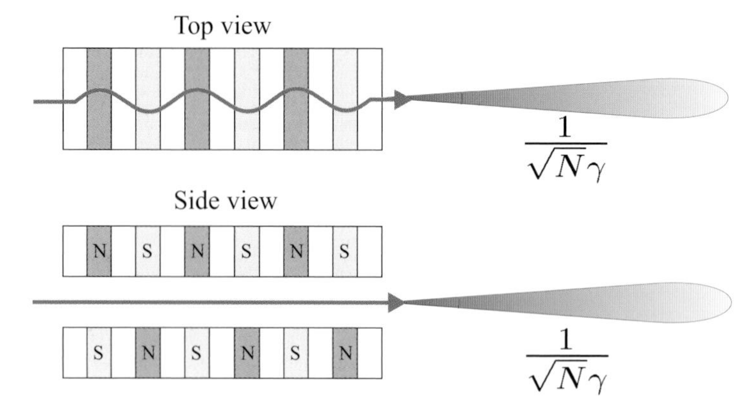

Figure 2.7: Radiation from a wiggler (top) and from an undulator (bottom). The difference in the performance of these devices arises from the difference in the maximum angle of the electron oscillations in the horizontal plane: K is around 20 for a wiggler, and 1 for an undulator. The consequence is that the radiation cone from an undulator is compressed by a factor of approximately $1/\sqrt{N}$ (Eq. (2.15)) relative to the natural opening angle of synchrotron radiation, $1/\gamma$. The number of periods N is typically around 50.

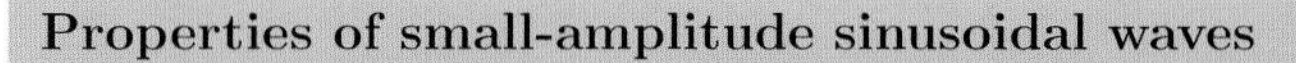

Properties of small-amplitude sinusoidal waves

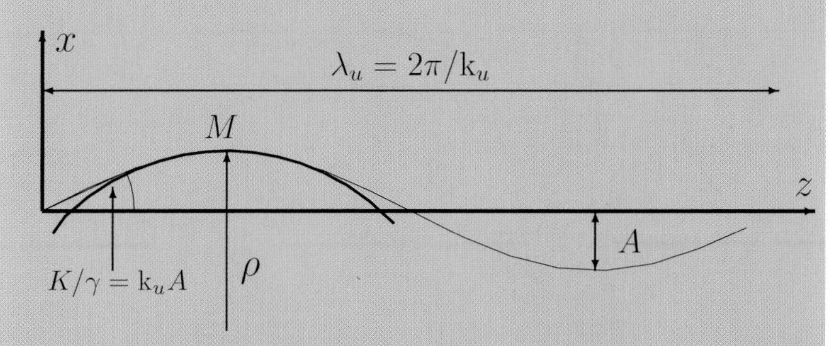

In the vicinity of M we can approximate the cosine wave by a circle of radius ρ, which for the amplitude $A \ll \lambda_u$ can be related to A and k_u from the following considerations:

Circle: $\quad x + (\rho - A) = \sqrt{\rho^2 - z^2} \quad \Rightarrow \quad x \approxeq A - \frac{1}{2}\frac{z^2}{\rho}$

Cosine path: $\quad x = A\cos(\mathrm{k}_u z) \quad \Rightarrow \quad x \approxeq A - \frac{A}{2}\mathrm{k}_u^2 z^2$

Identifying the two expressions leads to the result $\rho \approxeq (A\mathrm{k}_u^2)^{-1}$.

The electron path length S for one period of the undulator is evaluated as

$$S \cdot \lambda_u = \int ds \qquad = \int \sqrt{1 + (dx/dz)^2}dz$$
$$\approxeq \lambda_u[1 + (A\mathrm{k}_u)^2/4] \quad = \lambda_u[1 + K^2\gamma^{-2}/4]$$

we shall use the parameter K defined in this way to characterize the undulator. From Eq. (2.5) we have an expression for ρ in terms of the magnetic field, and from the drawing in the box on this page we have an alternative expression for ρ in terms of $K\gamma^{-1}$, so one finds K readily in terms of the maximum magnetic field B_0 in the undulator as:

$$K = \frac{e\mathrm{B}_0}{mc\mathrm{k}_u} = 0.934\,\lambda_u\,[\mathrm{cm}]\,\mathrm{B}_0\,[\mathrm{T}] \tag{2.9}$$

where $\mathrm{k}_u = 2\pi/\lambda_u$.

The fundamental wavelength, λ_1

An expression for the fundamental wavelength λ_1 in the undulator spectrum is now derived. It is a simple matter to find the relation between λ_1 and the undulator period λ_u. Consider as shown in Fig. 2.8 one undulation of the

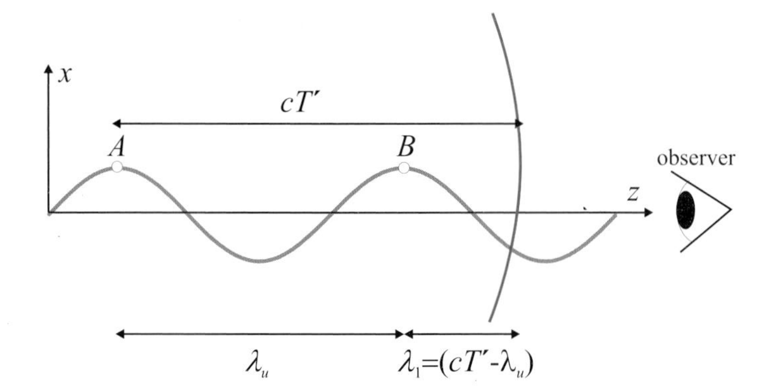

Figure 2.8: Constructive interference occurs when the wavefront emitted by the electron when it was at A is one wavelength λ_1 ahead of the wavefront emitted by the electron when it reaches B. This is then the fundamental wavelength emitted from the undulator.

electron path. At laboratory time $t' = 0$ the electron is at point A. The electron is one undulation further downstream at $t' = T'$. The signal from A is then at a position cT' and the condition for coherence is that $(cT' - \lambda_u)$ is one wavelength λ_1 (or a multiple thereof). The electron path length between A and B is a factor S larger than the period λ_u, so $T' = S\lambda_u/\text{v}$. In the box on the preceding page the path length is derived for one period in the limit of small amplitude oscillations. Altogether then we find

$$\lambda_1(\theta = 0) = \lambda_u(\frac{S}{\beta_e} - 1) \quad \xrightarrow[S=1+\gamma^{-2}K^2/4]{} \quad \frac{\lambda_u}{2\gamma^2}(1 + \frac{K^2}{2}) \tag{2.10}$$

Here it has been tacitly assumed that the direction of observation is on-axis – if the observation direction had been at angle θ with the undulator axis, then $(S/\beta_e - 1)$ should be substituted by $(S/\beta_e - \cos\theta)$ with the result

$$\lambda_1(\theta) = \lambda_u(\frac{S}{\beta_e} - \cos\theta) \quad \xrightarrow[S=1+\gamma^{-2}K^2/4]{} \quad \frac{\lambda_u}{2\gamma^2}(1 + \frac{K^2}{2} + (\gamma\,\theta)^2) \tag{2.11}$$

It has been shown above that γ^{-2} is of order 10^{-8}, so that with λ_u of order 1 cm, λ_1 becomes of order an Ångström, and is hence in the X-ray region. It is also important to note that the radiated, first-order wavelength λ_1 is tunable – by changing the magnetic field by varying the gap between the poles, one changes K according to Eq. (2.9), and thereby the wavelength in accordance with Eq. (2.10). Somewhat counter-intuitively a larger field produces a softer X-ray fundamental.

Higher harmonics

Let us now look at the time dependence of the transverse electron oscillations, both in terms of the laboratory time t' and in terms of the retarded time t. The fundamental differential equation relating t' to t has already been derived (Eq.

(2.4)). Here it is rewritten in terms of a unit vector $\mathbf{n}$ pointing towards the observer, and the instantaneous velocity vector $\boldsymbol{\beta}_e$:

$$\frac{dt}{dt'} = 1 - \mathbf{n} \cdot \boldsymbol{\beta}_e(t')$$

It is important to distinguish between the angle φ in the horizontal plane where the undulations occur, and the vertical angle ψ, the combination of which gives the total deviation angle θ. The geometry is shown in the top part of Fig. 2.9.

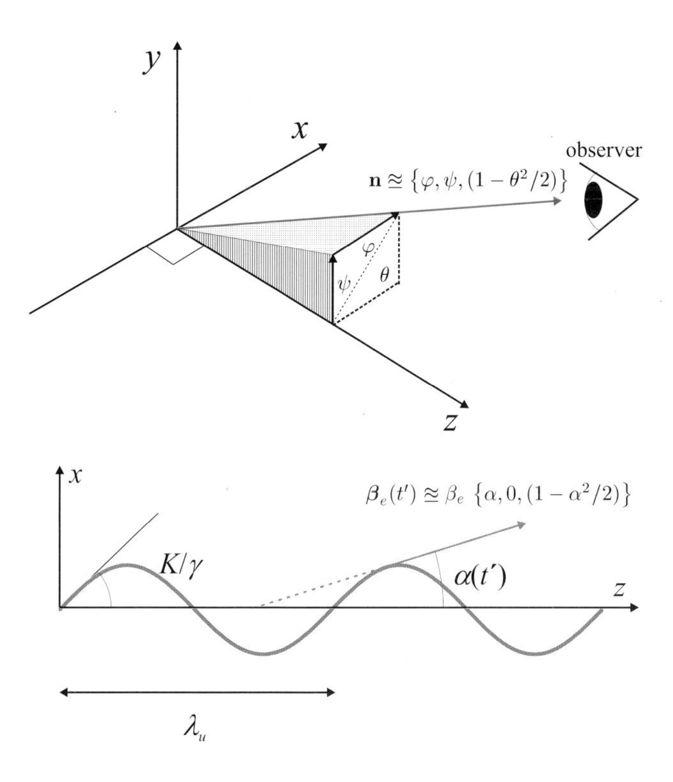

Figure 2.9: The electron undulations take place in the horizontal $x - z$ plane. The direction to the observer is at angle ψ from the horizontal plane and at a horizontal angle φ from the undulator axis. The resulting angle θ is then given by $\theta^2 = \psi^2 + \varphi^2$ and the unit vector $\mathbf{n}$ has the coordinates as indicated. The other vector of interest is the velocity vector $\beta_e(t')$. Its angle with the z-axis varies sinusoidally, or rather as a $\cos(\omega_u t')$, with a maximum value of K/γ.

Since $\mathbf{n}$ is a unit vector it has the coordinates

$$\mathbf{n} = \left\{\varphi, \psi, \sqrt{1-(\varphi^2+\psi^2)}\right\} \approxeq \{\varphi, \psi, (1-\theta^2/2)\}$$

The components of the velocity vector can be written in terms of the instantaneous angular deviation $\alpha(t')$ as

$$\boldsymbol{\beta}_e(t') = \beta_e \left\{\alpha, 0, \sqrt{1-\alpha^2}\right\} \approxeq \beta_e \ \{\alpha, 0, (1-\alpha^2/2)\}$$

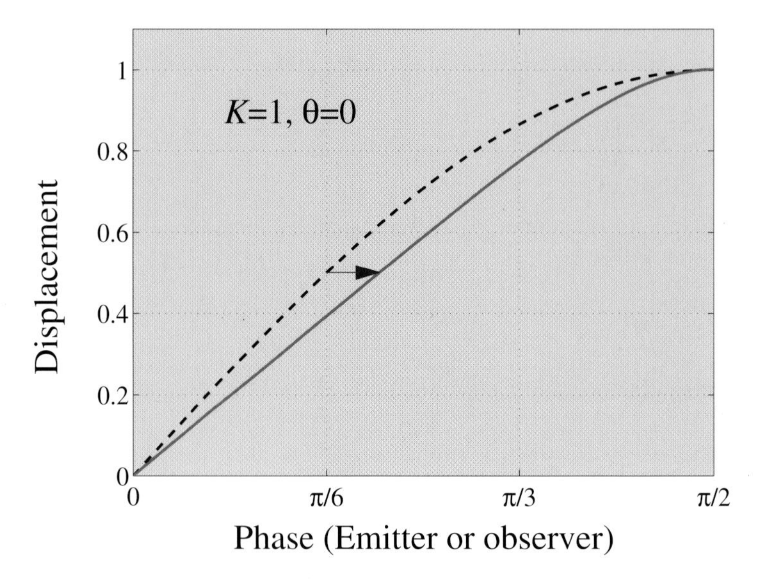

Figure 2.10: Construction to show the relationship between the displacement in the laboratory time frame of the emitting electron (dashed sinusoidal line) to that in frame of the observer (solid line), the retarded time frame. For example, the difference between the emitter, $\omega_u t'$, and observer, $\omega_1 t$, phases at the point $\omega_u t' = \pi/6$ is represented by the arrow (see Eq. (2.14)).

so that the differential equation for dt/dt' becomes

$$\begin{aligned} \frac{dt}{dt'} &= 1 - \mathbf{n} \cdot \boldsymbol{\beta}_e(t') \approxeq 1 - \beta_e \left[\alpha\,\varphi + (1 - \theta^2/2 - \alpha^2/2)\right] \\ &\approxeq 1 - (1 - \gamma^{-2}/2)\left[\alpha\,\varphi + (1 - \theta^2/2 - \alpha^2/2)\right] \\ &\approxeq \frac{1}{2}\left[\gamma^{-2} + \theta^2 + \alpha^2(t')\right] - \alpha(t')\,\varphi \end{aligned} \tag{2.12}$$

where $\beta_e = 1 - \gamma^{-2}/2$ from Eq. (2.3). The solution to this equation is

$$\begin{aligned} \omega_1 t \;\; &= \;\; \omega_u t' + \frac{K^2/4}{[1 + (\gamma\theta)^2 + K^2/2]} \sin(2\omega_u t') \\ &\quad - \frac{2K\gamma}{[1 + (\gamma\theta)^2 + K^2/2]}\,\varphi\,\sin(\omega_u t') \end{aligned} \tag{2.13}$$

The derivation is given in the box on the next page.

We can use the solution given in Eq. (2.13) to estimate quantitatively the content of higher harmonics in the undulator spectrum in the following way. The electron displacement varies sinusoidally in laboratory time, but the observed displacement will, in general, have a different time dependence. Only in the limit $K \to 0$ will t and t' be proportional, $\omega_1 t = \omega_u t'$, and only in this limit will the displacement also appear sinusoidal to the observer. Let us discuss a numerical example to understand how the observer in general experiences deviations from harmonic time variation. For simplicity the example is restricted to the case of on-axis radiation, i.e. $\theta = 0$, and a value of K=1 is chosen as the undulation

Solution of the differential equation (2.12)

With a sinusoidal path $x = (K\gamma^{-1})\mathrm{k}_u^{-1}\sin(\mathrm{k}_u z)$, which has the required maximal deviation angle $(dx/dz)_{\max} = (K\gamma^{-1})$, one finds that the general deviation angle α is

$$\alpha \approxeq \tan(\alpha) = \frac{dx}{dz} = (K\gamma^{-1})\cos(\mathrm{k}_u z) = (K\gamma^{-1})\cos(\omega_u t')$$

Therefore $\alpha^2/2$ in Eq. (2.12) can be expressed as

$$\frac{\alpha^2}{2} = \frac{1}{2}(K\gamma^{-1})^2\cos^2(\omega_u t') = \frac{1}{4}(K\gamma^{-1})^2[1+\cos(2\omega_u t')]$$

and the differential equation reads

$$\frac{dt}{dt'} = \frac{\gamma^{-2}}{2}[1+(\gamma\theta)^2+K^2/2] + \frac{(K\gamma^{-1})^2}{4}\cos(2\omega_u t') \\ - (K\gamma^{-1})\varphi\cos(\omega_u t')$$

This expression can be simplified by introducing the parameter χ, where

$$\chi = [1+(\gamma\theta)^2+K^2/2]$$

and multiplying both sides in the differential equation with $\omega_u dt' = d(\omega_u t')$ we obtain

$$\omega_u dt = \frac{\gamma^{-2}}{2}\chi d(\omega_u t') + \frac{\gamma^{-2}}{2}\frac{K^2}{2}\cos(2\omega_u t')d(\omega_u t') \\ - \frac{\gamma^{-2}}{2}(2K\gamma\varphi)\cos(\omega_u t')d(\omega_u t')$$

Further, by introducing the frequency ω_1 where

$$\omega_1 = \omega_u(2\gamma^2/\chi)$$

and multiplying by $2\gamma^2/\chi$ gives

$$d(\omega_1 t) = d(\omega_u t') + (K^2/2)\chi^{-1}\cos(2\omega_u t')d(\omega_u t') \\ - 2K\gamma\chi^{-1}\varphi\cos(\omega_u t')d(\omega_u t')$$

which is readily integrated to give

$$\omega_1 t = \omega_u t' + (K^2/4)\chi^{-1}\sin(2\omega_u t') - 2K\gamma\chi^{-1}\varphi\sin(\omega_u t')$$

parameter. The solution Eq. (2.13) then reads

$$\omega_1 t = \omega_u t' + \left(\frac{1}{4}\right)\left(\frac{2}{3}\right)\sin(2\omega_u t') = \omega_u t' + \left(\frac{1}{6}\right)\sin(2\omega_u t') \tag{2.14}$$

The displacement is proportional to $\sin(\omega_u t')$, and in arbitrary units the displacement is as shown by the dashed curve in Fig. 2.10, when plotted against the emitter phase $\omega_u t'$. When plotted against the observer phase $\omega_1 t$ the result is the solid curve. The two curves of course coincide when $\omega_1 t = \omega_u t'$, i.e. at $\omega_u t' = 0$ or at $\pi/2$, but differ at other points. For example, when $\omega_u t' = \pi/6$, the displacement is $\sin(\pi/6) = 1/2$, and $\omega_1 t$ is larger than $\omega_u t'$ by the amount $\sin(\pi/3)/6$, as indicated by the arrow. Hence by writing a simple computer program it is possible to generate the displacement as it appears to the observer which can then be resolved into Fourier components. The Fourier components of the apparent acceleration can then be calculated, since from elementary considerations it is expected that the apparent acceleration is proportional to the square of the frequency times the apparent displacement. Then as the amplitude of the observed radiation is proportional to the apparent acceleration, the observed undulator spectrum for any K and for any angle (φ, ψ) can be computed.

For an on-axis observer (θ=0) we plot in Fig. 2.11⋆, upper left, the transverse displacement for K =1, 2 and 5 both versus the laboratory time t' (or rather the emitter phase $\omega_u t'$), where we have a simple sinusoidal variation, and versus retarded time t (or rather the observer phase $\omega_1 t$), where we note that the displacement deviates more and more from a sinusoidal variation the larger the value of K. This implies immediately that the frequency spectrum is dominated by the first harmonic at low K, but as K increases the spectrum acquires a successively greater content of higher harmonics.

Another point of interest is the symmetry of the electrons displacement as seen from the on-axis observer: the curves are symmetric around $\pi/2$ and $3\,\pi/2$. This implies that all even harmonics vanish. This symmetry is lifted when the observer is off-axis. The upper right part of Fig. 2.11⋆ shows the electron movement as seen from the observer in the off-axis case with $\psi = 0, \varphi = \theta = \gamma^{-1}$, and for $K = 2$. Now it is apparent that the symmetry around $\pi/2$ and $3\pi/2$ is broken, and as a result even harmonics appear in the spectrum.

Monochromaticity and angular divergence

So far we have only discussed one oscillation in the undulator, although in deriving the coherence condition (Eq. (2.10)) it was tacitly assumed that if there was coherence between waves from A and B, then there was also coherence with the waves from all other oscillations in the undulator path. There is however only a finite number of periods, N, in the undulator. Suppose that we consider a wavelength, or a frequency for that matter, which has a relative deviation of ϵ from the coherence condition:

$$\omega = \omega_1(1+\epsilon)$$

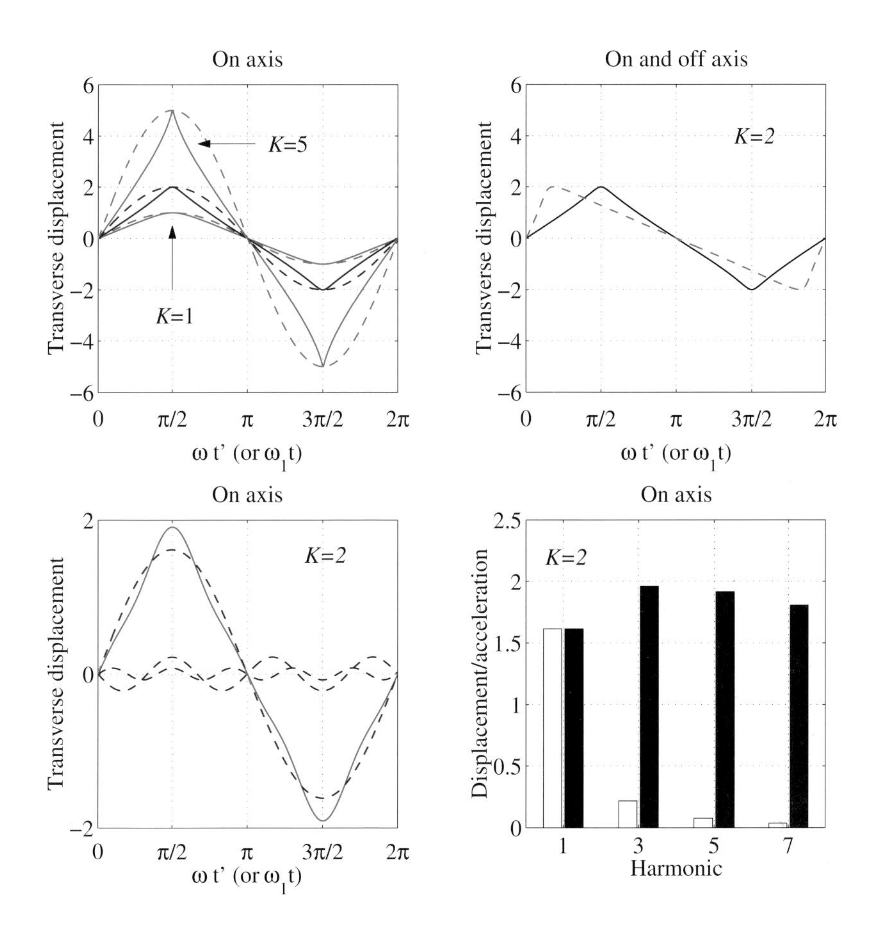

Figure 2.11: ⋆ Undulator characteristics. Top left: Sinusoidal displacement vs. time in the lab frame (dashed lines) for K=1, 2 and 5 are observed on axis as progressively more triangular shaped displacements (full lines). Top right: For $K = 2$ the curve as seen on-axis is shown as the solid line, but off-axis in the horizontal plane ($\psi = 0$) at $\varphi = 1/\gamma$ the shape is no longer symmetric around $\omega_1 t = \pi$ (dashed line) and the Fourier transform contains even harmonics. Bottom left: The on-axis K=2 curve is decomposed into first, third and fifth harmonics, the sum being shown as the full line. Bottom right: The composition of harmonics in the displacement curve for $K = 2$ is represented by the white boxes. The corresponding composition in acceleration, shown in black, is derived by multiplying by the square of harmonic number.

In summing up the amplitudes from all periods we obtain the result from one period multiplied by

$$S_{N,1} \equiv \sum_{m=1}^{N} \mathrm{e}^{i\,m2\pi\epsilon} = \mathrm{e}^{i\,2\pi\epsilon}\left(\frac{1-\mathrm{e}^{i\,2\pi\epsilon N}}{1-\mathrm{e}^{i\,2\pi\epsilon}}\right) = \mathrm{e}^{i(N+1)\pi\epsilon}\left(\frac{\sin(\pi N\epsilon)}{\sin(\pi\epsilon)}\right)$$

or

$$|S_{N,1}| = \frac{\sin(\pi N\epsilon)}{\sin(\pi\epsilon)}$$

The summation of phase factors is described in the box on page 51. This result

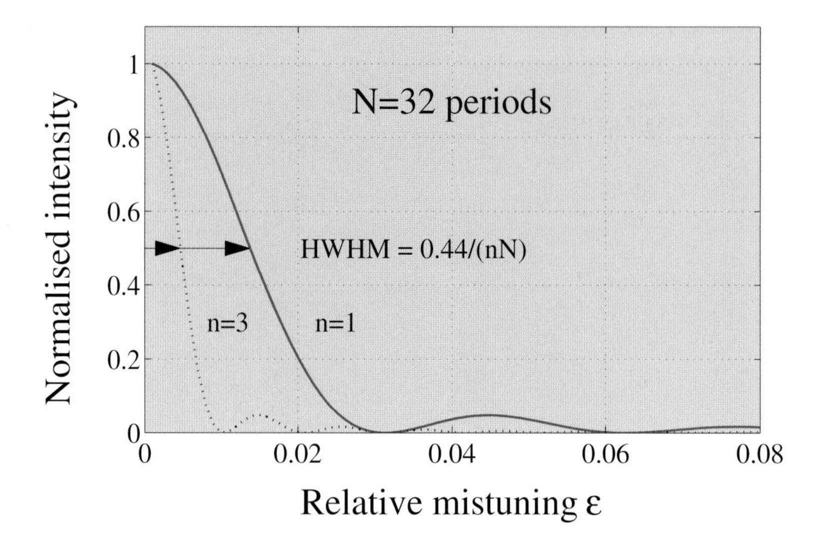

Figure 2.12: Monochromaticity of first and third harmonic for an undulator with N =32 periods as seen through an on-axis pin hole from a zero emittance source.

can be generalized to the n'th harmonic to give

$$|S_{N,n}| = \frac{\sin(\pi N n \epsilon)}{\sin(\pi n \epsilon)} \quad \text{with} \quad \omega = n\omega_1(1+\epsilon)$$

The intensity is proportional to $|S_{N,n}|^2$ and this is normalized by N^2 and plotted in Fig. 2.12 for $N = 32$ with n=1 and n=3. The full width at half maximum (FWHM) is around $0.88/nN$. In other words the monochromaticity $(\lambda/\Delta\lambda)$ in undulator radiation is directly proportional to the number of periods N and also to the higher harmonic number n.

An additional important feature of off-axis observation is the change in the coherence condition, which means that a finite θ (see Fig. 2.9) corresponds to an offset in wavelength. Quantitatively it follows from Eq. (2.10) and (2.11) that

$$\lambda_1(\theta) = \lambda_1(0)\left[1 + \frac{(\gamma\theta)^2}{1+K^2/2}\right] \equiv \lambda_1(0)\left[1+\epsilon_\theta\right]$$

so a certain value of θ corresponds to a relative offset ϵ_θ in wavelength. We saw above that a relative mistuning of the frequency or wavelength by ϵ implied a FWHM in the interference intensity of approximately $1/nN$. It can therefore be concluded that the FWHM in θ must fulfill the equation

$$\epsilon_\theta = \frac{(\gamma\theta_{\text{FWHM}})^2}{1+K^2/2} \approxeq \frac{1}{nN}$$

which can be rearranged to give

$$\boxed{\theta_{\text{FWHM}} \approxeq \frac{1}{\gamma}\sqrt{\frac{1+K^2/2}{nN}}} \tag{2.15}$$

Phase factor summation and the geometrical series

Throughout this book we shall be interested in evaluating sums over N phase factors of the form

$$S_N(x) = \sum_{n=0}^{N-1} e^{i\,2\pi n x}$$

with x a continuous variable. This is nothing other than the geometrical series

$$S_N = \sum_{n=0}^{N-1} k^n = 1 + k + k^2 + \cdots + k^{N-1} = \frac{1-k^N}{1-k}$$

The proof follows once it is realized that $S_N - S_{N-1} = k^{N-1}$ and $kS_{N-1} + 1 = S_N$. The sum is convergent in the limit $N \to \infty$ if and only if $|k| < 1$, for which

$$S_\infty = \frac{1}{1-k}$$

We can now evaluate the sum over phase factors as

$$\begin{aligned} S_N(x) &= \frac{1 - e^{i\,2\pi N x}}{1 - e^{i\,2\pi x}} = \frac{e^{-i\,\pi N x} - e^{i\,\pi N x}}{e^{-i\,\pi x} - e^{i\,\pi x}} \frac{e^{i\,\pi N x}}{e^{i\,\pi x}} \\ &= \frac{\sin(\pi N x)}{\sin(\pi x)} e^{i(N-1)\pi x} \end{aligned}$$

and below we plot its modulus squared.

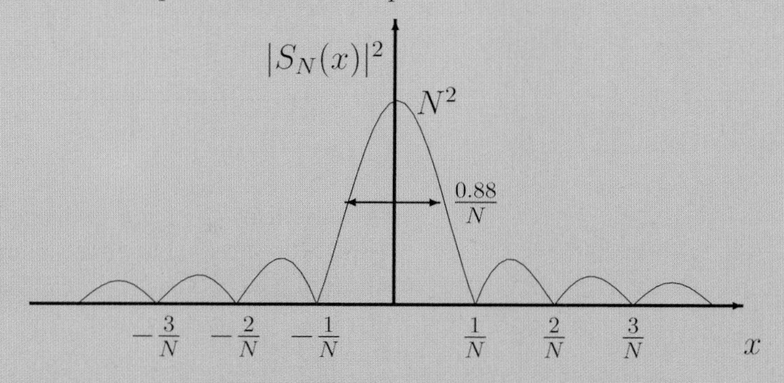

The full width at half of the maximum value (FWHM) is approximately $0.88/N$.

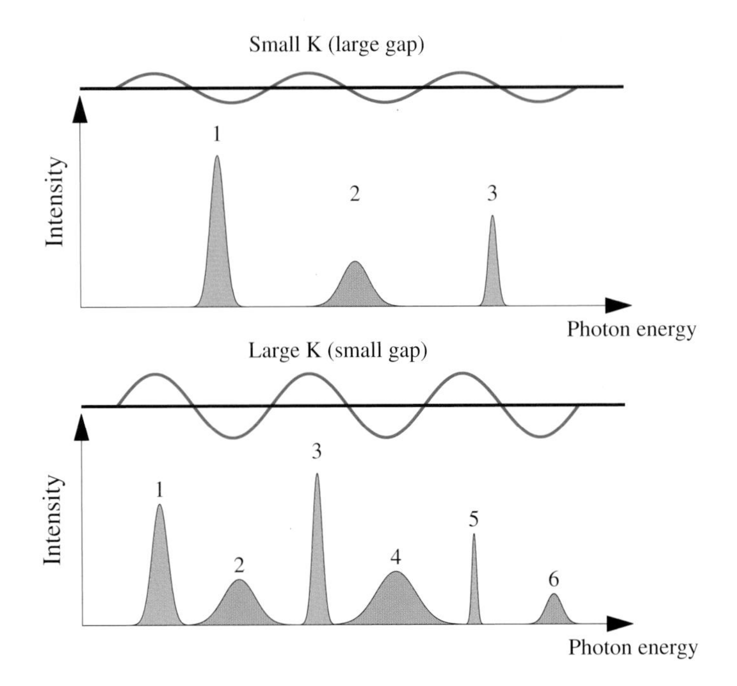

Figure 2.13: Schematic of the spectrum from an undulator. The energy of the harmonics can be tuned by K so that a larger gap, implying a lower field and thereby a smaller K, gives a higher energy.

Note that this is a substantial reduction compared to the natural divergence γ^{-1}, and that it is independent of the azimuthal angle relative to the undulator axis.

In comparing with experiments one has to consider the effect of the finite angular divergence of the *electron* beam. This can be added in quadrature to the intrinsic divergence as given by Eq. (2.15) in order to obtain the observed divergence. The electron beam divergence within the plane of undulation is usually different from that in the perpendicular direction, so the observable synchrotron radiation divergence will not be symmetric around the undulator axis.

Summary: undulator radiation

The salient features of undulator radiation can be summarized as follows:

1. The undulator is characterized by the K parameter (proportional to the peak magnetic field), the period λ_u, and the number of periods N.

2. The on-axis spectrum has a fundamental peak in wavelength given by Eq. (2.10), and has odd harmonics with a relative width of $1/nN$ (FWHM). The higher the K value, the higher is the relative proportion of the harmonics.

3. The intrinsic angular divergence of the fundamental (and the odd harmonics) is much smaller than γ^{-1} and is given by Eq. (2.15).

4. The finite divergence of the electron beam implies that the on-axis spectrum contains contributions from finite values of φ, and thus has intensity also at energies corresponding to the even harmonics.

These properties are shown schematically in Fig. 2.13. The ideal X-ray source is monochromatic with a tunable energy. The angular divergence of the beam should be small, preferably in all directions. The main power of the source beam should be in a quasi-monochromatic band, so that the heat load on the first optical elements in the beamline is not unduly high. The undulator beam from a third-generation synchrotron storage ring has all of these desired properties.

Helical undulators

The *linear* undulator that has been discussed so far in this section is the most common insertion device used at present day synchrotron facilities, but it is by no means the only one. In some applications it is useful to have an undulator with specific characteristics, for example one that is capable of producing circularly polarized instead of linearly polarized radiation. The *helical* undulator is such a device, and it is instructive to consider the difference between a linear and helical undulator in terms of the higher-harmonic content in their spectra.

The content of harmonics of undulator radiation can be understood qualitatively by considering the apparent electron acceleration that an on-axis observer experiences. First, recall the linear undulator. In the lab frame the electron performs a harmonic, sinusoidal motion. But it is not sinusoidal seen from the on-axis observer. Here the Doppler shift is a little different when the electron is at its maximum displacement (where its instantaneous direction is exactly towards the observer), compared to when it is passing the undulator axis (where its instantaneous direction makes the angle K/γ with the direction to the observer). Therefore the observer does not see a sinusoidal displacement versus time; he sees the sinusoidal curve distorted more towards a triangular shape, which of course can be resolved into Fourier components (see Fig. 2.11). By symmetry the on-axis observer will see odd harmonics only (in contrast to the off-axis observer who will also see even harmonics). Clearly the distortion from sinusoidal shape increases with increasing K, so the content of higher harmonics also increases with K. Now consider the helical undulator. In the lab frame the electron performs a spiral. Seen from the on-axis observer though, the electron trajectory describes a circle, and by the symmetry of the circle there are no points where the Doppler effect is more pronounced than at other points. In other words the observer sees the same Doppler shift all the time, and therefore no distortion of the circular motion. Thus the only difference between the lab frame, projected onto a plane perpendicular to the undulator axis, and the observer's retarded time frame, is the scale of timing: the observer sees the circular movement as being much faster. But there is no distortion, and hence there are no harmonics. As the electrons execute a circular path, the emitted radiation is circularly polarized.

2.5 Emittance and the diffraction limit

In the last section it was shown that an undulator is a highly brilliant source of radiation. The question naturally arises whether the brilliance of an undulator, or indeed any source, can be increased without limit. According to its definition in Eq. (2.1) the brilliance is inversely proportional to the square of the product of the linear source size and angular divergence. The product of source size and divergence is known as the *emittance*, ε, of a source. In this section the lower limit of the emittance of the photon beam is discussed. This turns out to be determined by the *convolution* of the emittance of the electron beam circulating in the storage ring, and the emittance of the photon beam for the passage of a single electron through the source path that is visible to the observer.

For a synchrotron storage ring, the electron beam emittance is a constant along the orbit around the ring. This is a consequence of Liouville's theorem, which states that for beams of particles the product of beam size and divergence is a constant. Although the product of source size and angular divergence is a constant, the two individual components may be manipulated by magnetic fields. It is therefore convenient to represent the electron beam emittance at a given position around the ring by a plot in phase space, with the spatial coordinate along the abscissa and the divergence along the ordinate, as shown in Fig. 2.14(a). Here z is the spatial coordinate in the vertical direction, and z' is the angular divergence in the same direction. In the figure the source size and divergence are written as σ_z and σ_z' respectively, and the contour representing the root-mean squared (r.m.s.) value of these quantities is shown as an ellipse. It follows from Liouville's theorem that the phase-space ellipse has a constant area around the orbit, which can be tilted by magnetic fields as indicated in the figure. Typical values for the electron beam parameters of an undulator at the ESRF in the vertical direction are $\sigma_z = 10.3\,\mu$m and $\sigma_z' = 3.8\,\mu$rad. The vertical emittance is thus $\varepsilon_z = \sigma_z \sigma_z' = 39$ pm rad. The ratio of the emittances in the vertical and horizontal directions is known as the *coupling*. For the ESRF the coupling is currently chosen to be 1 %, and the horizontal emittance is a factor of 100 times larger than in the vertical direction. In the future the coupling is likely to be reduced by a factor of between two to four.

The concept of a phase-space ellipse is also a convenient way to visualize the properties of the X-ray photon beam. Let us discuss this for an undulator source by considering the passage of a single electron. The angular divergence is given in Eq. (2.15), which for the first harmonic can be re-written as

$$\theta_{\text{FWHM}} \approxeq \frac{1}{\gamma}\sqrt{\frac{1+K^2/2}{N}} = \sqrt{2}\sqrt{\frac{\lambda_1}{L}}$$

where Eq. (2.10) has been used to relate the undulator length $L = N\lambda_u$ to the X-ray photon wavelength λ_1 of the first harmonic. Converting from FWHM to r.m.s. introduces a factor of $2\sqrt{2\ln 2} \approx 2.355$, but neglecting the difference between that and the factor of $\sqrt{2}$ in the equation above, leads to the expression for the r.m.s. photon beam divergence of

$$\sigma_r' \approx \sqrt{\frac{\lambda}{L}}$$

On the other hand, we know that the photon source size, σ_r, can never be smaller than the value set by the diffraction limit. Beyond the diffraction limit

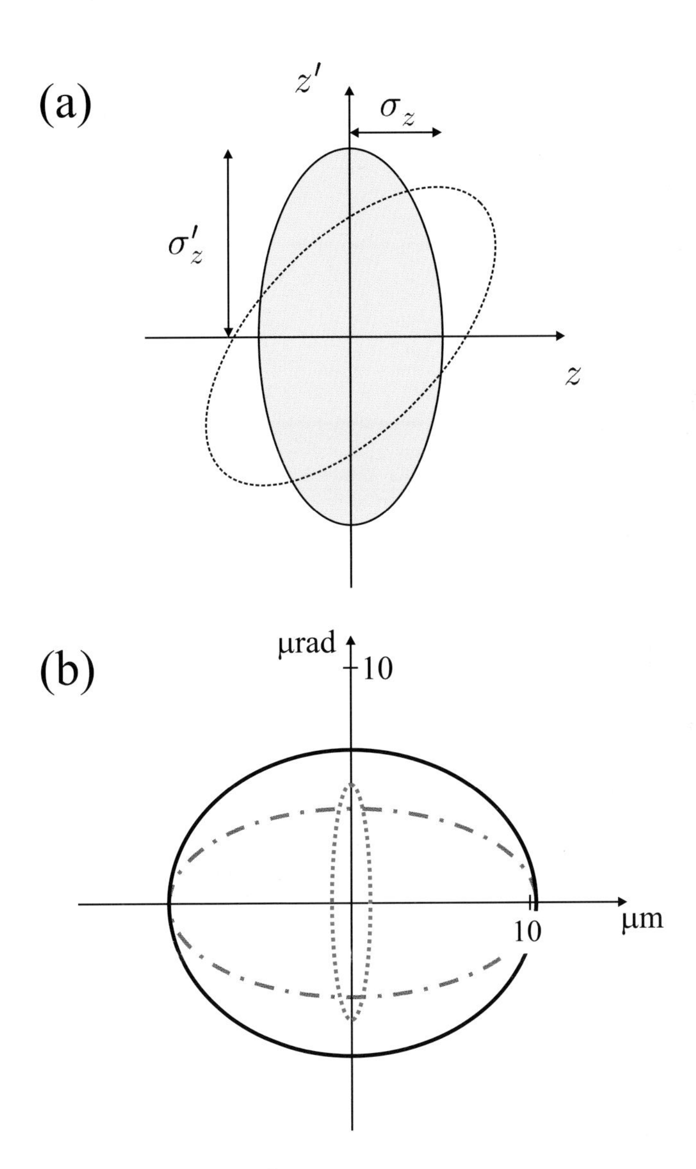

Figure 2.14: Phase-space representation of the emittance ellipse, where the abscissa is the spatial dimension, z, and the ordinate is the divergence, z'. The emittance is defined to be the product of source size and divergence. (a) For the electron beam in the vertical direction, z, the emittance is written as $\varepsilon = \sigma_z \sigma'_z$. It is a constant around the orbit of the ring, and may therefore be represented at different points on the orbit by ellipses of equal area. (b) The photon beam emittance from an undulator at the ESRF in the vertical direction. Dotted line: the diffraction limit of a 1 Å photon beam arising from the passage of a single electron through a 4 m undulator. Dashed-dotted line: electron beam parameters in the vertical direction for a bunch of electrons in the storage ring. Full line: the phase-space ellipse of the resulting photon beam.

any reduction in source size leads to an increase in source divergence, and *vice versa*. The condition for the diffraction limit can be obtained from Heisenberg's uncertainty relation. In the present context this is written as

$$\sigma_r \Delta \mathrm{p} \geq \frac{\hbar}{2}$$

where Δp is the uncertainty in the momentum of the photon. This in turn can be related to the angular divergence of the photon beam through

$$\begin{aligned}\Delta \mathrm{p} &= \hbar \Delta \mathrm{k} \\ &= \hbar\, \mathrm{k}\, \sigma'_r = \hbar \frac{2\pi}{\lambda} \sqrt{\frac{\lambda}{L}} \\ &= \hbar \frac{2\pi}{\sqrt{L\lambda}}\end{aligned}$$

It follows that the diffraction limited source size and angular divergence for an undulator are

$$\boxed{\sigma_r = \frac{\sqrt{L\lambda}}{4\pi} \quad \text{and} \quad \sigma'_r = \sqrt{\frac{\lambda}{L}}}$$

To take a definite example we consider an undulator at the ESRF, and show how the emittance of the electron beam and the diffraction limit of the photon beam combine to produce the final emittance of the X-ray beam. From the above, a 4 m long undulator at a wavelength of 1 Å has σ_r= 1.6 μm and σ'_r= 5 μrad. The source size and divergence of the electron beam in the vertical direction were stated earlier as σ_z= 10.3 μm and σ'_z= 3.8 μrad. These are shown graphically in Fig. 2.14(b). The phase-space ellipse for the X-ray photon due to passage of a single electron through the undulator is represented by the dotted line, while the ellipse from the passage of all the electrons in a bunch (approximately 10^{11} electrons) is shown by dashed-dotted line. The resulting photon beam pulse is the convolution of the two, as indicated by the solid line. From this it is clear that a further reduction in the electron beam size would be beneficial for the brilliance of the resulting photon beam, but a reduction in the electron beam divergence will not make much difference. Similar considerations apply to the horizontal direction.

If the electron beam phase-space ellipse could be made considerably smaller than the diffraction limit ellipse, the source would have full transverse coherence (Section 1.5). This is difficult to achieve at a synchrotron source for both the vertical and horizontal directions. For that it is necessary to consider a radically different type of source, the free-electron laser, which is described in the next section.

2.6 The free-electron laser

Although the undulator at a synchrotron source has many desirable properties it could nonetheless be improved upon considerably. The reason is that although the radiation from a single electron is coherent, in the sense that the radiation from one oscillation is in phase with that from the following ones, the radiation from different electrons is incoherent. This results from the fact that the electrons traverse the undulator in a bunch without any positional order, in other words as an electron gas. If somehow the electrons in the bunch (or macro-bunch) could be ordered spatially into smaller micro-bunches, with a separation equal to the X-ray wavelength, then the radiation from one micro-bunch would be in phase with that from all of the following micro-bunches. It follows that the charge, q, in a single micro-bunch would be much larger than e, and, since the micro-bunch is confined spatially within a distance shorter than the emitted wavelength, this charge can be considered as point like.

In an undulator the radiation field increases from zero at the entrance to its full value at the exit. An electron traversing through an undulator experiences the force from the magnetic lattice of the insertion device, and after some distance it begins to coherently feel the fields from the other electrons in the bunch. The interaction with the electric field is spatially modulated with a period equal to the X-ray wavelength, and will hence tend to modulate the electrons within a bunch into micro-bunches. Once this effect occurs it will, through boot-strapping, enhance itself, as the radiation field increases rapidly as the electrons move downstream. The mechanism is called Self Amplified Stimulated Emission, or SASE for short, and an undulator designed to exploit the SASE principle is known as a free-electron laser [Derbenev et al., 1982, Murphy and Pellegrini, 1985]. It has already been demonstrated in the infrared, and at the time of writing projects are underway to extend the wavelength range down to the UV and soft X-ray regime. Ultimately it may be possible to reach wavelengths around 1 Å in which case we will have a pulsed X-ray source which is completely coherent in the transverse plane, and with a brilliance many orders of magnitude higher than the radiation from third-generation storage rings (see Fig. 1.1). A schematical representation of the SASE effect is shown in Fig. 2.15.

The boot-strapping mechanism requires that the radiation field acting on the electrons is sufficiently strong, and the more localized the electron gas bunch is, the stronger will be the electric field. The electron gas density is therefore a decisive parameter for the SASE mechanism to function. Even in small-emittance, third-generation storage rings the electron density is not sufficiently high. This is partly because the horizontal beam width is too large, and in particular because the bunch length, of order 100 psec times c, or about 30 mm, is far too long. One solution to this problem is to use a linear accelerator, usually abbreviated to LINAC. In a LINAC it is possible to produce ultra small electron beams, of order 100 μm (FWHM) in diameter, with a very small angular divergence, of order 1 μrad. Most importantly, with recently developed electron guns and LINAC bunch compression devices one can obtain the required high electron density, and bunch times as low as 0.1 psec, corresponding to a bunch length of 30 μm. With such a compressed electron gas volume the SASE principle should work even for radiation in the hard X-ray regime.

Estimating the expected flux from a free-electron laser requires detailed numerical calculations, as the SASE mechanism is more complicated than our

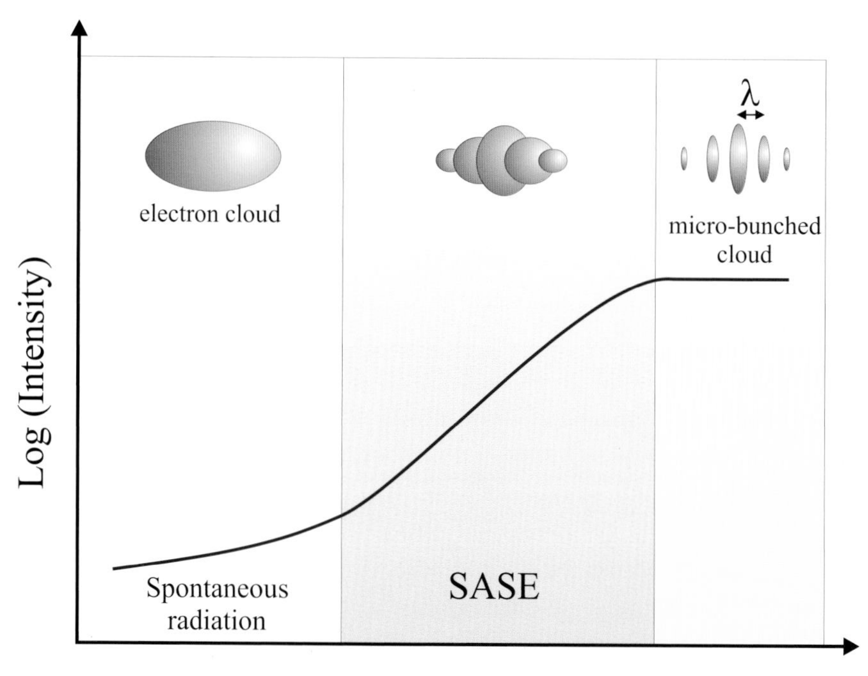

Figure 2.15: The intensity of the radiation emitted from an electron cloud as it traverses through a long undulator. For short distances along the undulator there is no correlation between the radiation emitted by the different electrons in the cloud. Each electron emits as a coherent source, and the intensity of the "spontaneous radiation" is proportional to the number of electrons in the cloud. Further downstream the electrons start to form micro-bunches and the SASE effect switches on, leading to a large increase in intensity. Eventually a train of micro-bunches forms with a spacing equal to the X-ray wavelength. Each micro-bunch can be regarded as a point-like charge, so that in the ideal case the intensity is now proportional to the *square* of the number of electrons in a microbunch.

description above suggests. As the electron bunch enters the undulator, the electrons are distributed like in a gas. There will be spontaneous fluctuations in the electron density within the bunch, and the region that happens to have a slightly higher density than the average will spontaneously be the seed for SASE growth of a linearly modulated electron density with a wavelength equal to the first harmonic (or possibly the third harmonic) X-ray wavelength of the undulator. This region will *not* grow to extend over the entire macro-bunch; its size will be limited by the distance where the photon field will be in phase with the electrons that created it. Numerical estimates indicate that the length of an ordered region will be some hundred wavelengths. Within the total length of the bunch there will be many such partly ordered regions, but the radiation from different regions will be incoherent. Superimposed on this SASE radiation, which is extremely brilliant, is the radiation from all the electrons that are not ordered in any way, and this is like the ordinary undulator radiation that was discussed in an earlier section.

Further reading

The Feymann Lectures on Physics, Vol. 1, Ch. 34, Richard P. Feynman, Robert B Leighton, and Matthew Sands (Addison-Wesley, 1977)

Brightness, Coherence and Propagation Characteristics of Synchrotron Radiation K.-J. Kim, Nucl. Instrum. Methods Phys. Res. **A246**, 71 (1986)

3. Refraction and reflection from interfaces

A ray of light propagating in air changes direction when it enters glass, water or other transparent materials. This is the basis for the classical optics of lenses. Quantitatively, the phenomenon is described by Snell's law. For visible wavelengths the refractive index n of most transparent materials has a value in the range between 1.2 and 2. The refractive index depends on the frequency ω of the light, so that blue light is refracted more than red light, etc.

The index of refraction for electromagnetic waves displays resonant behaviour at frequencies corresponding to electronic transitions in atoms and molecules. On the low frequency side of a resonance, n increases with ω, and this is known as normal dispersion. Immediately above the resonance frequency it decreases, and as more and more resonances are passed, the magnitude of the index of refraction decreases. X-ray frequencies are usually higher than all transition frequencies, perhaps with the exception of those involving the inner K- or maybe L-shell electrons. As a result in the X-ray region n turns out to be *less* than unity. As we shall see, this reflects the phase shift of π in the Thomson scattering of X-rays. Furthermore, it leads to the phenomenon of total external reflection from a flat, sharp interface: for incident glancing angles α below a certain critical angle α_c the ray will no longer penetrate into the material but will be totally reflected from it. The deviation of n from unity is tiny, so the critical angle is small. The reader might wonder how n can be less than unity, since the velocity in the material is c/n, and this would seem to imply that the speed of light is higher in the material than in vacuum. However, c/n is the phase velocity, not the group velocity. The latter, evaluated as $d\omega/d\mathrm{k}$, is indeed less than c.

In this chapter we shall see that the deviation of n from unity, δ, is related to the scattering properties of the medium. Each electron scatters the X-ray beam with the Thomson scattering amplitude r_0, and δ turns out to be proportional to the product of r_0 and the electron density ρ. With the explicit formula derived below, Eq. (3.1), one finds that δ is of order 10^{-5}. Snell's law evaluated at small glancing angles implies that the critical angle $\alpha_c = \sqrt{2\delta}$, and therefore is of order of a few milli-radians. Although this is small, it transpires that it is sufficient in practice to allow the production of highly reflecting X-ray mirrors, which can be shaped so as to focus an incident X-ray beam. The glancing angle geometry implies a long focal length, of order 10 meters, and rather long mirrors, since the "footprint" of the beam on the mirror is the beam height divided by the sine of the glancing angle. This problem can be overcome with a multilayer mirror, which has peaks in its reflectivity curve at angles well beyond the critical angle, allowing shorter mirrors to be used. The fact that δ is so small would seem to make the construction of refractive lenses in the X-ray region unfeasible. This turns out not to be the case, and we shall expand on this, and other aspects of

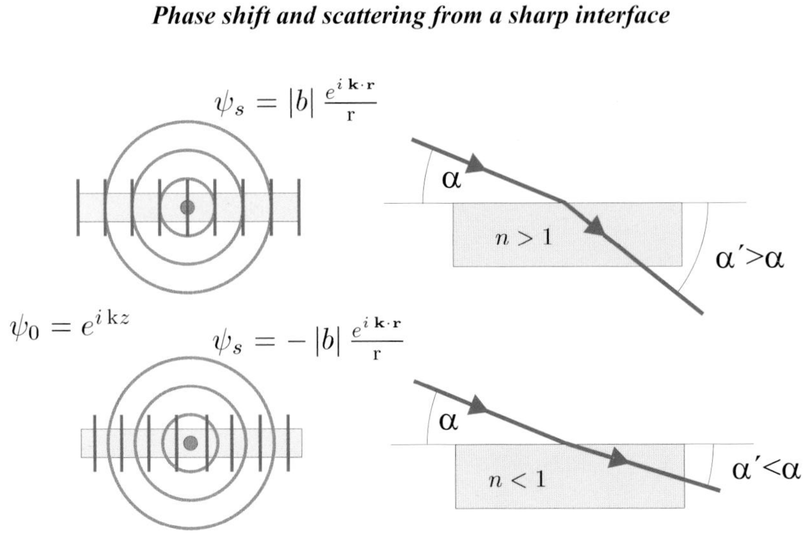

Phase shift and scattering from a sharp interface

Figure 3.1: The spherical wave ψ_s from a point scatterer can be in phase (top) or 180° out of phase with the incident plane wave ψ_0 (bottom). The refractive index is greater than unity in the first case, and smaller than unity in the second case.

X-ray optical elements, towards the end of this chapter.

In general X-ray reflectivity is a very powerful probe of the structure of interfaces. The reference interface is the sharp, flat interface for which the reflectivity is given by the Fresnel equations. Although these may be well known to the reader in the optical region, we shall derive them here and see how they are simplified in the X-ray region. Real interfaces are rarely sharp on a length scale of an Ångström, and neither are they ideally flat. Most importantly, one can determine deviations from the ideal sharp, flat interface by reflectivity studies. The ideally flat, but graded interface, and the ideally sharp, but roughened interface, will be considered in later sections, followed by examples of reflectivity experiments where these ideas have been an essential component of the models used to interpret the data. The interesting part of the spectrum of the reflected intensity versus incident glancing angles is often at reflectivities below 10^{-6}, so an intense beam from a synchrotron source is often of advantage.

3.1 Refraction and phase shift in scattering

In this first section absorption processes are neglected, and it is assumed that the interface between vacuum and the medium of interest is both flat and sharp. The medium has a homogeneous density of scatterers each giving rise to a spherical scattered wave of real amplitude b.

As shown in Fig. 3.1 there are two possibilities for the spherical wave emanating from a scattering centre: either it is in phase with the incident wave (top), or there is a phase shift of π in the scattering process (bottom). A phase shift cannot be detected in an ordinary scattering experiment, since the intensity is proportional to the absolute square of the scattering length b. But, as we shall prove in the next section, there is a distinct difference in the refraction at an interface. With no phase shift the refractive index n is larger than unity, and the ray is therefore refracted as shown in the top, right part of Fig. 3.1. With a phase shift of π, the index of refraction becomes smaller than unity and the phenomena of *total external reflection* occurs at sufficiently small glancing angles α. The argument does not depend on the kind of radiation considered – for example, it also applies to beams of neutrons. This is the reason why the nomenclature of b has been used for the amplitude of the spherical wave, since the nuclear scattering length for neutrons is commonly denoted by b. Furthermore, the sign of b varies from nucleus to nucleus. A well known example is the deuteron and the proton, which have nuclear scattering lengths of opposite sign. However, here we are mainly interested in X-rays, and in that case the scattering length for each electron is r_0, and as was shown in Chapter 1 there is a phase shift of π between the incident and scattered waves. The refraction of neutrons is discussed further in Appendix F.

Below we derive the equation relating the index of refraction, n, to the scattering properties of the medium given by the number density of electrons, ρ, and the scattering amplitude per electron, r_0. The equation is

$$\boxed{n = 1 - \delta} \tag{3.1}$$

with

$$\boxed{\delta = \frac{2\pi\,\rho\,r_0}{\mathrm{k}^2}} \tag{3.2}$$

and where the wavelength of the radiation, λ, is related to the wavevector, k, by $\mathrm{k} = 2\pi/\lambda$. The electron density ρ in condensed matter is of order 1 electron/Å^3. This means that with r_0= 2.82× 10^{-5} Å and k around 4 Å^{-1}, δ is of the order of 10^{-6}. This is very much smaller than unity and explains why refraction phenomena in the X-ray region are not completely trivial to observe[1]. Snell's law relates the glancing angles α and α' defined in Fig. 3.1 to each other through the equation

$$\boxed{\cos\alpha = n\,\cos\alpha'}$$

which is also derived below. The critical angle α=α_c for total external reflection is obtained by setting α'=0°, and by expanding the cosines to yield

$$\alpha_c = \sqrt{2\delta} = \frac{\sqrt{4\pi\,\rho\,r_0}}{\mathrm{k}} \tag{3.3}$$

Using the values of the typical parameters given above, α_c is of order one milliradian.

[1]W.C. Röntgen had a suspicion that X-rays were waves, and attempted to look for refraction phenomena, but without success.

3.2 Refractive index and scattering length density

In order to derive the relationships stated in Eqs. (3.1) and (3.2) between the refractive index of a material and its scattering properties we consider in Fig. 3.2 a plane wave at normal incidence to a thin plate. The presence of the plate is sensed at the observation point P by a change in the wave ψ^P_{tot} compared to the situation without a plate. For X-rays ψ^P_{tot} describes the electric field, for neutrons it is the Schrödinger wavefunction. The notation of ψ is used to emphasize the similarity of X-ray and neutron optics. The derivation is simplified by considering normal incidence, since then the wave does not change direction when entering the material, and for a thin plate any phase difference between waves scattered from the front or the back of the plate can be neglected. There are then two equivalent descriptions: on the one hand a refractive description where the presence of the plate is taken into account by a phase difference of $(n-1)\,\mathrm{k}\,\Delta$; and on the other hand a scattering description, where the wave at P is a superposition of the infinitesimal spherical waves emanating from each scattering centre in the plate, plus of course the incident wave. For the refractive description the total wave at P is

$$\psi^P_s = \psi^P_0\,\mathrm{e}^{i(n-1)\mathrm{k}\Delta} \approxeq \psi^P_0[1+i(n-1)\mathrm{k}\Delta] \tag{3.4}$$

A quantitative evaluation of the scattering description requires a little work. First of all, the incident plane wave is approximated by a train of spherical waves coming from a very distant source point S. For convenience, let the observation point P be the point symmetric to S on the other side of the plate. Next, let us consider the waves scattered from an element in the plate, lying within the plane of the drawing at a distance x from the axis. The distance to the source and to the observation point from this element, R, is a little longer than the closest distance R_0.

By expansion we find $R = (R_0^2+x^2)^{1/2} \approxeq R_0\left[1+x^2/(2R_0^2)\right]$, and the phase difference compared to the direct path from S to P is $2\mathrm{k}x^2/(2R_0)$, where the first factor of 2 accounts for equal phase lags on the source and observation sides of the plate. A similar expression is obtained for an element at coordinate $(0,y)$, so that the phase difference, $\phi(x,y)$, for rays emanating from the element at (x,y) is

$$\mathrm{e}^{i\,\phi(x,y)} = \mathrm{e}^{i\,(x^2+y^2)\mathrm{k}/R_0} = \mathrm{e}^{i\,x^2\mathrm{k}/R_0}\mathrm{e}^{i\,y^2\mathrm{k}/R_0}$$

The number of scattering centres in that element is $\rho\,\Delta\,dxdy$, and each wave has the scattering amplitude r_0 with a phase shift of π. Altogether, the contribution $d\psi^P_s$ to the scattered wave at P from the volume element at (x,y) is

$$d\psi^P_s \approxeq \left(\frac{\mathrm{e}^{i\,\mathrm{k}R_0}}{R_0}\right)(\rho\Delta\,dxdy)\left(-b\frac{\mathrm{e}^{i\,\mathrm{k}R_0}}{R_0}\right)\mathrm{e}^{i\,\phi(x,y)}$$

as indicated in Fig. 3.2. The subscript "s" stands for "scattered" to remind ourselves that, after integration over all elements in the plate, we have to add the scattered wave to the incident wave in order to obtain the total. The scattered

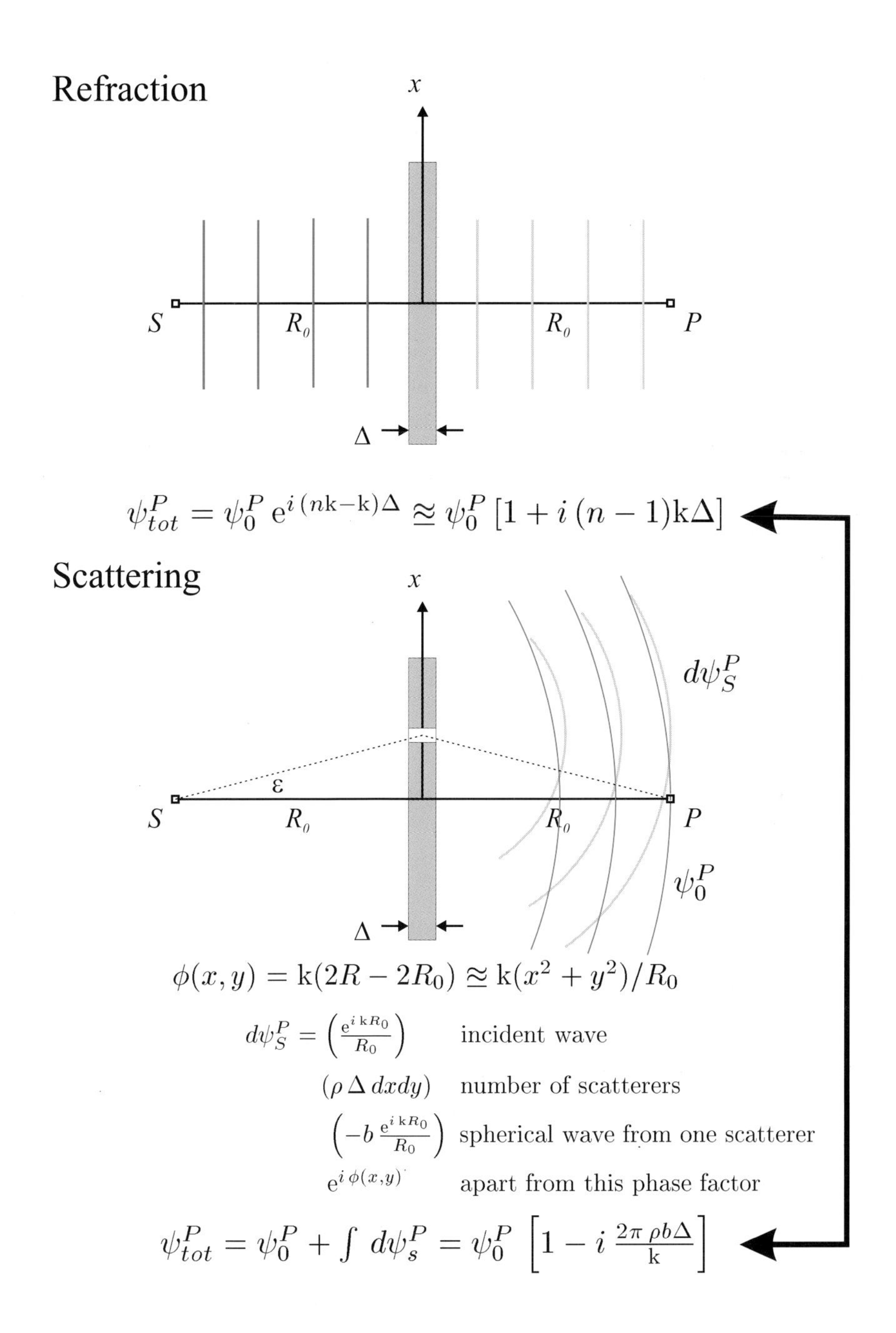

Figure 3.2: The refractive description (top) implies that the thin plate introduces a small phase shift in the wave observed at point P. In the scattering description (bottom) the incident plane wave is approximated by a point source far away, and the perturbation of the plate is derived as a superposition of scattered spherical waves. (Absorption has been neglected in both treatments). The two descriptions are equivalent and lead to the relation between the refractive index and the scattering properties.

Evaluation of the integral in Eq. (3.5)

Here the integral

$$I = \lim_{L\to\infty} \int_{-L}^{L} \mathrm{e}^{i\,(\mathrm{k}/R_0)\,x^2}\, dx$$

is evaluated. This is achieved by using a theorem from the mathematics of complex functions: *The integral of a function in the complex plane between two points A and B is independent of the path from A to B.*

Therefore instead of going from point A at $-L$ on the real axis to point B at $+L$ on the real axis, we go first from A along the circular path to point P, then from P to the symmetric point P′, and finally along another circular segment from P′ to B, as shown in the Fig. 3.3. On the line PP′ the complex variable z is given in polar coordinates by

$$z = r\,\mathrm{e}^{i\,\pi/4} = r\,\tfrac{(1+i)}{\sqrt{2}} \quad\Rightarrow\quad dz = \tfrac{(1+i)}{\sqrt{2}}\, dr$$

The integrand then becomes $\mathrm{e}^{i\,(\mathrm{k}/R_0)\,z^2} \longrightarrow \mathrm{e}^{-(\mathrm{k}/R_0)\,r^2}$ since $z^2 = i\,r^2$. The integral is then evaluated as

$$\lim_{L\to\infty} \int_{\mathrm{P}}^{\mathrm{P}'} \cdots = \lim_{L\to\infty} \int_{-L}^{+L} \tfrac{(1+i)}{\sqrt{2}}\, \mathrm{e}^{-(\mathrm{k}/R_0)\,r^2}\, dr = \tfrac{(1+i)}{\sqrt{2}}\, \sqrt{\pi}\sqrt{\tfrac{R_0}{\mathrm{k}}}$$

where use has been made of the standard integral:

$$\int_{-\infty}^{\infty} \mathrm{e}^{-r^2}\, dr = \sqrt{\pi}$$

Along the arcs A to P and P′ to B the complex variable z is written in polar coordinates as $z = (L, \theta) \Rightarrow dz = L\, d\theta$. The contribution to I from these arc segments vanishes since

$$\lim_{L\to\infty} \int_{\mathrm{A}}^{\mathrm{P}} \cdots \equiv \lim_{L\to\infty} \int_{\mathrm{P}'}^{\mathrm{B}} \cdots = \lim_{L\to\infty} \int_{\theta=\pi/4}^{\theta=0} \mathrm{e}^{-(\mathrm{k}/R_0)\,L^2}\, L\, d\theta$$

$$= -\lim_{L\to\infty} \mathrm{e}^{-(\mathrm{k}/R_0)\,L^2}\, L\, \left(\tfrac{\pi}{4}\right) \longrightarrow 0$$

Including also the y integration, the square of the integral is

$$I^2 = (1+i)^2 \frac{\pi R_0}{2\mathrm{k}} = i\left(\frac{\pi R_0}{\mathrm{k}}\right)$$

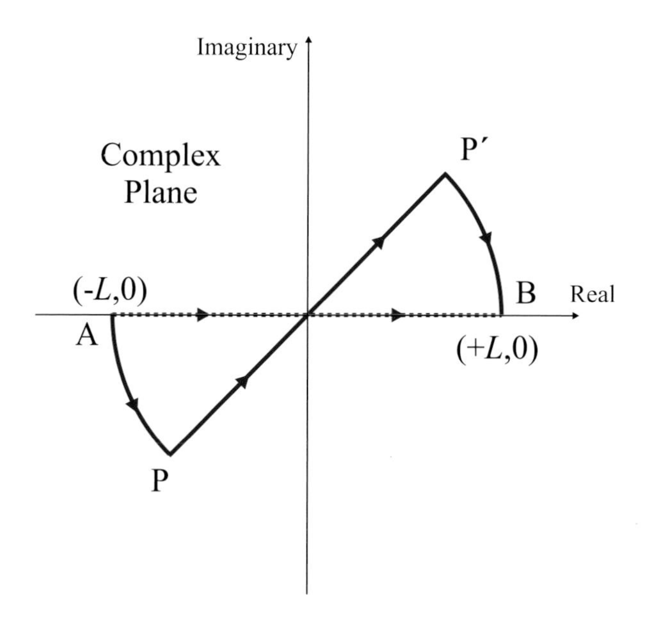

Figure 3.3: Evaluation of the integral in Eq. (3.5) is facilitated by integrating along the path A-P-P′-B in the complex plane instead of along the real axis from $(-L,0)$ to $(+L,0)$.

wave at P is found by integrating the above to yield

$$\begin{aligned}\psi_s^P = \int d\psi_s^P &= -\rho b\Delta\left(\frac{\mathrm{e}^{i\,2\mathrm{k}R_0}}{R_0^2}\right)\int_{-\infty}^{\infty}\mathrm{e}^{i\,\phi(x,y)}\,dxdy\\ &= -\rho b\Delta\left(\frac{\mathrm{e}^{i\,2\mathrm{k}R_0}}{R_0^2}\right)I^2\end{aligned} \tag{3.5}$$

The integration over (x,y) is described in the box on the preceding page, and leads to the result

$$I^2 = \int_{-\infty}^{\infty}\mathrm{e}^{i\,\phi(x,y)}\,dxdy = i\left(\frac{\pi R_0}{\mathrm{k}}\right)$$

The incident wave at P, a distance of $2R_0$ from the source S, is

$$\psi_0^P = \frac{\mathrm{e}^{i\,\mathrm{k}2R_0}}{2R_0}$$

and the total wave at the observation point P becomes

$$\psi_{tot}^P = \psi_0^P + \psi_s^P = \psi_0^P\left[1 - i\,\frac{2\pi\,\rho\,b\,\Delta}{\mathrm{k}}\right] \tag{3.6}$$

When the expression for ψ_{tot}^P from the scattering picture is identified with ψ_{tot}^P in the refractive description (compare the top and bottom panels in Fig. 3.2) one arrives at the result stated in Eq. (3.1).

Our discussion of the forward scattered wave shown in Fig. 3.2 can be extended to include a thin plate composed of atoms, instead of the uniform distribution of electrons that has been considered so far. All that needs to be done is to replace the electron number density ρ in Eq. (3.6) by the product of the atomic number density, ρ_a, and the atomic scattering factor $f(\mathbf{Q})$ (see Eq. (1.9)). The expression for ψ_{tot}^P given in Eq. (3.6) can then be written as

$$\psi_{tot}^P = \psi_0^P \left[1 - i \, \frac{2\pi \, \rho_a f^0(0) \, r_0 \, \Delta}{\mathrm{k}} \right] \tag{3.7}$$

where for the forward direction $\mathbf{Q} = 0$, and the dispersion corrections to $f(\mathbf{Q})$ have been neglected. Later we shall consider the diffraction from atomic planes where the angle of incidence θ is not necessarily 90°. This is taken into account by replacing Δ by $\Delta / \sin\theta$. To emphasize that the effect of a thin plate is to introduce a phase shift of the forward scattered wave the above is rewritten as

$$\psi_{tot}^P = \psi_0^P \left[1 - i \, g_0 \right] \approxeq \psi_0^P \, \mathrm{e}^{-i \, g_0} \tag{3.8}$$

where g_0 is the phase shift given by

$$\boxed{g_0 = \frac{\lambda \rho_a f^0(0) r_0 \Delta}{\sin\theta}} \tag{3.9}$$

The factor of $\sin\theta$ has been introduced to allow for the change in thickness of material traversed as the incident angle is changed. In terms of the atomic density and atomic scattering length δ becomes

$$\delta = \frac{2\pi \, \rho_a \, f^0(0) \, r_0}{\mathrm{k}^2} \tag{3.10}$$

3.3 Refractive index including absorption

Suppose now that in addition to scattering, absorption processes also take place in the medium. Absorption implies that the beam is attenuated in the material with a characteristic 1/e length which is denoted by μ^{-1}, where μ is known as the absorption coefficient. It is important to note that this length refers to the *intensity* attenuation, and not to the *amplitude* attenuation. After traversing a distance z in the material the intensity is attenuated by a factor $\mathrm{e}^{-\mu z}$, but the amplitude only by a factor of $\mathrm{e}^{-\mu z/2}$.

In the refractive description of a wave incident at normal angles to a plate (see Fig. 3.4) the wavevector changes from k in vacuum to nk in the medium. If the refractive index n is now allowed to be a complex number, $n = 1 - \delta + i\beta$, then the wave propagating in the medium is

$$\mathrm{e}^{i \, n \mathrm{k} z} = \mathrm{e}^{i \, (1-\delta) \mathrm{k} \, z} \, \mathrm{e}^{-\beta \mathrm{k} \, z}$$

From this equation for the amplitude it can be inferred that $\beta \, \mathrm{k} = \mu/2$, or

$$\boxed{n \equiv 1 - \delta + i \, \beta} \tag{3.11}$$

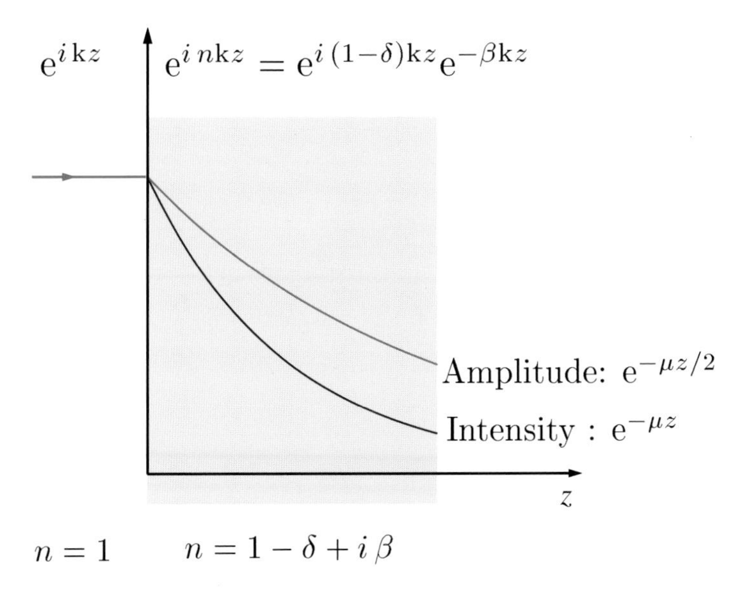

Figure 3.4: A plane wave at normal incidence to a plate with absorption length $1/\mu$. The absorption is formally equivalent to an imaginary part of the refractive index.

with

$$\boxed{\delta = \frac{2\pi\,\rho_a f^0(0)\,r_0}{\mathrm{k}^2}} \tag{3.12}$$

and

$$\boxed{\beta = \frac{\mu}{2\,\mathrm{k}}} \tag{3.13}$$

An alternative approach is to write the atomic scattering length $f(\mathbf{Q})$ as a complex number by including the dispersion corrections (see Chapter 7). The atomic scattering length is then $f(\mathbf{Q}){=}f^0(\mathbf{Q}){+}f'{+}if''$, and the refractive index becomes

$$n \equiv 1 - \frac{2\pi\,\rho_a\,r_0}{\mathrm{k}^2}\left\{f^0(0) + f' + i\,f''\right\}$$

with

$$-\left(\frac{2\pi\,\rho_a\,r_0}{\mathrm{k}^2}\right) f'' = \beta$$

Using Eq. (1.14) and Eq. (3.13) this can be rearranged by writing

$$f'' = -\left(\frac{\mathrm{k}^2}{2\pi\rho_a r_0}\right)\beta = -\left(\frac{\mathrm{k}^2}{2\pi\rho_a r_0}\right)\frac{\mu}{2\mathrm{k}} \tag{3.14}$$

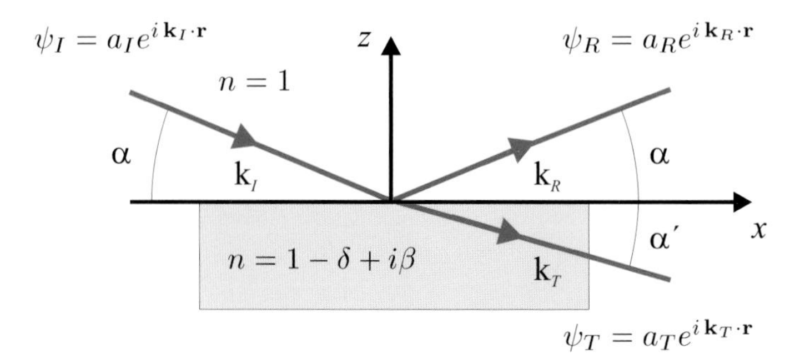

Figure 3.5: Snell's law and the Fresnel equations can be derived by requiring continuity at the interface of the wave and its derivative.

to read

$$\boxed{f'' = -\left(\frac{\mathrm{k}}{4\pi r_0}\right)\sigma_a} \tag{3.15}$$

Thus the absorption cross-section, σ_a, is proportional to the imaginary part of the atomic scattering length, f'', in the forward direction. This result is sometimes known as the optical theorem. It should be noted that f'' is negative since σ_a is a positive real number, and that in other texts the sign convention is sometimes such that f'' is positive.

3.4 Snell's law and the Fresnel equations in the X-ray region

In the X-ray wavelength region, both δ and β are very much smaller than unity. It follows that when considering refraction and reflection phenomena we can limit ourselves to small angles and take advantage of the appropriate expansions.

The incident wavevector is $\mathbf{k}_I$, and the amplitude is a_I, as indicated in Fig. 3.5. Similarly the reflected and the transmitted wavevectors (at angle α') are $\mathbf{k}_R$ and $\mathbf{k}_T$, respectively, and the amplitudes are a_R and a_T. Snell's law and the Fresnel equations are derived by imposing the boundary conditions that the wave and its derivative at the interface $z = 0$ must be continuous. These require that the amplitudes are related by

$$a_I + a_R = a_T \tag{3.16}$$

and

$$a_I\mathbf{k}_I + a_R\mathbf{k}_R = a_T\mathbf{k}_T \tag{3.17}$$

The wavenumber in vacuum is denoted by $\mathrm{k}=|\mathbf{k}_I|=|\mathbf{k}_R|$ and in the material it is $n\mathrm{k}=|\mathbf{k}_T|$. Taking components of $\mathbf{k}$ parallel and perpendicular to the surface yields respectively

$$a_I \mathrm{k} \cos\alpha + a_R \mathrm{k} \cos\alpha = a_T (n\mathrm{k}) \cos\alpha' \tag{3.18}$$

$$-(a_I - a_R) \mathrm{k} \sin\alpha = -a_T (n\mathrm{k}) \sin\alpha' \tag{3.19}$$

From Eq. (3.16) together with the projection parallel to the interface (Eq. (3.18)) one readily derives Snell's law:

$$\boxed{\cos\alpha = n \, \cos\alpha'} \tag{3.20}$$

As α and α' are small the cosines can be expanded to yield

$$\begin{aligned} \alpha^2 &= \alpha'^2 + 2\delta - 2\,i\,\beta \\ &= \alpha'^2 + \alpha_c^2 - 2\,i\,\beta \end{aligned} \tag{3.21}$$

where the refractive index n has been taken from Eq. (3.11), and Eq. (3.3) has been used to relate δ to the critical angle, α_c, for total reflection.

From Eq. (3.16) together with the projection perpendicular to the interface (Eq. (3.19)) it follows that

$$\frac{a_I - a_R}{a_I + a_R} = n \frac{\sin\alpha'}{\sin\alpha} \approx \frac{\alpha'}{\alpha} \tag{3.22}$$

from which the Fresnel equations are derived as

$$\boxed{\begin{aligned} r &\equiv \frac{a_R}{a_I} = \frac{\alpha - \alpha'}{\alpha + \alpha'} \\ t &\equiv \frac{a_T}{a_I} = \frac{2\alpha}{\alpha + \alpha'} \end{aligned}} \tag{3.23}$$

Here the *amplitude* reflectivity, r, and transmittivity, t, have been introduced. The corresponding *intensity* reflectivity (transmittivity), denoted by the capital letter R (T), is the absolute square of the amplitude reflectivity (transmittivity).

Note that α' is a complex number to be derived from Eq. (3.21) for a given incidence angle α. By decomposing α' into its real and imaginary parts

$$\alpha' \equiv Re(\alpha') + i\, Im(\alpha')$$

it can be seen that the transmitted wave falls off with increasing depth into the material as

$$a_T \mathrm{e}^{i\,(\mathrm{k}\alpha')z} = a_T \mathrm{e}^{i\,\mathrm{k}\,Re(\alpha')z} \mathrm{e}^{-\mathrm{k}\,Im(\alpha')z}$$

The *intensity* therefore falls off with a 1/e penetration depth Λ given by

$$\boxed{\Lambda = \frac{1}{2\mathrm{k}\, Im(\alpha')}.} \tag{3.24}$$

The results for r, t and Λ depend on several parameters: the incident angle α, the density and absorption in the medium, as well as the wavevector. In order to get an overview of this multi-parameter problem, it is convenient to use suitable units. The normalization unit for angles is the critical angle α_c. However, in connection with diffraction and reflection phenomena, the wavevector transfers are more useful than angular variables:

$$\mathrm{Q} \equiv 2\,\mathrm{k}\sin\alpha \approxeq 2\,\mathrm{k}\,\alpha \quad ; \quad \mathrm{Q}_c \equiv 2\,\mathrm{k}\sin\alpha_c \approxeq 2\,\mathrm{k}\,\alpha_c \tag{3.25}$$

and in particular their dimensionless counterparts

$$q \equiv \frac{\mathrm{Q}}{\mathrm{Q}_c} \approxeq \left(\frac{2\,\mathrm{k}}{\mathrm{Q}_c}\right)\alpha \quad ; \quad q' \equiv \frac{\mathrm{Q}'}{\mathrm{Q}_c} \approxeq \left(\frac{2\,\mathrm{k}}{\mathrm{Q}_c}\right)\alpha'$$

Equation (3.21) can then be rewritten in terms of the dimensionless wavevectors q and q' by multiplying both sides of the equation by $(2\mathrm{k}/\mathrm{Q}_c)^2$ to yield

$$\boxed{q^2 = q'^2 + 1 - 2\,i\,b_\mu} \tag{3.26}$$

where from Eq. (3.13) the parameter b_μ is related to the absorption coefficient μ through

$$b_\mu = \left(\frac{2\,\mathrm{k}}{\mathrm{Q}_c}\right)^2 \beta = \left(\frac{4\,\mathrm{k}^2}{\mathrm{Q}_c^2}\right)\frac{\mu}{2\mathrm{k}} = \frac{2\mathrm{k}}{\mathrm{Q}_c^2}\,\mu \tag{3.27}$$

The wavevector Q_c at the critical angle is

$$\mathrm{Q}_c = 2\mathrm{k}\alpha_c = 2\mathrm{k}\sqrt{2\delta} = 4\sqrt{\pi\,\rho\,r_0\left(1 + \frac{f'}{Z}\right)} \tag{3.28}$$

as can be seen from Eq. (3.3) and Eq. (3.12). For completeness the dispersion correction f' to f^0 has been included in the expression for Q_c (see Chapter 7 for a complete discussion of the dispersion corrections).

Calculation of the reflectivity, transmittivity and penetration depth proceeds as follows. For the material in question values for the absorption length μ^{-1} (at the particular X-ray wavelength being used), the electron density ρ, and possibly the dispersion correction f', are obtained from standard sources, such as the International Tables of Crystallography. From these numbers the quantity b_μ is calculated. The complex number q' can then be derived from Eq. (3.26), and thereby the complex amplitude reflectivity (transmittivity) from the wavevector form of Eq. (3.23):

$$\boxed{r(q) = \frac{q - q'}{q + q'} \quad ; \quad t(q) = \frac{2q}{q + q'} \quad ; \quad \Lambda(q) = \frac{1}{\mathrm{Q}_c Im(q')}} \tag{3.29}$$

Let us consider the solutions to some limiting cases, recalling that $b_\mu \ll 1$ in all cases.

	Z	Molar density (g/mole)	Mass density (g/cm^3)	ρ (e/Å^3)	Q_c (1/Å)	$\mu \times 10^6$ (1/Å)	b_μ
C	6	12.01	2.26	0.680	0.031	0.104	0.0009
Si	14	28.09	2.33	0.699	0.032	1.399	0.0115
Ge	32	72.59	5.32	1.412	0.045	3.752	0.0153
Ag	47	107.87	10.50	2.755	0.063	22.128	0.0462
W	74	183.85	19.30	4.678	0.081	33.235	0.0409
Au	79	196.97	19.32	4.666	0.081	40.108	0.0495

Table 3.1: Parameters required for reflectivity calculations for selected elements: electron density, ρ; critical wavevector, Q_c; linear absorption coefficient, μ, at λ=1.54051 Å.

$q \gg 1$ The solution to Eq. (3.26) yields $Re(q') \approxeq q$ and $Im(q') \approxeq b_\mu/q$. From Eq. (3.29), $r(q)$ can be written as $r(q) = (q^2 - q'^2)/(q + q')^2$ so in the considered limit $r(q) \approxeq (2q)^{-2}$, i.e. the reflected wave is in phase with the incident wave. The intensity reflectivity falls off as $R(q) \approxeq (2q)^{-4}$, there is almost complete transmission, and the penetration depth is $\alpha\mu^{-1}$.

$q \ll 1$ In this case q' is almost completely imaginary with $Im(q') \approxeq 1$ and $r(q) \approxeq -1$, i.e. the reflected wave is out of phase with the incident wave, so the transmitted wave becomes very weak. It propagates along the surface with a minimal penetration depth of $1/Q_c$, independent of α as long as $\alpha \ll \alpha_c$. Due to the small penetration depth, it is called an *evanescent* wave.

$q = 1$ From Eq. (3.26) one finds $q' = \sqrt{b_\mu}(1 + i)$. The penetration depth is $b_\mu^{-1/2}$ times larger than the asymptotic value of $1/Q_c$. Since $b_\mu \ll 1$ the amplitude reflectivity is close to +1 so the reflected wave is in phase with the incident wave. This implies that the evanescent amplitude is almost twice that of the incident wave.

An overview of the different quantities versus scattering vector or incident angle is given in Fig. 3.6⋆. In Table 3.1 the parameters needed to compute the reflectivity for several elements are given.

It should be emphasised here that the Fresnel reflectivity is *specular*. This means that the reflected intensity is confined to the plane spanned by the incident wavevector and the interface normal, and that within this plane the angle of the reflected beam equals the angle of the incident beam. Non-specular reflectivity is produced by rough surfaces, as we shall explain later in this chapter.

3.5 Reflection from a homogeneous slab

In this section the reflectivity from a slab of finite thickness is considered. This is shown schematically in Fig. 3.7 where it is compared with the case of the infinitely thick medium described in the previous section.

Consider first Fig. 3.7(a) where a wave propagating in medium 0 with refractive index 1 is incident on an infinitely thick medium of refractive index n.

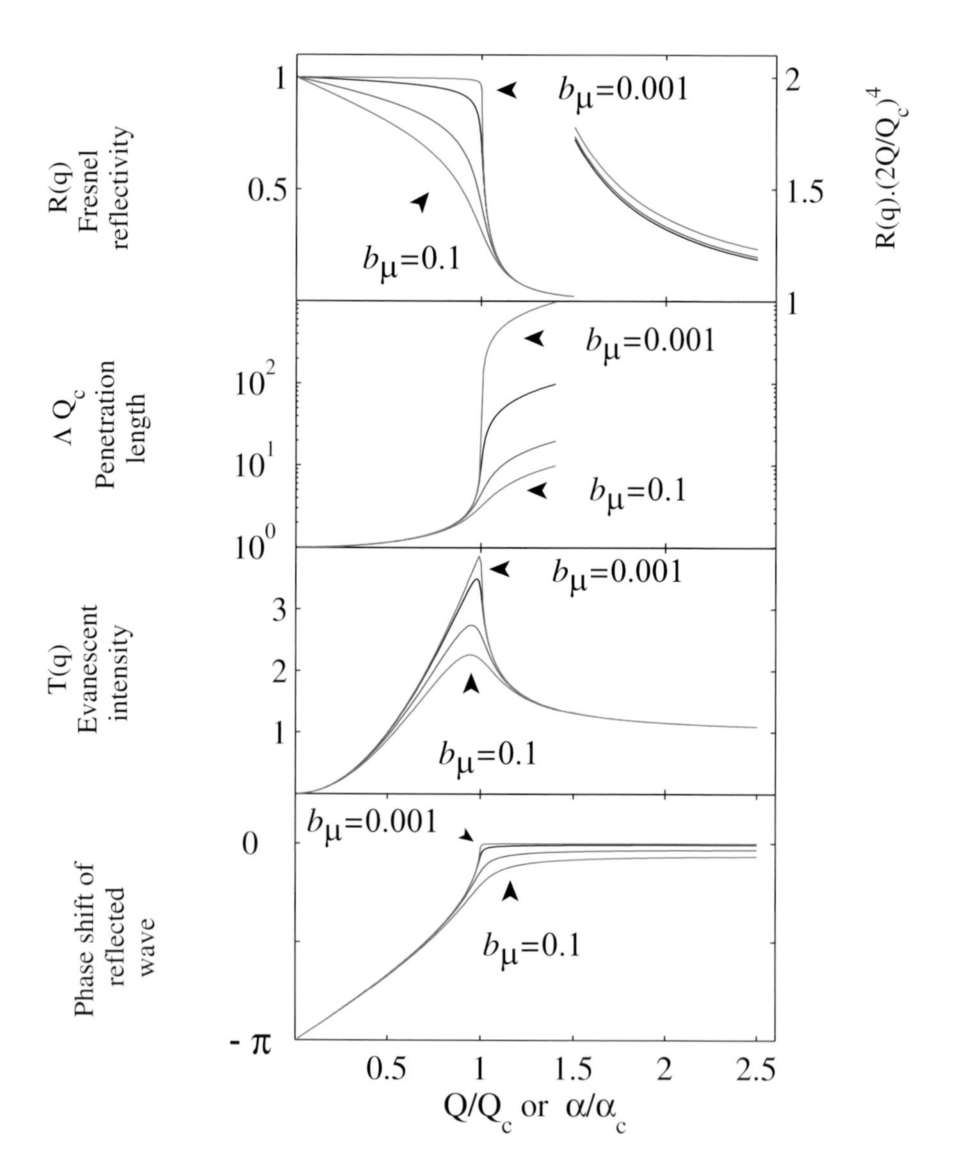

Figure 3.6: ⋆ The intensity reflectivity $R(q)$, the penetration depth $\Lambda\,\mathrm{Q}_c$, the intensity transmittivity $T(q)$ and the phase of the reflected wave, all versus Q/Q_c or α/α_c. In each case a family of curves is given corresponding to different values of the (small) parameter b_μ. The values of $b_\mu (= 2\mu\mathrm{k}/\mathrm{Q}_c^2)$ used were: 0.001, 0.01, 0.05, 0.1. The right hand side of the figure gives the asymptotic behaviour, scaled to the expressions in the text for $q \gg 1$, which by definition approaches unity as $q \gg 1$.

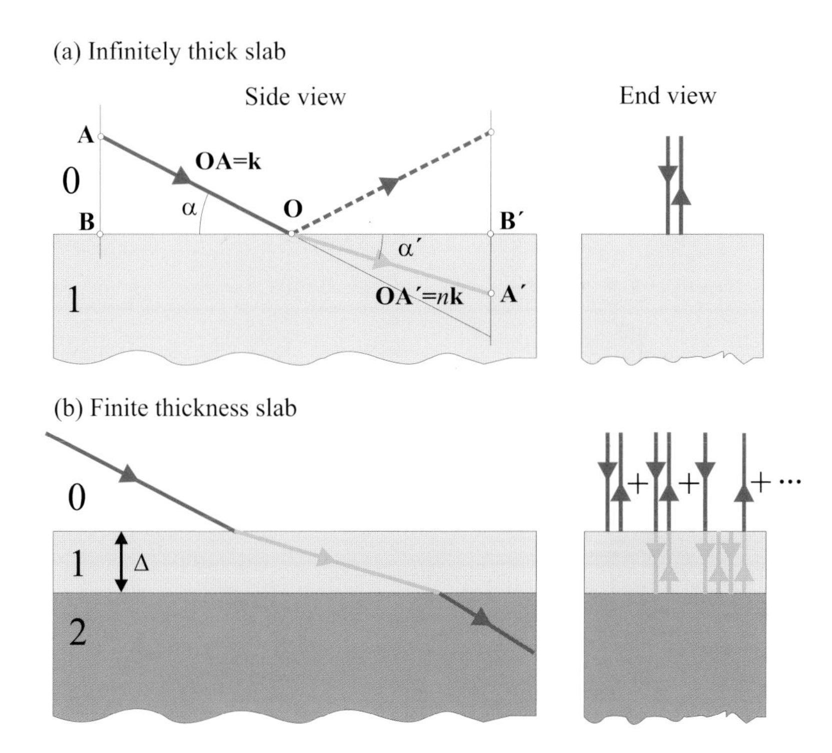

Figure 3.7: Reflection and transmission from a slab of infinite (a) and finite (b) thickness. The finite slab is of thickness Δ and the total reflectivity is the sum of the infinite number of reflections, as indicated in the right panel of (b).

This part of the figure illustrates how Snell's law can be inferred directly from the boundary condition that the waves are continuous at the interface. The incident plane wave, $e^{i\,\mathbf{k}\cdot\mathbf{r}}$, with wavevector **OA**=**k**, can be decomposed into two plane waves with wavevectors along, k_x, and normal, k_z, to the interface: $e^{i\,\mathbf{k}\cdot\mathbf{r}} = e^{i\,\mathrm{k}_x x} e^{i\,\mathrm{k}_z z}$. Continuity at the interface implies that the component of k_x cannot change in going from medium 0 to medium 1, i.e. the waves propagating along x in medium 0 and medium 1 must necessarily have the same wavelength if a continuous transition is to be made on crossing the interface at any arbitrary point. The wavevector of the transmitted wave in medium 1 must therefore terminate on the vertical line through B$'$. The termination point A$'$ is determined by the condition that OA$'$=nk. Snell's law then follows immediately.

Next consider a slab of finite thickness shown in Fig. 3.7(b). The side view depicts only the transmitted wavevectors across the two interfaces from medium 0 to 1, and from 1 to 2, whereas the right panel shows the z components of the wavevectors. In contrast to the case of the infinite slab there is now an infinite series of possible reflections, and the first three of these are drawn in the figure:

1. Reflection at interface 0 to 1, amplitude r_{01}.
2. Transmission at interface 0 to 1, t_{01}, then reflection at interface 1 to 2, r_{12}, followed by transmission at interface 1 to 0, t_{10}. In adding this wave to the above it is necessary to include the phase factor $p^2 = e^{i\,Q\Delta}$.
3. Transmission at interface 0 to 1, t_{01}, then reflection at interface 1 to 2,

r_{12}, followed by reflection at interface 1 to 0, r_{01}, then another reflection at interface 1 to 2, r_{12}, finally followed by transmission 1 to 0, t_{10}. The total phase factor for this wave is p^4.

The total amplitude reflectivity is therefore:

$$\begin{aligned} r_{\text{slab}} &= r_{01} + t_{01}t_{10}r_{12}\,p^2 + t_{01}t_{10}r_{10}r_{12}^2\,p^4 + t_{01}t_{10}r_{10}^2r_{12}^3\,p^6 \cdots \\ &= r_{01} + t_{01}t_{10}r_{12}\,p^2\left\{1 + r_{10}r_{12}\,p^2 + r_{10}^2r_{12}^2\,p^4 \cdots\right\} \\ &= r_{01} + t_{01}t_{10}r_{12}\,p^2 \sum_{m=0}^{\infty}(r_{10}r_{12}p^2)^m \end{aligned}$$

This is a geometric series which may be summed as described on page 51 to give

$$r_{\text{slab}} = r_{01} + t_{01}t_{10}r_{12}\,p^2\,\frac{1}{1 - r_{10}r_{12}p^2}$$

This expression may be simplified using the Fresnel equations (Eq. (3.23)). Using the notation in Fig. 3.7 we have

$$r_{01} = \frac{\mathrm{Q}_0 - \mathrm{Q}_1}{\mathrm{Q}_0 + \mathrm{Q}_1} \quad \text{and} \quad t_{01} = \frac{2\mathrm{Q}_0}{\mathrm{Q}_0 + \mathrm{Q}_1}$$

which in turn imply that

$$r_{01} = -r_{10}$$

and

$$r_{01}^2 + t_{01}t_{10} = \frac{(\mathrm{Q}_0 - \mathrm{Q}_1)^2}{(\mathrm{Q}_0 + \mathrm{Q}_1)^2} + \frac{2\mathrm{Q}_0\,2\mathrm{Q}_1}{(\mathrm{Q}_0 + \mathrm{Q}_1)^2} = \frac{(\mathrm{Q}_0 + \mathrm{Q}_1)^2}{(\mathrm{Q}_0 + \mathrm{Q}_1)^2} = 1$$

When these expressions are inserted into the equation for r_{slab} one obtains

$$\boxed{r_{\text{slab}} = \frac{r_{01} + r_{12}p^2}{1 + r_{01}r_{12}p^2}} \tag{3.30}$$

The phase factor, p^2, of the rays reflected from the top and bottom faces of the slab is $\mathrm{e}^{i\,\mathrm{Q}_1\Delta}$, where $\mathrm{Q}_1 = 2\mathrm{k}_1 \sin\alpha_1$.

For simplicity it is further assumed that the media on either side of the slab are the same, or in other words that $r_{01} = -r_{12}$. In this case the slab reflectivity becomes

$$r_{\text{slab}} = \frac{r_{01}(1 - p^2)}{1 - r_{01}^2 p^2} \tag{3.31}$$

The intensity reflectivity given by this formula is plotted in Fig. 3.8⋆, and displays oscillations known as Kiessig fringes [Kiessig, 1931] due to the interference of waves reflected from the top and bottom interfaces. The peaks in the oscillations correspond to the waves scattering in phase, and the dips to them scattering out of phase. In the figure Δ has been chosen to be equal to $10 \times 2\pi$ Å, which makes it clear that the oscillations occur with a period of $2\pi/\Delta$ in Q.

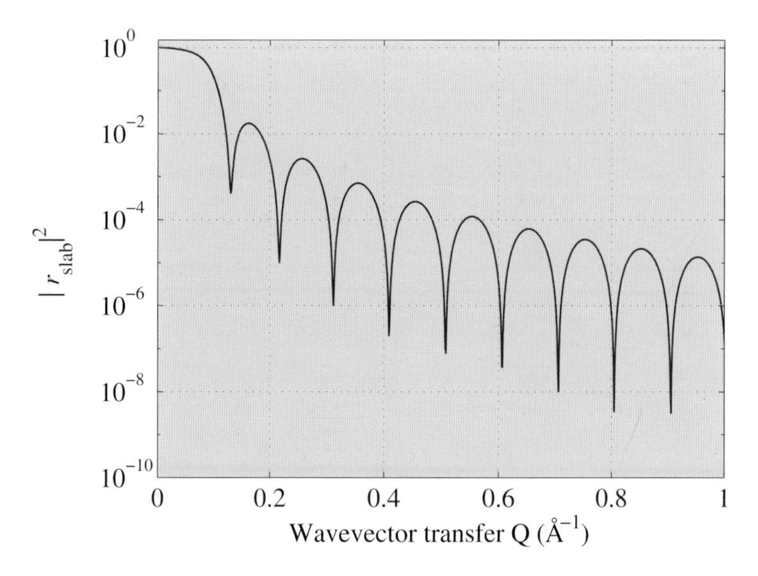

Figure 3.8: ⋆ Kiessig fringes from a homogeneous slab of Tungsten. Solid curve: the calculated reflectivity $|r_{\text{slab}}|^2$ for a slab of thickness $10 \times 2\pi$ Å. The density of the film is 4.678 electrons per Å^3, Table 3.1.

The expression for the reflectivity of a slab is exact, but it is useful to consider the limiting case of a thin slab. It is also assumed that the angles are sufficiently large that refraction effects can be neglected, with the result that $|r_{01}| \ll 1$. In this case the slab reflectivity becomes

$$r_{\text{slab}} = \frac{r_{01}(1-p^2)}{1-r_{01}^2 p^2} \approxeq r_{01}(1-p^2) \approxeq r_{01}(1-\mathrm{e}^{i\,\mathrm{Q}\Delta}) \approxeq \left(\frac{\mathrm{Q}_c}{2\mathrm{Q}}\right)^2 (1-\mathrm{e}^{i\,\mathrm{Q}\Delta})$$

This can be recast in a form that will be of use to us in the section on multilayers by rewriting it as

$$\begin{aligned} r_{\text{slab}} &= -\frac{16\pi\rho r_0}{4\mathrm{Q}^2}\,\mathrm{e}^{i\,\mathrm{Q}\Delta/2}\left(\mathrm{e}^{i\,\mathrm{Q}\Delta/2} - \mathrm{e}^{-i\,\mathrm{Q}\Delta/2}\right) \\ &= \left(\frac{16\pi\rho r_0\Delta}{2\mathrm{Q}}\right)\frac{\mathrm{e}^{i\,\mathrm{Q}\Delta/2}}{2(\mathrm{Q}\Delta/2)}(-i)\frac{\left(\mathrm{e}^{i\,\mathrm{Q}\Delta/2} - \mathrm{e}^{-i\,\mathrm{Q}\Delta/2}\right)}{2i} \\ &= -i\left(\frac{4\pi\rho r_0\Delta}{\mathrm{Q}}\right)\left(\frac{\sin(\mathrm{Q}\Delta/2)}{\mathrm{Q}\Delta/2}\right)\mathrm{e}^{i\,\mathrm{Q}\Delta/2} \end{aligned} \tag{3.32}$$

Second, in addition to neglecting refraction effects it is assumed that the slab is thin, i.e. $\mathrm{Q}\Delta \ll 1$, and the reflectivity is then

$$r_{\text{thin slab}} \approxeq -i\frac{4\pi\rho r_0\Delta}{\mathrm{Q}} \equiv -i\frac{\mathrm{Q}_c^2\Delta}{4\mathrm{Q}} = -i\frac{4\pi\rho r_0\Delta}{2\mathrm{k}\sin\alpha} = -i\frac{\lambda\rho r_0\Delta}{\sin\alpha} \tag{3.33}$$

This expression is valid for angles well away from the critical angle, where the reflectivity is weak, and both multiple reflections and refraction effects can be neglected. This is referred to as the region of kinematical reflectivity.

An alternatively way to derive the reflectivity from a thin film is to use the following heuristic argument. The amplitude of the reflected wave must be proportional to the density of electrons, ρ, as well as to the scattering length r_0, and to the thickness of sample traversed which is equal to $\Delta/\sin\alpha$. However, the product of these three variables has the dimension of inverse length, whereas the reflectivity is a dimensionless number. The only length remaining in the problem is the X-ray wavelength. Thus from a dimensional analysis the reflectivity from a thin slab is $r_{\text{thin slab}} = C\,(\rho r_0 \lambda\Delta/\sin\alpha)$, where C is a constant to be determined. The value of C can be found by imagining an infinitely thick medium as being formed from an infinite stack of thin slabs. In other words, by integrating the expression for $r_{\text{thin slab}}$ from 0 to ∞, taking into account of course the appropriate phase factor, we should obtain the Fresnel reflectivity, r_F:

$$\begin{aligned} r_F = C \int_0^\infty \left(\frac{\rho r_0 \lambda}{\sin\alpha}\right) \mathrm{e}^{i\,\mathrm{Q}z}\, dz &= C\left(\frac{\rho r_0 \lambda}{\sin\alpha}\right)\left(\frac{1}{i\mathrm{Q}}\right)\int_0^\infty \mathrm{e}^{i\,\mathrm{Q}z}\, d(i\mathrm{Q}z) \\ = -iC\left(\frac{\rho r_0 2\pi}{\mathrm{Qk}\sin\alpha}\right)\left[\mathrm{e}^{i\,\mathrm{Q}z}\right]_0^\infty &= iC\left(\frac{\mathrm{Q}_c}{2\mathrm{Q}}\right)^2 \end{aligned}$$

where the expression given in Eq. (3.28) for Q_c has been used. However, the Fresnel amplitude reflectivity is $r_F \approx (\mathrm{Q}_c/2\mathrm{Q})^2$ in the limit that $\alpha \gg \alpha'$, which implies that $C = -i$. Thus the heuristic argument taken together with this determination of C is in accordance with Eq. 3.33.

3.6 Specular reflection from multilayers

The scattering from multilayer structures has assumed particular significance in recent years. Modern growth techniques allow materials to be designed and fabricated at the atomic or molecular level. Many technologically important materials are now produced in this way, as under favourable conditions it is possible to tailor make materials with desired physical properties. One particularly interesting and useful class of materials is the multilayer or superlattice. This is a system grown by depositing one material on top of another in a repetitive sequence as shown in Fig. 3.9.

Materials that are used to fabricate multilayers range from metallic or semiconducting elements, through to complex molecules such as are found in a Langmuir layer. In most cases a specific growth technique has been developed to produce the multilayer system of interest. What is common to all of these systems is that there is a need to characterize the resulting structure. X-ray and neutron reflectivity turns out to be an excellent tool for this task, as the contrast in scattering density between the two materials gives rise to scattering. However, from now on we consider X-ray reflectivity only, for which the scattering length density is simply ρr_0.

The most general approach is to extend what has been developed so far for the single slab, so as to obtain an expression that is valid at all scattering angles. It is more instructive, however, to start by considering the kinematical reflectivity, where multiple reflections and refraction are assumed to be small. The resulting formulae are then valid only at angles well away from the critical angle, but have the advantage that the connection between the equations and the electron density profile is more transparent. The mathematical implication

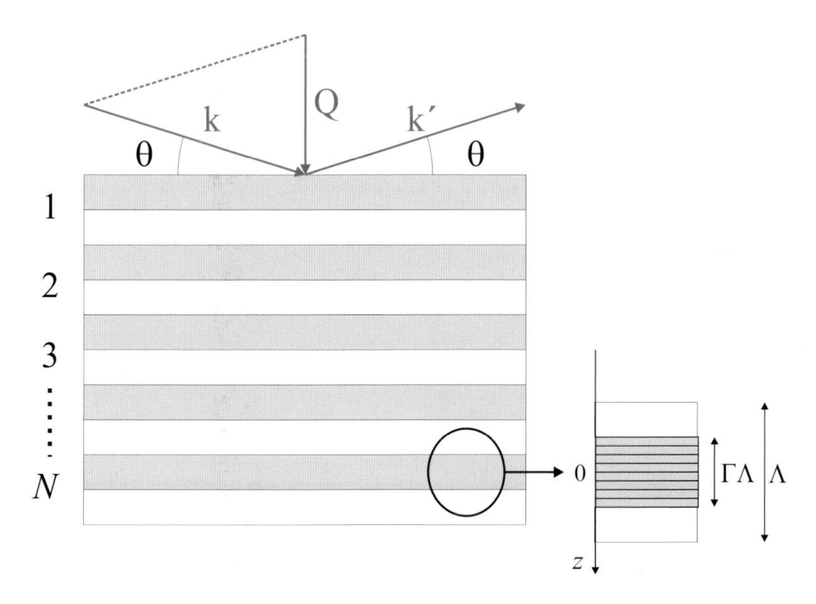

Figure 3.9: Schematic of a multilayer which here is a stack of bilayers. Each bilayer, as shown to the right has a homogeneous high electron density region of thickness $\Gamma\Lambda$, and a low density region. The total thickness of a bilayer is Λ.

of restricting ourselves to the kinematical region is that we can deduce the amplitude reflectivity as a superposition of reflected waves from infinitesimal sheets, taking into account of course the phase factor e^{iQz} for the sheet at depth z. Unsurprisingly, the equations are very similar to those used to describe the scattering of light from an optical diffraction grating.

Kinematical approximation

For convenience we imagine the structure of the multilayer as being composed of N repetitions of a single bilayer of thickness Λ formed from one layer of material A followed by B, as shown in Fig. 3.9. No assumption is made about the detailed structure of A or B, so that the formulae apply equally well to amorphous or crystalline materials: all that matters is that there is an electron density contrast between A and B. Having decomposed the multilayer into a sum of bilayers it is then straightforward to write down an expression for the reflectivity. First the scattering amplitude from a single bilayer is calculated, and then a sum made over the N bilayers, making suitable allowance for the difference in phase factors for the waves scattered from each bilayer. Here it is assumed that the interfaces are flat, and the wavevector $\mathbf{Q}$ is parallel to the surface normal, so the reflectivity is *specular*, and the problem of summing the phases is then one dimensional.

If r_1 is the reflectivity from a single bilayer, then the reflectivity from N bilayers comprising the multilayer is

$$r_N(\zeta) = \sum_{\nu=0}^{N-1} r_1(\zeta)\, e^{i\,2\pi\zeta\nu} e^{-\beta\nu} = r_1(\zeta)\, \frac{1 - e^{i\,2\pi\zeta N} e^{-\beta N}}{1 - e^{i\,2\pi\zeta} e^{-\beta}} \tag{3.34}$$

where ζ is defined by $Q = 2\pi\zeta/\Lambda$, and β is the average absorption per bilayer. The bilayer reflectivity, r_1, is evaluated using the expression for the reflectivity of a thin slab given in Eq. (3.33). To apply this result to the bilayer two modifications need to be made. First, the electron density of the slab must be replaced by the difference in electron densities between A and B, where it is assumed that $\rho_{\mathrm{A}} > \rho_{\mathrm{B}}$. Second, as usual it is necessary to allow for the change in phase of waves reflected from different depths in the bilayer. To do so we imagine the high density material A to comprise a fraction Γ of the bilayer, and to be subdivided into thin sheets, each of which has the reflectivity of a thin slab, but with ρ replaced by $\rho_{\mathrm{AB}} = \rho_{\mathrm{A}} - \rho_{\mathrm{B}}$.

From Eq. (3.33) the amplitude reflectivity from one bilayer may then be written as

$$\begin{aligned} r_1(\zeta) &= -i\frac{\lambda r_0 \rho_{\mathrm{AB}}}{\sin\theta} \int_{-\Gamma\Lambda/2}^{+\Gamma\Lambda/2} \mathrm{e}^{i\,2\pi\zeta z/\Lambda}\, dz \\ &= 4\pi r_0 \rho_{\mathrm{AB}} \frac{1}{i\,Q} \int_{-\Gamma\Lambda/2}^{+\Gamma\Lambda/2} \mathrm{e}^{i\,2\pi\zeta z/\Lambda}\, dz \\ &= -2ir_0\rho_{\mathrm{AB}} \left(\frac{\Lambda^2\Gamma}{\zeta}\right) \frac{\sin(\pi\Gamma\zeta)}{\pi\Gamma\zeta} \end{aligned} \tag{3.35}$$

To evaluate the absorption parameter β for a bilayer we note that the incident X-ray has a path length $\Lambda/\sin\theta$ in the bilayer, of which a fraction Γ is through A and a fraction $(1-\Gamma)$ through B. Remembering that the absorption coefficient μ refers to intensity and not amplitude, the amplitude absorption for a bilayer is $\mathrm{e}^{-\beta}$ with

$$\begin{aligned} \beta &= 2\left[\left(\frac{\mu_{\mathrm{A}}}{2}\right)\left(\frac{\Gamma\Lambda}{\sin\theta}\right) + \left(\frac{\mu_{\mathrm{B}}}{2}\right)\left(\frac{(1-\Gamma)\Lambda}{\sin\theta}\right)\right] \\ &= \frac{\Lambda}{\sin\theta}\left[\mu_{\mathrm{A}}\Gamma + \mu_{\mathrm{B}}(1-\Gamma)\right] \end{aligned} \tag{3.36}$$

where the factor of 2 in the first line allows for the path length of both the incident and reflected beam.

In Fig. 3.10(a)$\star$ the reflectivity curve of a multilayer is shown. This serves to illustrate how the different factors in Eq. (3.34) combine to produce the resulting curve. A specific example has been chosen of a multilayer formed from 10 bilayers, where each bilayer has 10 Å of W and 40 Å of Si. (This or similar types of multilayers are useful optical components in X-ray beamlines, as we shall see in Section 3.10.) The main peaks in the reflectivity occur when ζ is an integer, as the denominator in Eq. (3.34) is then zero (at least if β may be assumed to be negligible). These correspond to the principal diffraction maxima from a diffraction grating. In between the principal maxima there are auxiliary maxima due to oscillations in the numerator. In real optical components for X-ray applications the number of bilayers is normally much larger than 10, so that the spacing between the auxiliary maxima becomes small and the reflectivity of the first principal maxima tends to 100%.

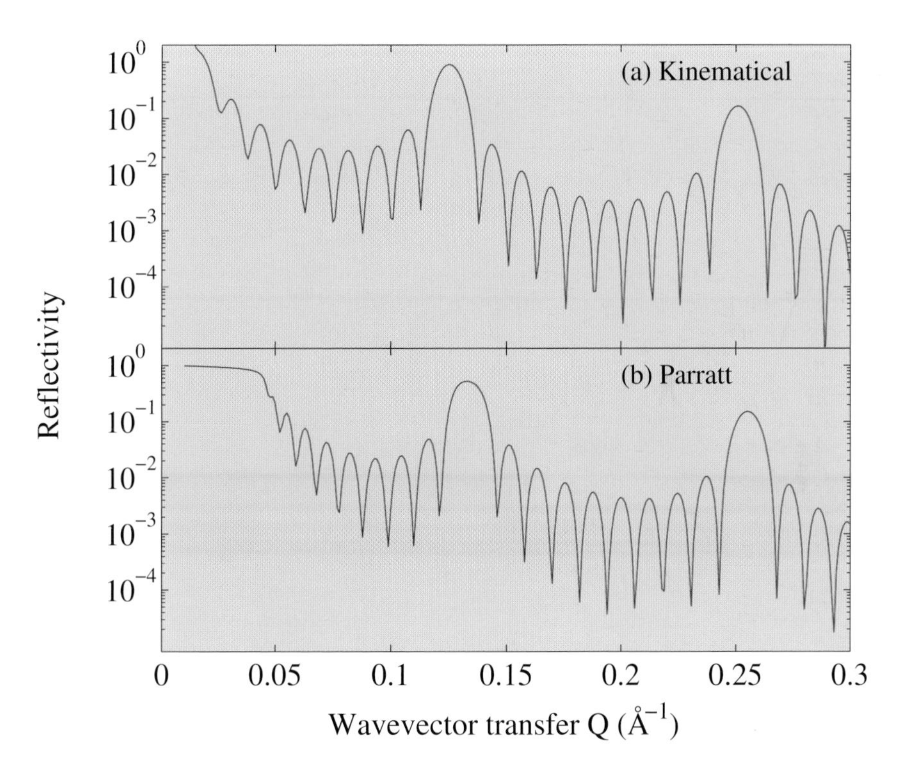

Figure 3.10: ⋆ Specular reflectivity from a W/Si multilayer: 10 bilayers each being 10 Å (amorphous) W on 40 Å (amorphous) Si. (a) Kinematical reflectivity. (b) Reflectivity curve calculated using Parratt's method. The parameters used in the calculation were taken from Table 3.1.

Parratt's exact recursive method

A method to extend the exact result for a single slab (Eq. (3.30)) to the case of a stratified medium has been described by Parratt [Parratt, 1954]. The medium is imagined as being composed of N strata, or layers, sitting on top of an infinitely thick substrate. By definition the N'th layer sits directly on the substrate. Each layer in the stack has a refractive index $n_j = 1 - \delta_j + i\beta_j$ and is of thickness Δ_j. It follows from Fig. 3.7 that the z component of the wavevector, $\mathrm{k}_{z,j}$, in the slab labelled j is determined from the total wavevector $\mathrm{k}_j = n_j\mathrm{k}$ and the x component, $\mathrm{k}_{x,j}$, which is conserved through all layers so $\mathrm{k}_{x,j} = \mathrm{k}_x$ for all j. The value of $\mathrm{k}_{z,j}$ is found from

$$\mathrm{k}_{z,j}^2 = (n_j\mathrm{k})^2 - \mathrm{k}_x^2 = (1 - \delta_j + i\,\beta_j)^2\mathrm{k}^2 - \mathrm{k}_x^2 \approxeq \mathrm{k}_z^2 - 2\delta_j\mathrm{k}^2 + i\,2\beta_j\mathrm{k}^2$$

Noting that $\mathrm{Q}_j = 2\mathrm{k}_j \sin\alpha_j = 2\mathrm{k}_{z,j}$, the wavevector transfer in the j'th layer is

$$\mathrm{Q}_j = \sqrt{\mathrm{Q}^2 - 8\mathrm{k}^2\delta_j + i\,8\mathrm{k}^2\beta_j}$$

In the absence of multiple reflections, the reflectivity (Eq. (3.29)) of each interface is obtained from the Fresnel relation

$$r'_{j,j+1} = \frac{\mathrm{Q}_j - \mathrm{Q}_{j+1}}{\mathrm{Q}_j + \mathrm{Q}_{j+1}}$$

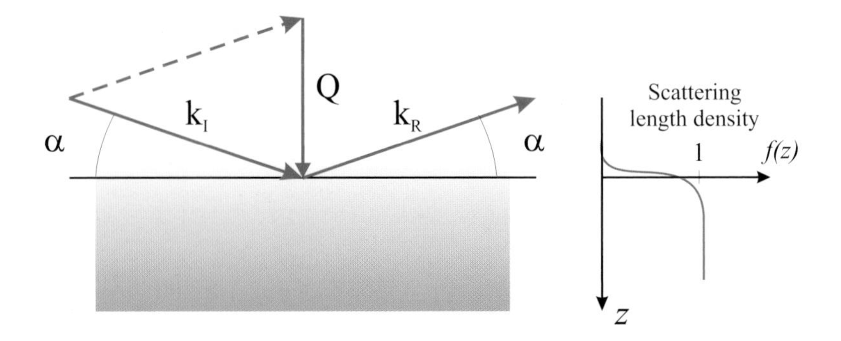

Figure 3.11: A flat interface with a graded density given by the shape function $f(z)$, normalized to unity at large z. The Fourier transform of its derivative, $\phi(\mathrm{Q})$, can be considered to be the form factor of the density variation across the interface.

where the prime is used to denote a reflectivity amplitude that does not include multiple scattering effects.

The first step is to calculate the reflectivity from the interface between the bottom of the N'th layer and the substrate. As the substrate is infinitely thick there are no multiple reflections to consider and

$$r'_{N,\infty} = \frac{\mathrm{Q}_N - \mathrm{Q}_\infty}{\mathrm{Q}_N + \mathrm{Q}_\infty}$$

The reflectivity from the top of the N'th layer is then evaluated using Eq. (3.30) as

$$r_{N-1,N} = \frac{r'_{N-1,N} + r'_{N,\infty} p_N^2}{1 + r'_{N-1,N} r'_{N,\infty} p_N^2}$$

which allows for the multiple scattering and refraction in the N'th layer, and where p_N^2 is the phase factor $\mathrm{e}^{i\,\Delta_N \mathrm{Q}_N}$, or in general $p_j^2 = \mathrm{e}^{i\,\Delta_j \mathrm{Q}_j}$. It follows that the reflectivity from the next interface up in the stack is

$$r_{N-2,N-1} = \frac{r'_{N-2,N-1} + r_{N-1,N} p_{N-1}^2}{1 + r'_{N-2,N-1} r_{N-1,N} p_{N-1}^2}$$

and it is clear that the process can be continued recursively until the total reflectivity amplitude, $r_{0,1}$, at the interface between the vacuum and first layer is obtained.

The reflectivity from the same W/Si multilayer discussed above has been calculated using Parratt's method and is plotted in Fig. 3.10(b)$\star$. A comparison of the two curves shows that as expected there is little difference at high values of Q where the kinematical approximation is valid, but close to the critical wavevector $\mathrm{Q}_c \approx 0.04$ Å^{-1} the approximation fails completely.

3.7 Reflectivity from a graded interface

So far we have considered the reflectivity from systems that have sharp, flat interfaces. Many interesting systems cannot be described in this way, and hence it is necessary to extend the formalism to include graded interfaces. For the sake of simplicity we shall limit ourselves to the kinematical region, where Q is much larger than Q_c. As with the example of reflectivity from a multilayer, this means that the reflectivity is derived by considering the contribution from a thin slab at a depth z, and then summing up all the contributions from the graded interface, making allowance for the change in phase e^{iQz}. The density profile of the interface is given by the function $f(z)$, which is defined so that $f(z) \rightarrow 1$ as $z \rightarrow \infty$ as shown in Fig. 3.11. From Eq. (3.33) the contribution to the reflectivity from an infinitesimal thin slab at depth z is

$$\delta r(Q) = -i\left(\frac{Q_c^2}{4Q}\right) f(z)\, dz \tag{3.37}$$

The amplitude reflectivity for the superposition of infinitesimal layers is thus

$$\begin{aligned} r(Q) &= -i\left(\frac{Q_c^2}{4Q}\right)\int_0^\infty f(z)\,e^{iQz}dz \\ &= i\frac{1}{iQ}\left(\frac{Q_c^2}{4Q}\right)\int_0^\infty f'(z)\,e^{iQz}dz \\ &= r_F(Q)\,\phi(Q) \end{aligned} \tag{3.38}$$

where $r_F(Q)$ is the Fresnel reflectivity, and $\phi(Q)$ is defined by

$$\phi(Q) = \int_0^\infty f'(z)\,e^{iQz}dz$$

In the second line of Eq. (3.38) we have used partial integration, and in the third we have used the expression from Eq. (3.29) for the Fresnel reflectivity of a sharp interface in the limit $q \gg 1$. The measured reflectivity is the intensity reflectivity, obtained as the absolute square of $r(Q)$. The *master formula* for the intensity reflectivity of a graded interface is therefore

$$\boxed{\frac{R(Q)}{R_F(Q)} = \left|\int_0^\infty \left(\frac{df}{dz}\right) e^{iQz}\, dz\right|^2} \tag{3.39}$$

or expressed in words: the ratio between the actual reflectivity and that for an ideal sharp interface is the absolute square of the Fourier transform of the normalized gradient of the density across the interface. In the box on page 85 and in Appendix E the reader is reminded of the definition of the Fourier transform.

The master formula is particularly useful as it allows analytical expressions to be used for the density gradient at an interface. One commonly used function in this context is the error function

$$f(z) = \mathrm{erf}\left(\frac{z}{\sqrt{2}\sigma}\right) \tag{3.40}$$

where σ is a measure of the width of the graded region. The derivative of the error function is a Gaussian

$$\frac{df(z)}{dz} = \frac{1}{\sqrt{2\pi\sigma^2}} \mathrm{e}^{-\frac{1}{2}\left(\frac{z}{\sigma}\right)^2} \tag{3.41}$$

and the Fourier transform of a Gaussian is another Gaussian, $\mathrm{e}^{-Q^2\sigma^2/2}$ (see Appendix E). The intensity reflectivity for this model may then be written in the compact form

$$\boxed{R(Q) = R_F(Q)\,\mathrm{e}^{-Q^2\sigma^2}} \tag{3.42}$$

3.8 Rough interfaces and surfaces

Real interfaces are rarely, if ever, perfectly flat or uniformly graded. Instead it is expected that the height of an interface has a degree of randomness, in other words the interface is rough. In this section it is explained how the presence of roughness gives rise to distinctive features in the X-ray reflectivity. In keeping with the approach adopted in the preceding sections the reflectivity from a rough interface is treated within the kinematical approximation, where the scattering is assumed to be weak so that multiple reflections may be neglected. This approach has the advantage that it is possible to understand the effects of roughness by comparing directly the results for a rough interface with the limiting form of the Fresnel reflectivity from a flat interface, which varies as $(Q_c/2Q_z)^4$ at high angles.

The formalism in this section differs from what has gone before in that the interface is now described by a statistical distribution [Wong, 1985, Sinha et al., 1988, Cowley, 1994]. In addition the heights of the interface (or surface) at different points on the rough interface are correlated in a way that is characteristic of the particular type of roughness. One important consequence of the existence of these correlations is that the reflectivity is no longer necessarily strictly specular, as is the case for the Fresnel reflectivity from a sharp interface or from a graded but flat interface. Instead it develops a diffuse component, which is also referred to as the off-specular reflectivity.

Figure 3.12(a) illustrates the reflection of an X-ray beam from a rough surface. A beam of intensity I_0 is incident at a glancing angle θ_1, and the reflected beam is observed at a glancing exit angle θ_2. The incident beam illuminates a volume V (indicated by the darker shading) to a depth determined by the absorption coefficient. Within the kinematical approximation the reflected amplitude of the beam is calculated by summing all of the beams scattered from volume elements $d\mathbf{r}$ within V, taking into account the appropriate phase factors. The scattering amplitude is

$$r_V = -r_0 \int_V (\rho d\mathbf{r})\,\mathrm{e}^{i\,\mathbf{Q}\cdot\mathbf{r}} \tag{3.43}$$

Here r_0 is the Thomson scattering length of a single electron, $(\rho d\mathbf{r})$ is the number of electrons in a volume element centred at position $\mathbf{r}$, and the last term in the

Fourier transforms and the Convolution Theorem

Fourier transforms occur naturally and ubiquitously in the mathematical description of scattering. The reason is that the scattering amplitude from an extended body often appears as a Fourier transform. Here we remind the reader of a few important results. The Fourier transform of the one dimensional function $f(x)$ is defined by

$$F(\mathrm{q}) = \int_{-\infty}^{\infty} f(x) \mathrm{e}^{i\,\mathrm{q}x}\, dx$$

and the inverse transform is

$$f(x) = \frac{1}{2\pi} \int_{-\infty}^{\infty} F(q) \mathrm{e}^{-i\,\mathrm{q}x}\, d\mathrm{q}.$$

These results are easily generalized to higher dimensions.

One particularly useful result in the context of scattering is the Convolution Theorem. This states that the Fourier transform of the convolution of two functions $f(x)$ and $g(x)$ is equal to the product of the two individual Fourier transforms $F(\mathrm{q})$ and $G(\mathrm{q})$. The convolution or folding integral $h(x)$ of two functions $f(x)$ and $g(x)$ is defined by

$$h(x) = \int_{-\infty}^{\infty} f(x_1)\, g(x_1 - x)\, dx_1$$

Its Fourier transform is

$$\begin{aligned} H(\mathrm{q}) &= \int_{-\infty}^{\infty} h(x)\, \mathrm{e}^{i\,\mathrm{q}x}\, dx \\ &= \int_{-\infty}^{\infty} f(x_1)\, \mathrm{e}^{i\,\mathrm{q}x_1}\, dx_1 \int_{-\infty}^{\infty} g(x_1 - x)\, \mathrm{e}^{-i\,\mathrm{q}(x_1 - x)}\, dx \\ &= F(\mathrm{q})\, G(\mathrm{q}) \end{aligned}$$

The great utility of this result is that in many scattering problems the object of interest may be described mathematically as the convolution of two component functions in real space. One important example is a crystal lattice, for which the density may be viewed as the convolution of a lattice function and a function that describes what sits at each lattice point. The scattering amplitude is proportional to the Fourier transform of the density, and hence from the convolution theorem is equal to the product of the Fourier transforms of the component functions. If the latter are known, as is often the case, then the scattering amplitude may be obtained almost by inspection.

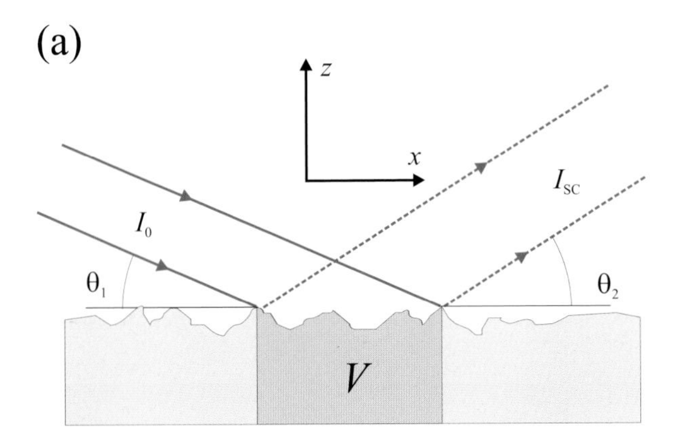

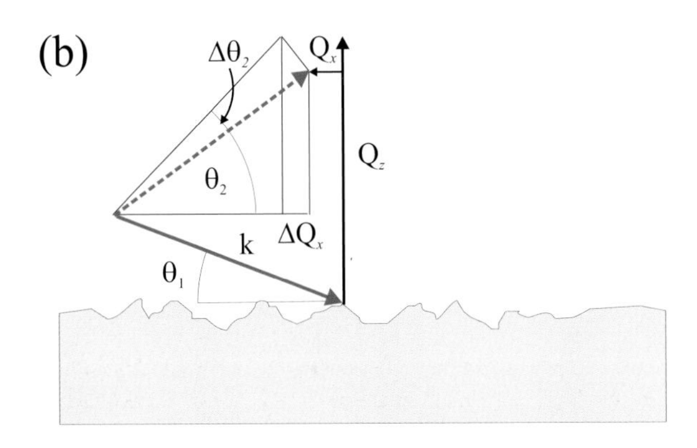

Figure 3.12: (a) Scattering from a rough surface. (b) Definition of ΔQ_x.

integrand is the phase factor. The volume integral can be transformed to a surface integral using Gauss' theorem, which states that

$$\int_V (\nabla \cdot \mathbf{C})\, d\mathbf{r} = \int_S \mathbf{C} \cdot d\mathbf{S}$$

where $\mathbf{C}$ is a vector field, S refers to the surface, and $d\mathbf{S}$ is normal to the surface at position (x, y) and has a magnitude equal to $dxdy$.

Gauss' theorem may be applied to transform Eq. (3.43) to a surface integral in the following way. Let $\mathbf{C}$ be the unit vector $\hat{\mathbf{z}}$ along the z axis multiplied by the function $\mathrm{e}^{i\mathbf{Q}\cdot\mathbf{r}}/(i\mathrm{Q}_z)$. The divergence of $\mathbf{C}$ is then $\nabla \cdot \mathbf{C}=\mathrm{e}^{i\mathbf{Q}\cdot\mathbf{r}}/(i\mathrm{Q}_z)$ $\times(i\mathrm{Q}_z)=\mathrm{e}^{i\mathbf{Q}\cdot\mathbf{r}}$, which is the integrand in Eq. (3.43). Expressed as a surface integral the scattering amplitude becomes

$$\begin{aligned} r_S &= -r_0 \int_V (\rho d\mathbf{r})\, \mathrm{e}^{i\,\mathbf{Q}\cdot\mathbf{r}} \\ &= -r_0\rho \left(\frac{1}{i\mathrm{Q}_z}\right) \int_S \mathrm{e}^{i\,\mathbf{Q}\cdot\mathbf{r}}\, \hat{\mathbf{z}} \cdot d\mathbf{S} \end{aligned}$$

The dot product $\hat{\mathbf{z}} \cdot d\mathbf{S}$ is the the area element of the rough surface projected onto the $x - y$ plane, $\hat{\mathbf{z}} \cdot d\mathbf{S} = dxdy$, so that

$$r_S = -r_0\rho \left(\frac{1}{i\mathrm{Q}_z}\right) \int_S \mathrm{e}^{i\,\mathbf{Q}\cdot\mathbf{r}}\, dxdy$$

It should be noted that the rough surface is not the entire surface enclosing the volume V, as Gauss' formula assumes. However, the lower surface of V does not contribute, since the depth of V can be chosen such that, by the time the beam has penetrated to the lower surface, absorption reduces the beam intensity effectively to zero.

To proceed let the height variation of the rough surface be given by the function $h(x, y)$. Then the scalar product of $\mathbf{Q}$ and $\mathbf{r}$ is $\mathbf{Q} \cdot \mathbf{r} = \mathrm{Q}_z h(x, y)$ $+(\mathrm{Q}_x x + \mathrm{Q}_y y)$, so that the scattering amplitude from the surface is simply

$$r_S = -r_0\rho \left(\frac{1}{i\mathrm{Q}_z}\right) \int_S \mathrm{e}^{i\,\mathrm{Q}_z h(x,y)} e^{i(\mathrm{Q}_x x + \mathrm{Q}_y y)}\, dxdy$$

The differential scattering cross-section, $(d\sigma/d\Omega)$, is the absolute square of the scattering amplitude (see Appendix A):

$$\left(\frac{d\sigma}{d\Omega}\right) = \left(\frac{r_0\rho}{\mathrm{Q}_z}\right)^2 \int \mathrm{e}^{i\mathrm{Q}_z[h(x,y)-h(x',y')]}\, e^{i\mathrm{Q}_x(x-x')} e^{i\mathrm{Q}_y(y-y')}\, dxdx'dydy'$$

It is now assumed that the difference in heights, $h(x, y) - h(x', y')$, depends only on the relative difference in position $(x - x', y - y')$. The 4-dimensional integral above then reduces to the product of 2 two-dimensional integrals, one of which is simply $\int dxdy = A_0/\sin\theta_1$, the illuminated surface area, and we obtain

$$\left(\frac{d\sigma}{d\Omega}\right) = \left(\frac{r_0\rho}{\mathrm{Q}_z}\right)^2 \left(\frac{A_0}{\sin\theta_1}\right) \int \langle \mathrm{e}^{i\mathrm{Q}_z[h(0,0)-h(x,y)]} \rangle e^{i(\mathrm{Q}_x x + \mathrm{Q}_y y)}\, dxdy$$

The angular brackets indicate an ensemble average: for a fixed (x', y') one evaluates the average value of the function for all possible choices of the origin within the illuminated area. (We note in passing that the right hand side of the formula has the correct dimension of area.) One further assumption is now introduced, namely that the statistics of the height variations are Gaussian, with the consequence that the cross-section may be written as

$$\boxed{\left(\frac{d\sigma}{d\Omega}\right) = \left(\frac{r_0\rho}{\mathrm{Q}_z}\right)^2 \left(\frac{A_0}{\sin\theta_1}\right) \int \mathrm{e}^{-\mathrm{Q}_z^2\langle[h(0,0)-h(x,y)]^2\rangle/2}\, e^{i(\mathrm{Q}_x x + \mathrm{Q}_y y)}\, dxdy} \tag{3.44}$$

This follows from the Baker-Hausdorff theorem, which is proved in Appendix D. In the following the reflectivity is calculated for different models of the function $g(x, y)$ describing the ensemble average of height differences, where

$$g(x, y) = \langle [h(0, 0) - h(x, y)]^2 \rangle \tag{3.45}$$

3.8.1 The limiting case of Fresnel reflectivity

It is instructive to first check Eq. (3.44) against the kinematical form of the Fresnel reflectivity from a flat interface. To do so we set $h(x,y) = 0$ for all (x,y) with the result that

$$\left(\frac{d\sigma}{d\Omega}\right)_{\text{Fresnel}} = \left(\frac{r_0\rho}{\mathrm{Q}_z}\right)^2 \left(\frac{A_0}{\sin\theta_1}\right) \int \mathrm{e}^{i(\mathrm{Q}_x x + \mathrm{Q}_y y)}\, dx dy \tag{3.46}$$

According to the definition of the Fourier transform (see box on page 85) it can be seen that if $F(q) = 2\pi\delta(q)$ then $f(x) = (1/2\pi)\int F(q)\mathrm{e}^{-iqx}dq = 1$, and since also by definition $F(q) = \int f(x)\mathrm{e}^{iqx}dx = \int 1\,\mathrm{e}^{iqx}dx$, the double integral above is equal to $(2\pi)^2\delta(\mathrm{Q}_x)\delta(\mathrm{Q}_y)$, and thus

$$\left(\frac{d\sigma}{d\Omega}\right)_{\text{Fresnel}} = \left(\frac{2\pi r_0\rho}{\mathrm{Q}_z}\right)^2 \left(\frac{A_0}{\sin\theta_1}\right) \delta(\mathrm{Q}_x)\delta(\mathrm{Q}_y)$$

In order to make a connection between the cross-section, which has been derived here, and the formula for the intensity reflectivity that was derived earlier, we recall that the scattered intensity is related to the differential cross-section through

$$I_{\text{sc}} = \left(\frac{I_0}{A_0}\right)\left(\frac{d\sigma}{d\Omega}\right)\Delta\Omega$$

(see Appendix A). The element of solid angle $\Delta\Omega$ is evaluated with the help of Fig. 3.12(b), which shows the relationship between the angular variables (θ_1, θ_2) and the wavevector variables $(\Delta\mathrm{Q}_x, \Delta\mathrm{Q}_y)$. It is clear from this figure that $\mathrm{k}\Delta\theta_2\sin\theta_2 = \Delta\mathrm{Q}_x$, and as the y axis is perpendicular to the plane of the paper $\mathrm{k}\Delta\varphi = \Delta\mathrm{Q}_y$. Then since $\Delta\Omega = \Delta\theta_2\Delta\varphi$ the expression for the intensity becomes

$$I_{\text{sc}} = \left(\frac{I_0}{A_0}\right)\left(\frac{d\sigma}{d\Omega}\right)\frac{\Delta Q_x \Delta Q_y}{\mathrm{k}^2\sin\theta_2}$$

Inserting now the Fresnel scattering cross-section it can be seen that the delta functions in Q_x and Q_y imply that the Fresnel reflectivity is confined to the specular direction, $\theta_1 = \theta_2$. Furthermore we note that $\mathrm{k}^2\sin\theta_1\sin\theta_2 = (Q_z/2)^2$, and recall from Eq. (3.28) that $2\pi r_0\rho = \mathrm{Q}_c^2/8$, where the term f'/Z due to the dispersion correction has been neglected. Collecting all of these factors together the intensity reflectivity is

$$R(\mathrm{Q}_z) = \frac{I_{\text{sc}}}{I_0} = \left(\frac{\mathrm{Q}_c^2/8}{\mathrm{Q}_z}\right)^2 \left(\frac{1}{\mathrm{Q}_z/2}\right)^2 = \left(\frac{\mathrm{Q}_c}{2\mathrm{Q}_z}\right)^4$$

which is the expected form of the Fresnel reflectivity in the kinematical limit.

3.8.2 Uncorrelated surfaces

It is now assumed that the heights at different points (x,y) vary without any correlation: the height at (x',y') is independent of the height at (x,y) no matter how close (x,y) is to (x',y'). For points close to each other this is clearly an unphysical assumption, but it is instructive to carry out the analysis of this model anyway. For an uncorrelated surface the ensemble average of height differences Eq. (3.44) is

$$\langle[h(0,0)-h(x,y)]^2\rangle = 2\langle h^2\rangle - 2\langle h(0,0)\rangle\langle h(x,y)\rangle = 2\langle h^2\rangle$$

where the average value of h is defined to coincide with $z=0$. The cross-section then has the form

$$\left(\frac{d\sigma}{d\Omega}\right) = \left(\frac{r_0\rho}{Q_z}\right)^2 \left(\frac{A_0}{\sin\theta_1}\right) e^{-Q_z^2\sigma^2} \int e^{i(Q_x x+Q_y y)}\, dx dy \qquad (3.47)$$

which from Eq. (3.46) may be re-expressed as

$$\boxed{\left(\frac{d\sigma}{d\Omega}\right) = \left(\frac{d\sigma}{d\Omega}\right)_{\text{Fresnel}} e^{-Q_z^2\sigma^2}} \qquad (3.48)$$

where $\sigma = \sqrt{\langle h^2\rangle}$ is the rms roughness. From this the following points may be concluded

1. Fluctuations in height due to roughness diminish the Fresnel reflectivity. The $\mathbf{Q}$ dependence of this reduction, given in Eq. (3.48), is very much like the Debye-Waller factor we shall discuss in connection with thermal vibrations of atoms in a crystal in Chapter 4.

2. Since the height fluctuations are uncorrelated, the scattering is confined to the specular direction, as for the perfectly sharp interface

3. The result is identical to Eq. (3.42) for the particular example of a graded, but flat, surface. This illustrates that different models may yield the same reflectivity curve. In other words, reflectivity experiments cannot *uniquely* reveal the true nature of an interface.

3.8.3 Correlated surfaces

Our starting point is again Eq. (3.44). The difference from the previous section is that now the height fluctuations are correlated. It is further assumed that the correlations are isotropic in the plane of the surface (or interface), or in other words that $g(x,y)$ depends only on $\mathrm{r} = |\mathbf{r}| = \sqrt{x^2+y^2}$. For correlated surfaces it is possible to distinguish between two different cases, depending on the behaviour of $g(x,y)$ in the limit that $\mathrm{r}\to\infty$.

The first case to consider is when $g(x,y)$ is given by

$$g(x,y) = \langle[h(0,0)-h(x,y)]^2\rangle = \mathcal{A}\mathrm{r}^{2h} \qquad (3.49)$$

In this case the height fluctuations develop without limit as $\mathrm{r}\to\infty$. This type of roughness is displayed by fractal surfaces, and the exponent h determines the morphology of the surface: if $h \ll 1$ the surface is jagged, while as $h\to 1$ it becomes smoother. To evaluate the reflectivity in this case we simplify the mathematics by setting $y = 0$ in Eq. (3.44). This is justified if the resolution in the Q_y direction is very broad, as the intensity is proportional to the integral

$$\int_{-\infty}^{\infty} \mathrm{e}^{i\,\mathrm{Q}_y y}\, d\mathrm{Q}_y \propto \delta(y)$$

The problem then reduces to evaluating a one-dimensional integral, and as $g(x, y)$ depends on $|x|$ it is a symmetric function so that Eq. (3.44) becomes

$$\left(\frac{d\sigma}{d\Omega}\right) = \left(\frac{r_0\rho}{\mathrm{Q}_z}\right)^2 \left(\frac{A_0}{\sin\theta_1}\right) \int_0^{\infty} \mathrm{e}^{-\mathcal{A}\mathrm{Q}_z^2|x|^{2h}/2} \cos(\mathrm{Q}_x x)\, dx \tag{3.50}$$

In general the integral must be evaluated numerically, except for when $h = 1/2$ or $h = 1$, where it may be evaluated analytically using the results given in Appendix E to give

$$h = \frac{1}{2} \quad \Rightarrow \quad \left(\frac{d\sigma}{d\Omega}\right) = \left(\frac{A_0 r_0^2 \rho^2}{2\sin\theta_1}\right) \frac{\mathcal{A}}{\left(\mathrm{Q}_x^2 + (\mathcal{A}/2)^2\,\mathrm{Q}_z^4\right)} \tag{3.51}$$

$$h = 1 \quad \Rightarrow \quad \left(\frac{d\sigma}{d\Omega}\right) = \left(\frac{2\sqrt{\pi}A_0 r_0^2 \rho^2}{\sin\theta_1}\right) \frac{1}{\mathrm{Q}_z^4}\, \mathrm{e}^{-\frac{1}{2}\left(\frac{\mathrm{Q}_x^2}{\mathcal{A}\mathrm{Q}_z^2}\right)} \tag{3.52}$$

The first has a Lorentzian lineshape as a function of Q_x with a half width of $\mathcal{A}\mathrm{Q}_z^2/2$, while the second has a Gaussian lineshape with the variance of $\mathcal{A}\mathrm{Q}_z^2$. It is clear that the reflectivity from a surface where the height correlations are unbounded is completely diffuse, that is it lacks a delta function component in x (or y). This is contrast to the earlier cases of flat or uncorrelated surfaces, which have purely specular reflectivities. To illustrate this we show in Fig. 3.13 the reflectivity calculated using Eq. (3.51). Scans of Q_x at different fixed values of Q_z display a Lorentzian lineshape with a width that is proportional to Q_z^2.

The second case to consider is where the height fluctuations remain finite as $\mathrm{r} \to \infty$. This is best explored by writing

$$\begin{aligned} g(x, y) &= \langle [h(0,0) - h(x,y)]^2 \rangle = 2\langle h^2 \rangle - 2\langle h(0,0)h(x,y) \rangle \\ &= 2\sigma^2 - 2\,C(x, y) \end{aligned} \tag{3.53}$$

where $C(x, y) = \langle h(0,0)h(x,y) \rangle$ is known as the height-height correlation function. For example, if $C(x, y) = \sigma^2 \mathrm{e}^{-(r/\xi)^{2h}}$ then it can be seen that for $\mathrm{r} \ll \xi$, $g(x, y) \propto \mathrm{r}^{2h}$, and that as $\mathrm{r} \to \infty$, $g(x, y) \to 2\sigma^2$ as required. From Eqs. (3.44) and (3.53) the cross-section becomes

$$\left(\frac{d\sigma}{d\Omega}\right) = \left(\frac{r_0\rho}{\mathrm{Q}_z}\right)^2 \left(\frac{A_0}{\sin\theta_1}\right) \mathrm{e}^{-\mathrm{Q}_z^2\sigma^2} \int \mathrm{e}^{\mathrm{Q}_z^2 C(x,y)} \mathrm{e}^{i(\mathrm{Q}_x x + \mathrm{Q}_y y)}\, dx dy \tag{3.54}$$

By re-writing it in the form

$$\left(\frac{r_0\rho}{\mathrm{Q}_z}\right)^2 \left(\frac{A_0}{\sin\theta_1}\right) \mathrm{e}^{-\mathrm{Q}_z^2\sigma^2} \int \left[\mathrm{e}^{\mathrm{Q}_z^2 C(x,y)} - 1 + 1\right] \mathrm{e}^{i(\mathrm{Q}_x x + \mathrm{Q}_y y)}\, dx dy \tag{3.55}$$

it is possible to separate it into a specular and diffuse (or off-specular) term, since the last term in the square brackets is evidently the specular reflectivity

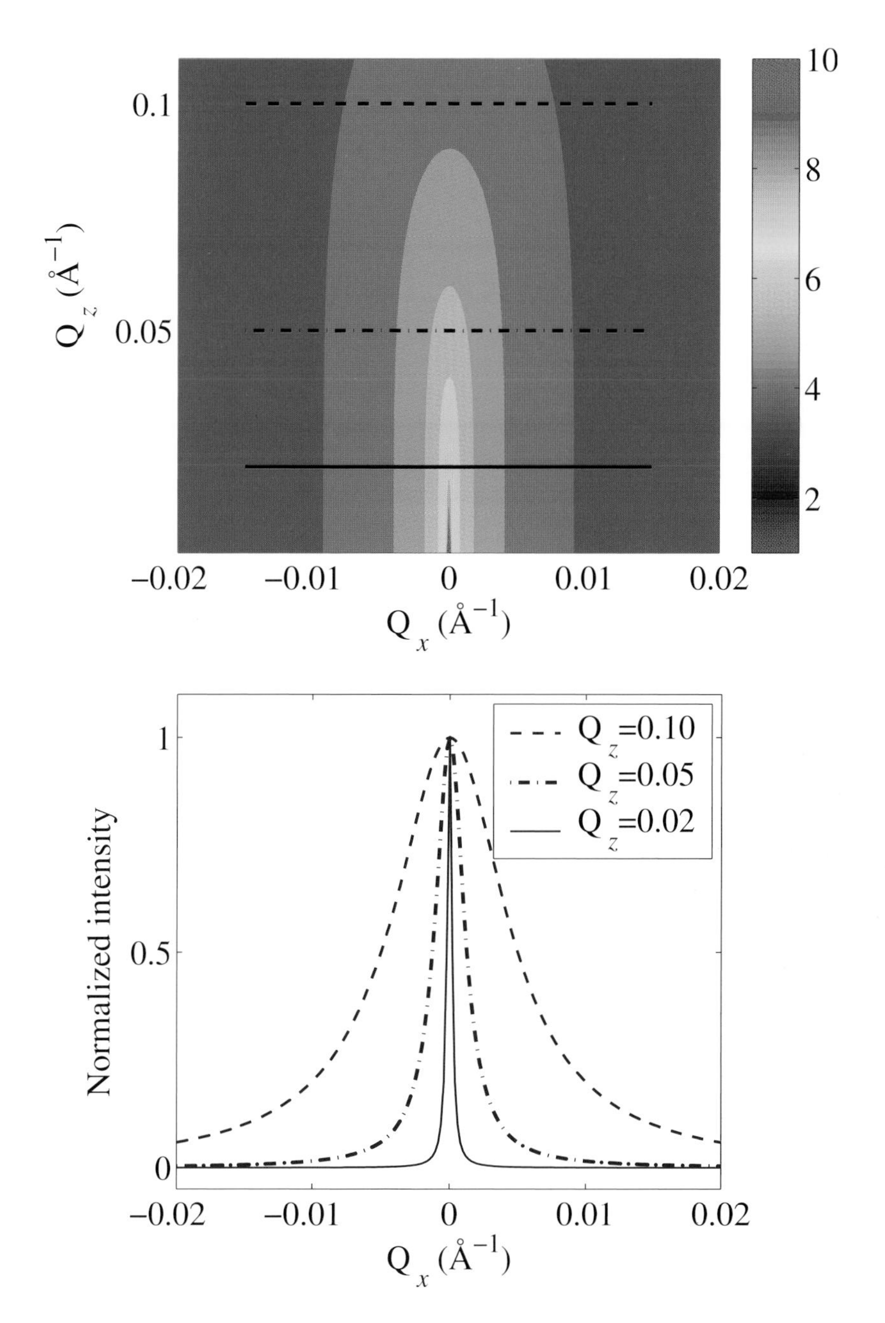

Figure 3.13: Top: the diffuse scattering from a rough surface described by $g(x, y) = \mathcal{A}r^{2h}$ with $h = 1/2$ (see Eq. (3.51)). The coordinate system is such that Q_z is perpendicular to the surface, and Q_x lies in the surface plane. The intensity has been plotted on a logarithmic scale. Bottom: Q_x scans at different values of Q_z indicated by the dashed lines in the top part of the figure. For clarity, each of the scans has been normalized to unity. The lineshape is seen to be Lorentzian, with a width that broadens as a function of Q_z.

from an *uncorrelated* surface given by Eq. (3.47). The total cross-section may then be written in the form

$$\boxed{\left(\frac{d\sigma}{d\Omega}\right) = \left(\frac{d\sigma}{d\Omega}\right)_{\text{Fresnel}} e^{-Q_z^2\sigma^2} + \left(\frac{d\sigma}{d\Omega}\right)_{\text{diffuse}}} \tag{3.56}$$

where the diffuse component is now given by

$$\left(\frac{d\sigma}{d\Omega}\right)_{\text{diffuse}} = \left(\frac{r_0\rho}{Q_z}\right)^2 \left(\frac{A_0}{\sin\theta_1}\right) e^{-Q_z^2\sigma^2} F_{\text{diffuse}}(\mathbf{Q})$$

with

$$F_{\text{diffuse}}(\mathbf{Q}) \equiv \int \left[e^{Q_z^2 C(x,y)} - 1\right] e^{i(Q_x x + Q_y y)}\, dx dy$$

The scattering from a surface where the height fluctuations are bounded therefore consists of two components. As a function of Q_x (or Q_y) the scattering has a sharp specular component superimposed on a diffuse component. In an experiment the ratio of the two will depend on the instrumental resolution. This takes us beyond the scope of this introduction, but it should be clear that X-ray reflectivity is a useful probe of the correlations displayed by rough surfaces.

3.9 Examples of reflectivity studies

Two examples of reflectivity studies are given. In the first the *specular* reflectivity from a Langmuir layer is considered. Although such a layer is a heterogeneous structure formed from complex organic molecules it turns out that X-ray reflectivity is an excellent tool for characterizing the overall morphology of such a layer. The second example concerns the reflectivity from liquid crystals. By studying the specular and *off-specular* reflectivities together it is shown how it is possible to understand the detailed nature of the critical fluctuations associated with the phase transitions in these systems.

3.9.1 Langmuir layers

X-ray reflectivity can be used to study heterogeneous structures with one or more atomic or molecular layers on a substrate. One such example is a so-called Langmuir layer. These are composed of insoluble amphiphilic molecules, with a hydrophilic chemical group at one end, and a hydrophobic chemical group at the other. When dissolved in a volatile solvent, for example chloroform, a drop of the resulting solution can be spread onto a water surface. The solvent quickly evaporates, and with the appropriate concentration one is left with a monolayer of amphiphilic molecules on top of the water surface – a Langmuir layer.

An example is shown in Fig. 3.14⋆. The hydrophobic part is a hydro-carbon chain $(CH_2)_nCH_3$ terminating in a methyl group, and the hydrophilic part is carboxylic acid COOH. When the pH of the subphase is increased by adding a base like NH_3 or Na(OH), the carboxylic acid is charged to COO^-. If in addition the subphase contains positive and negative ions from a salt, in the

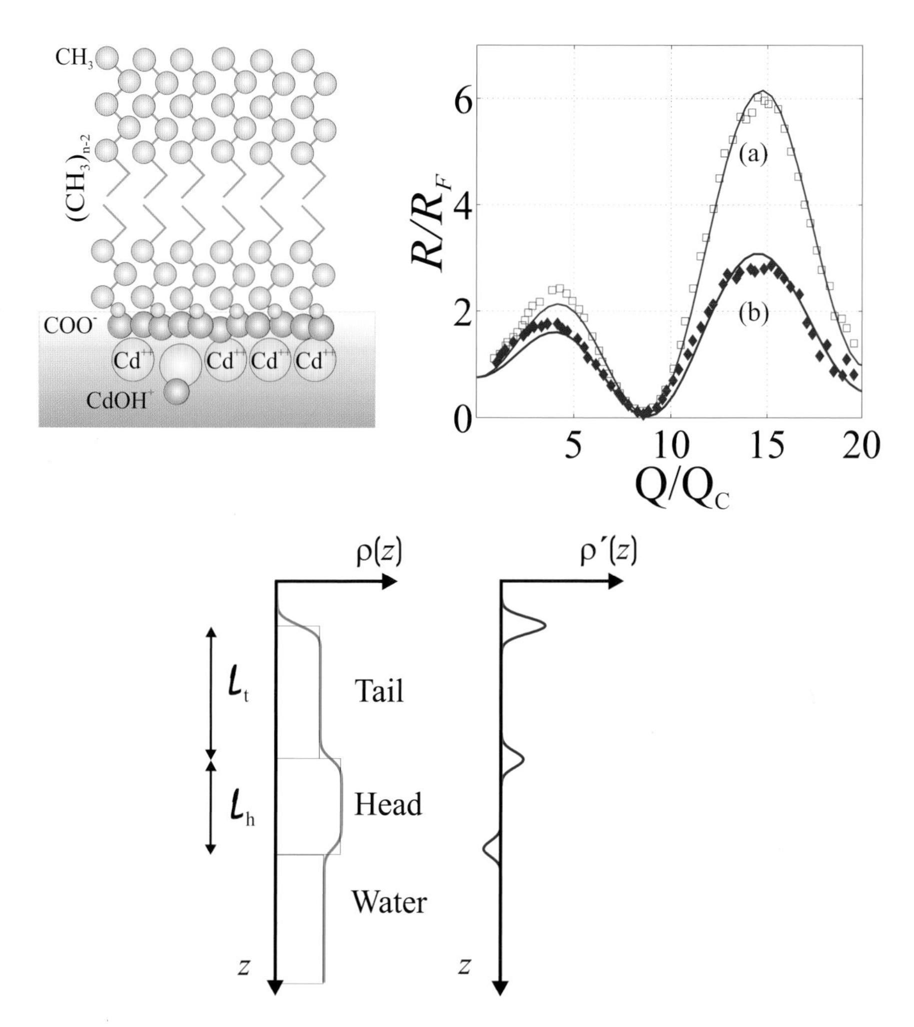

Figure 3.14: ⋆ Top left: Langmuir layer of arachidic acid (n=20) on a salt solution of $CdCl_2$. Top right: the measured reflectivity data, normalized to the Fresnel reflectivity, and plotted as a function of Q/Q_c, where Q_c=0.0217 Å^{-1} is the critical wavevector of water. Curve (a) and (b) corresponds to pH adjustments with NH_3 and Na(OH), respectively [Leveiller et al., 1994]. The dramatic difference shows that in the first case monovalent $Cd(OH)^+$ ions are bound to the monovalent COO^- head group in approximately a 1:1 ratio, whereas in the second case divalent Cd^{++} ions are bound in approximately the ratio 1:2. Bottom: two-box model of the density variation across the interfaces of a Langmuir layer on water. Each interface is smeared by a common parameter σ. Parameters deduced from fits to data: (a) $\rho_h/\rho_w = 2.28$, $\rho_t/\rho_w = 1.08$, $\ell_h = 6.2$ Å, $\ell_t = 22.0$ Å, $\sigma = 1.36$ Å; (b) $\rho_h/\rho_w = 3.35$, $\rho_t/\rho_w = 1.01$, $\ell_h = 2.7$ Å, $\ell_t = 23.4$ Å, $\sigma = 2.74$ Å. The monolayer coverage was 75% in both cases.

present case $CdCl_2$, the positive ions will be attracted to the negatively charged COO^- head groups.

We shall now discuss the modelling of the reflectivity data given in the upper right panel. As a first approximation the electron density of the molecules shown in the upper left panel can be modelled as a series of boxes: the upper box of the hydrocarbon tail is of length ℓ_t with a density ρ_t; then comes the head group, represented by a shorter box of length ℓ_h, but with a higher density ρ_h; finally a semi-infinite box corresponding to the water subphase of density ρ_w. The reflected waves from the different interfaces have different phases. Choosing the origin to be in the middle of the head-box, the phase from the upper interface of the tail box is $\phi_1 = \mathrm{Q}(\ell_t + \ell_h/2)$, the phase form the upper interface of the head-box is $\phi_2 = \mathrm{Q}\ell_h/2$, and $-\phi_2$ from the lowest interface. We can readily allow for imperfectly sharp interfaces between the different boxes using the master formula of Eq. (3.39), and by assuming that the change in density between the different boxes can be represented adequately by an error function (see Eq. (3.42)). The density and its derivative are shown schematically in the lower panel of Fig. 3.14⋆. The Fourier transform of the density gradient then assumes the simple form

$$\begin{aligned}\phi(\mathrm{Q}) &= \int \frac{\rho'(z)}{\rho_w}\, \mathrm{e}^{i\mathrm{Q}z}\, dz \\ &= \mathrm{e}^{-\mathrm{Q}^2\sigma^2/2}\,\frac{\left[\rho_t \mathrm{e}^{-i\,\phi_1} + (\rho_h - \rho_t)\mathrm{e}^{-i\,\phi_2} - (\rho_h - \rho_w)\mathrm{e}^{i\,\phi_2}\right]}{\rho_w}\end{aligned}$$

For this two-box model five parameters can be obtained by fitting it to the data; two for each box (length and density) plus a common smearing parameter. However, one can constrain the model from a knowledge of the molecular chemistry. For example, if there were no counter ions in the subphase, then the total number of electrons in the tail and head group together is known from the chemical formula CH_3-$(CH_2)_{18}$-COO^-. Least-squares fits of $|\phi(\mathrm{Q})|^2 \equiv R(\mathrm{Q})/R_F(\mathrm{Q})$ to the data are represented by the solid lines in the plot of the reflectivity data, and the values of the parameters deduced from the fits are given in the figure caption.

Dramatic differences in the reflectivity curve are apparent when the pH is altered. In (a) it was adjusted by adding NH_3, which resulted in singly charged $Cd(OH)^+$ ions being attracted to the COO^- head group in a 1:1 ratio. In (b) the base was Na(OH) and in this case doubly charged Cd^{++} ions were attracted in the ratio 1:2.

3.9.2 Free surface of liquid crystals

Liquid crystals consist of long molecules with a typical length-to-diameter ratio of five to one. In the upper part of Fig. 3.15 we show an example of a liquid crystal molecule, for brevity labelled nCB, consisting of a hydrocarbon chain C_nH_{2n+1} terminating in two aromatic rings. Two such molecules pair head-to-head, and this entity is considered as one rod-shaped building block of the structures that form at different temperatures. One must differentiate between the positional and the orientational order of such a building block. In the isotropic phase (I) both the position and orientation of the molecules are disordered, whereas in the nematic phase (N) the positions are disordered,

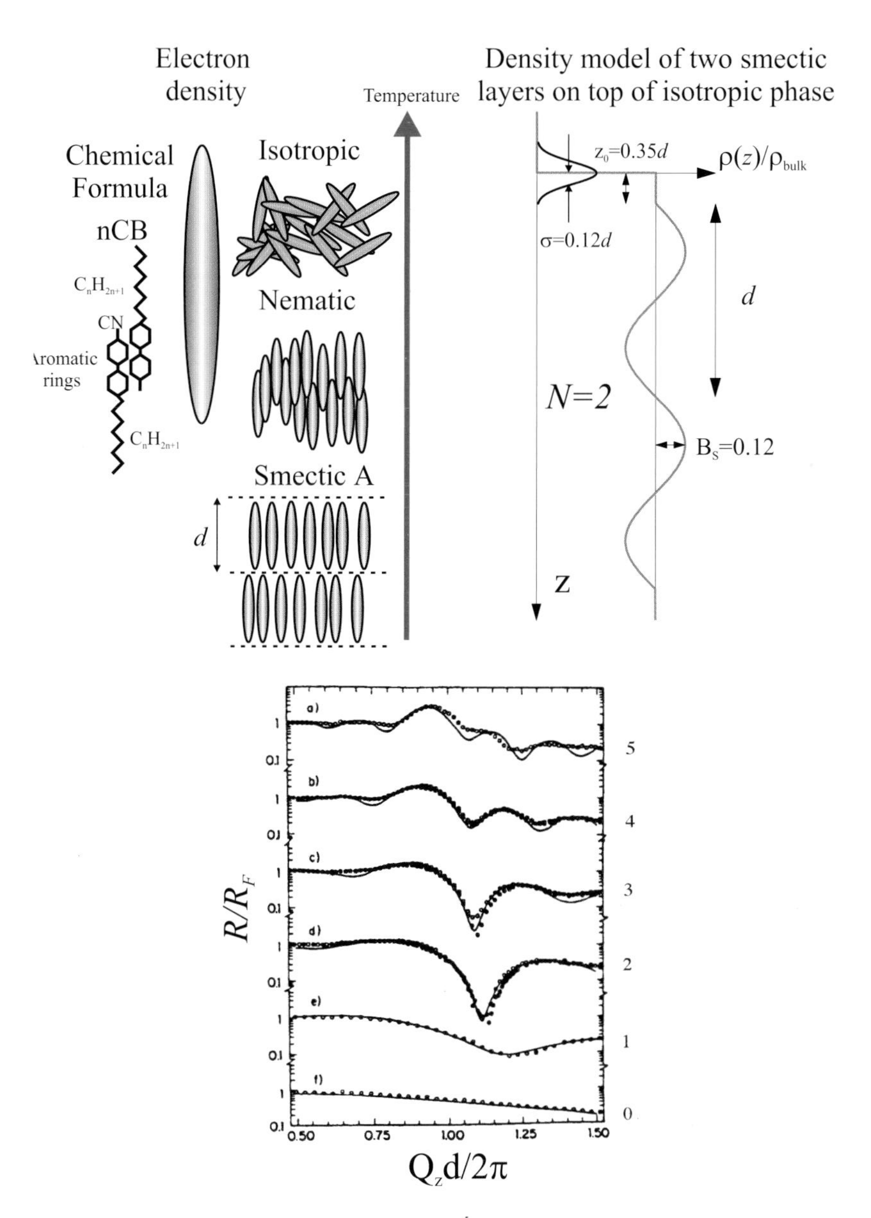

Figure 3.15: Liquid crystals consist of long, rod-shaped molecules (top left). Different phases are characterized by both positional and orientational order of the molecules. In the isotropic phase both of these are disordered. In the nematic phase the position of the molecules is still random, but they all have a common average orientation, the so-called director field. In the smectic-A phase the position of the molecules along the director field is ordered in layers with a repeat distance d but within a given layer the positions are random – it is like a crystal in one direction and like a liquid in the two perpendicular directions. A density model of two smectic layers on top of the isotropic phase is shown top right. This model was used to interpret reflectivity data (lower panel) from the molecule 12CB, which with decreasing temperature goes from the isotropic phase directly to the smectic-A phase [Ocko et al., 1986]. As the transition is approached a distinct number of layers form at the surface. In the lower figure, f) corresponds to zero layers, e) to one layer, etc.

but all molecules have a particular average direction. For the smectic-A phase (SmA) the common orientation is maintained, and in addition the molecules are ordered in layers perpendicular to their long axis with a well-defined repetition distance between the layers, but with positional disorder within the same layer. The smectic-A phase is a particularly interesting object from a structural point of view as it is like a solid crystal in one direction, and, in the plane perpendicular to it, it is like a liquid. Different sequences of transitions between these phases may occur. With decreasing temperature one may find I→SmA, or I→N→SmA. Here the I→N transition is first order, but the N→SmA transition may be discontinuous or continuous. In the latter case critical fluctuations of short-range ordered SmA regions in the N matrix become more and more extended as the N→SmA transition temperature is approached.

In the bottom part of Fig. 3.15 we show data at different temperatures in the isotropic phase for 12CB. This compound has no nematic phase but goes directly from the I phase to the SmA phase in a first-order transition. The reflectivity data shows that as the transition is approached by decreasing the temperature a discrete number of layers build up: the bottom curve corresponds to no layering, the next curve corresponds to one layer, the next to two layers and so on. Modelling of two layers is indicated in the upper right panel of Fig. 3.15. The smectic-A layering is a modulation of the density and may be modelled by a sine curve, which in the example shown has two repetitions, each of length d. Adjustable parameters used in fitting to the data are the amplitude B_S (best value is 0.12 of the bulk density), the phase displacement of the sine-curve with respect to the surface (best value $0.35d$) , and also a smearing of the surface (best r.m.s. value is $0.12d$). In fact, all of these parameters are the same for all of the full lines in the plot, except the number of layers N, i.e. the number of periods of the sine wave density.

All it takes to obtain a nematic phase between the I phase and the SmA phase is a shortening of the aliphatic tail of the nCB molecules from n=12 to say n=8. The surface layering is now quite different and so is the scattering and reflectivity data shown in Fig. 3.16.

Let us now consider the scattering from the schematic model shown in the bottom part of Fig. 3.16, and see that this is indeed consistent with the data shown in the top part of the figure. The surface layering has a well defined lattice spacing d, so as usual it must give a peak in the reflectivity curve when $Q_z=2\pi/d$. If the layer structure is extended very far in lateral directions, it means that the corresponding scattering must be very confined in reciprocal space, i.e. surface layering as shown must show up in the specular reflecting direction, and be modulated with a peak at $Q_0=2\pi/d$, c.f. the filled circles in the left data- panel. A scan in the same direction, but slightly displaced from the specular line, looks very different (open circles in the left data-panel). The intensity is much lower and the peak is symmetric, although with about the same width as the specular peak. Finally a scan in the lateral direction at fixed $Q_z=2\pi/d$ (right hand data-panel) shows a superposition of a very narrow peak, corresponding to the surface layering as already discussed, on top of a much broader peak which must reflect the bulk SmA clusters occurring in the nematic matrix. It is remarkable that one can in this case so clearly separate scattering from the bulk from scattering near the surface. The width of the central peak in the Q_x-scan is resolution limited and proves that the layering is perfect over macroscopic distances. That the width of the surface Q_z-scan

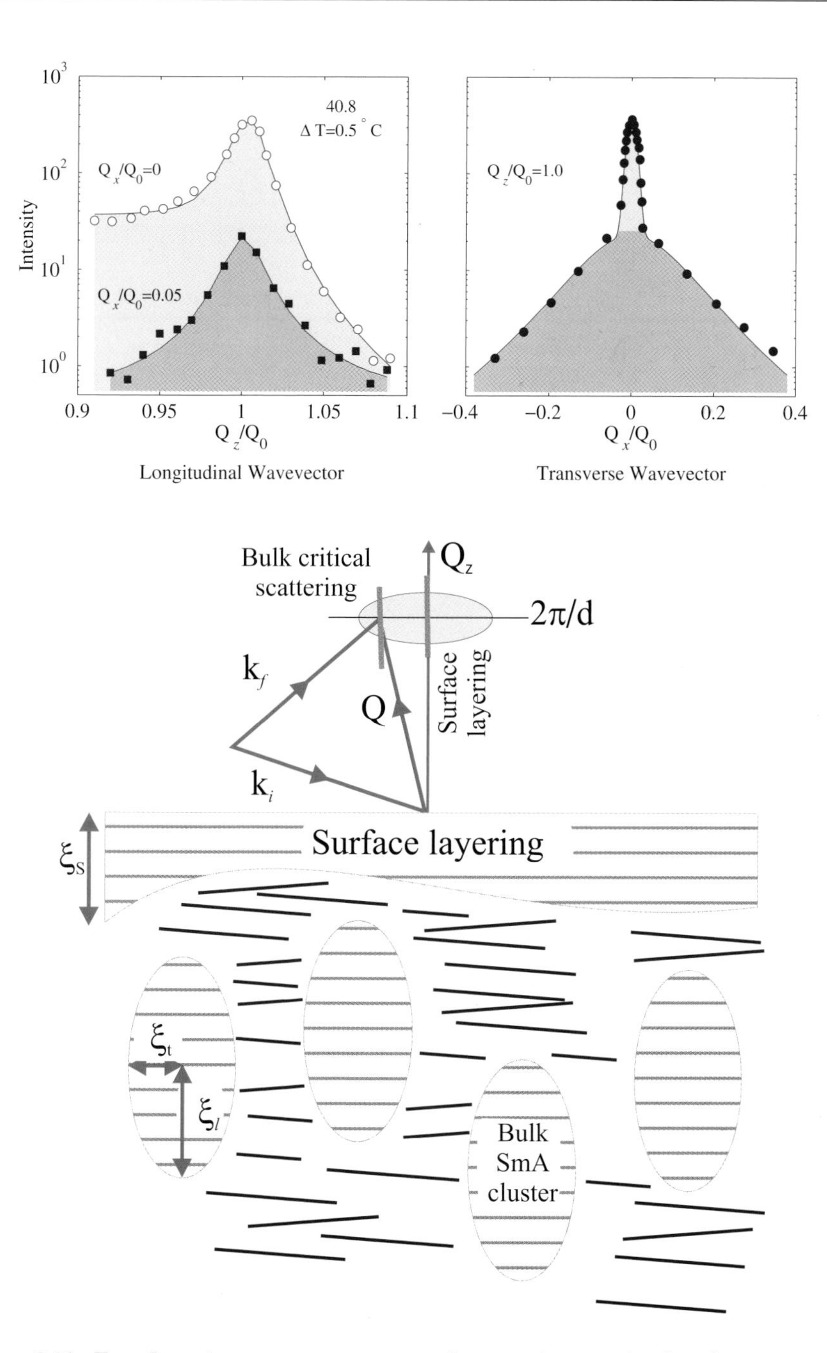

Figure 3.16: Top: Intensity vs. wavevector transfer near the second-order phase transition N → SmA of the liquid crystal 8CB in a free-surface geometry. The left panel shows the longitudinal scans indicated by the two green lines in the scattering diagram, whereas the right panel is the transverse scan indicated by the blue line. Q_0 is $2\pi/d$, d being the lattice plane spacing [Pershan et al., 1987]. Middle: reciprocal space. Critical scattering from the bulk SmA clusters are indicated by the shaded ellipses. Scattering from the surface layering is confined to the Q_z-axis, peaking at $Q_z = Q_0$. Bottom: schematic model for the surface layering used to interpret the data. Horizontal lines indicate planes of SmA molecules. The two Q_z-scans show that the surface penetration depth ξ_s equals the bulk longitudinal correlation length ξ_l of the critical fluctuations.

coincides with that of the bulk Q_z-scan tells us that the penetration of the surface layering is identical to the extent of the critical fluctuations – another remarkable feature that was found at all temperatures in the nematic phase. The reader is referred to the original research article to explain why this is so.

3.10 X-ray optics

3.10.1 Refractive X-ray optics

The ability to manipulate beams of visible light with lenses is of fundamental importance for ordinary optics. Optical lenses made from glass or plastic work so well since they have a refractive index that deviates considerably from unity, and this produces a significant change in the direction of light propagation at the air–lens interface. In addition they are transparent and hardly any losses take place in transmitting the beam through the lens. As we have seen, refraction at an interface also occurs for X-rays, but there are two basic differences from the case of visible light: the deviation in the index of refraction from unity is tiny, of order 10^{-5}; and the refractive index is less than one, not greater than one as it is for visible light. The latter implies that the shape of a converging X-ray lens must be the same as that of a diverging optical lens.

One consequence of the refractive index in the X-ray region being close to unity is that the focal length of a single lens is of order 100 m, which is impractically long for most applications. However, with a series of single lenses the combined focal length may be reduced to a more manageable size [Snigirev et al., 1996]. In synchrotron radiation research, beam lines are typically tens of meters, and a focal length of a similar magnitude may still be useful if it can provide an image spot which is significantly smaller than the source size. A comparative illustration of converging lenses for visible light and for X-rays is given in Fig. 3.17. In the following section the ideal shape of the air–lens interface needed to achieve focusing is discussed, and the formula for the focal length is derived for a given material and a given X-ray wavelength.

The ideal interface

In Fig. 3.18 we consider an X-ray beam entering a material at normal incidence, and exiting the interface at some angle, $\alpha(x)$, determined by the thickness of material, $h(x)$, where x is the distance from the central ray. The problem to be solved is to find the ideal form of $h(x)$ to achieve focusing. On entering the material the X-ray wavelength becomes slightly longer than in air, since the refractive index is a little bit smaller than unity by the amount δ. In Fig. 3.18(a) a comparison is made of two situations: to the left the X-ray beam continues in air, while to the right it has entered the material. The first question to address is how far the X-ray has to travel in the material before the two beams are back in phase again. We denote this distance by Λ. It is given by the condition that it is N wavelengths in the material, but $N+1$ wavelengths in air, i.e.

$$\Lambda = (N+1)\lambda_0 = N\lambda = N(1+\delta)\lambda_0 \qquad (3.57)$$

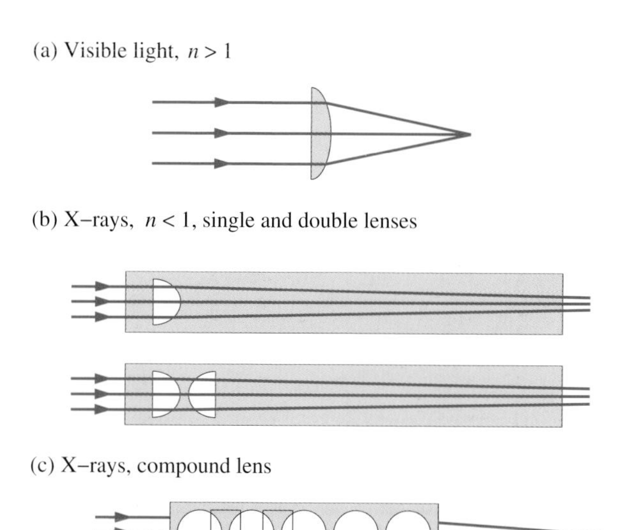

Figure 3.17: A comparison of converging lenses for visible light (a) and X-rays (b and c). The shaded areas represent material, and the white areas cavities. In (c), the rectangles indicate double lenses as shown in the bottom part of (b). The shape of a *converging* X-ray lens, is like that of a *diverging* lens for visible light.

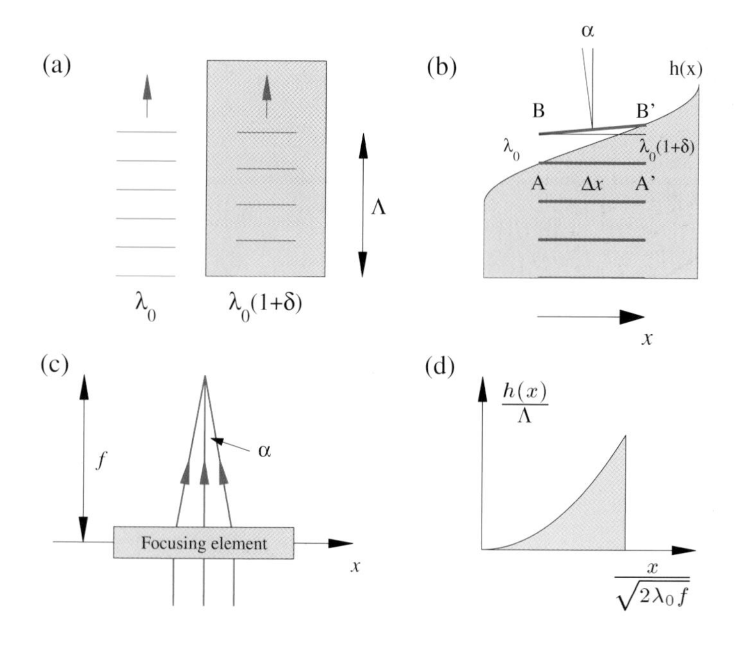

Figure 3.18: (a) Waves in vacuum and in a material are in phase after a distance Λ. (b) Just as the wave exits the material at A, the phases at A and A' are the same. After propagating one more wavelength the wave crest is at B in vacuum and at B' in the material. The direction of propagation is perpendicular to the line BB', and hence makes an angle α with the incident beam given by $\lambda_0\delta/\Delta x$. (c) The condition for focusing is that $\alpha(x) = x/f$, where f is the focal length. (d) The canonical parabolic shape of a focusing refractive lens.

which implies that $N\delta = 1$. We can thus write an approximate expression for Λ as

$$\Lambda \approxeq N\lambda_0 = \frac{\lambda_0}{\delta} \tag{3.58}$$

Using the formula for δ given in Eq. (3.2) we find

$$\boxed{\Lambda = \frac{2\pi}{\lambda_0 r_0 \rho}} \tag{3.59}$$

As usual, the electron number density is denoted by ρ, r_0 is the Thomson scattering length of an electron, and λ_0 is the X-ray wavelength in air. (Here dispersion corrections have been ignored.) For an order-of-magnitude estimate, note that λ_0 is around 1 Å, r_0 is 2.82 10^{-5} Å, and ρ is around 2 electrons per $Å^3$, so that Λ is of order 10^5 Å or 10 μm.

In Fig. 3.18(b) we consider the direction of the exit ray in leaving the material around the lateral position x. Exactly at x the wave has propagated one wavelength, λ_0, in going from point A at the interface to point B in air. From point A' within the material at $x + \Delta x$, where the wave is in phase with that in point A, Δx is chosen so that the wave has also propagated one wavelength $\lambda = \lambda_0(1 + \delta)$ to point B'. In other words, Δx is determined by

$$\lambda_0(1 + \delta) = h'(x)\Delta x \tag{3.60}$$

or

$$\Delta x \approx \frac{\lambda_0}{h'(x)} \tag{3.61}$$

Here $h'(x)$ denotes the derivative of the interface profile $h(x)$. The phases at B and B' are the same, since both points are one wavelength ahead from the equiphase points A and A'. The direction of propagation is therefore perpendicular to BB' and makes an angle with respect to the direction of the incident beam equal to $\alpha(x)=\lambda_0\delta/\Delta x$. This can be rewritten as

$$\alpha(x) = h'(x)\delta = h'(x)\frac{\lambda_0}{\Lambda} \tag{3.62}$$

using the expression for Δx given above together with Eq. (3.58).

We now impose the condition of focusing shown in Fig. 3.18(c): for all x the exit rays must intersect at the focal point, a distance f from the interface. Since the angular deflection of the exit ray is small, $\alpha(x) \ll 1$, and the focal length is large relative to the height of the interface profile, $f \gg h(x)$, we see that

$$\alpha(x) = \frac{x}{f} \tag{3.63}$$

The two equations for $\alpha(x)$ imply that $\lambda_0 h'(x)/\Lambda = x/f$, which by integration yields

$$\boxed{\frac{h(x)}{\Lambda} = \left[\frac{x}{\sqrt{2\lambda_0 f}}\right]^2} \tag{3.64}$$

In other words, by scaling the interface profile $h(x)$ with the "2π- length" Λ, and the lateral coordinate x with the combination of the X-ray wavelength λ_0 and the focal length f, we find a parabolic, universal interface shape, as illustrated in Fig. 3.18(d).

This approach can be applied to find the focal length of the array of X-ray lenses depicted in the bottom of Fig. 3.17. For a single lens the focal length is found from Eqs. (3.64) and (3.58) to be

$$f = \frac{1}{2\delta} \frac{x^2}{h(x)} \tag{3.65}$$

The focal length of N identical lenses is $1/N$ times that of a single lens, and if the ideal parabola is approximated by a circle of radius R, the focal length becomes

$$f_N \approx \frac{R}{2N\delta} \tag{3.66}$$

By way of example, consider an array of lenses formed by drilling a series of 30 holes of 2 mm diameter in Beryllium (Z=4), where the centre spacing of the holes is 2.1 mm. For a photon energy of 10 keV, δ=3.41 $\times$ 10^{-6} and the focal length is f_N = 4.9 m. While this may seem a little long, it is in fact well matched to the length of a typical synchrotron beamline, and arrays of lenses of the type illustrated in Fig. 3.17 have been used to successfully focus X-rays [Snigirev et al., 1996]. One potential problem with this type of lens is absorption. For Beryllium the absorption coefficient at 10 keV is $\mu = 1/(9589\,\mu\text{m})$, and the transmission is $\exp(-31 \times 0.1 \times 10^{-3}/9589 \times 10^{-6})$, or some 72%, more than sufficient for most applications.

The Fresnel zone plate

In the left part of Fig. 3.19 the canonical shape of a focusing X-ray lens has again been plotted. To illustrate the point to be made, we have explicitly indicated by the horizontal lines the depths where the central ray through $x = 0$ and the ray inside the material are in phase: that is at depths Λ, 2Λ, 3Λ..., since here the phase difference is 2π, 4π, 6π, etc. Recalling that the height $h(x)$ equals Λ when the lateral coordinate is $\sqrt{2\lambda_0 f}$, we see that even at a depth of 100Λ, which of order 1000 μm (or 1 mm), the lateral coordinate is only $10\sqrt{2\lambda_0 f}$, or some 100 μm. Not only would it be difficult to manufacture such a lens, but it would also be extremely inefficient, as the X-ray absorption through 1 mm of material could be considerable.[2] However, it is possible to produce a structure with the same optical performance as the standard lens with much lower absorption once it is realised that all of the material indicated in grey could be omitted, as all it does (apart from to absorb) is to provide a phase shift of a multiple of 2π. It follows that the structure shown in the upper right part of the figure would refract the incident beam in just the same way, but now the absorption never exceeds that occurring in a depth of Λ. This structure is known as a Fresnel zone plate. In recent years considerable effort has been devoted to producing Fresnel zone plates that work at X-ray wavelengths [see, for example, Yun, 1999, and references therein].

[2]It is quite likely that if the purpose is to obtain a high throughput in a small spot one would be better off by simply using an aperture instead of such a tiny lens.

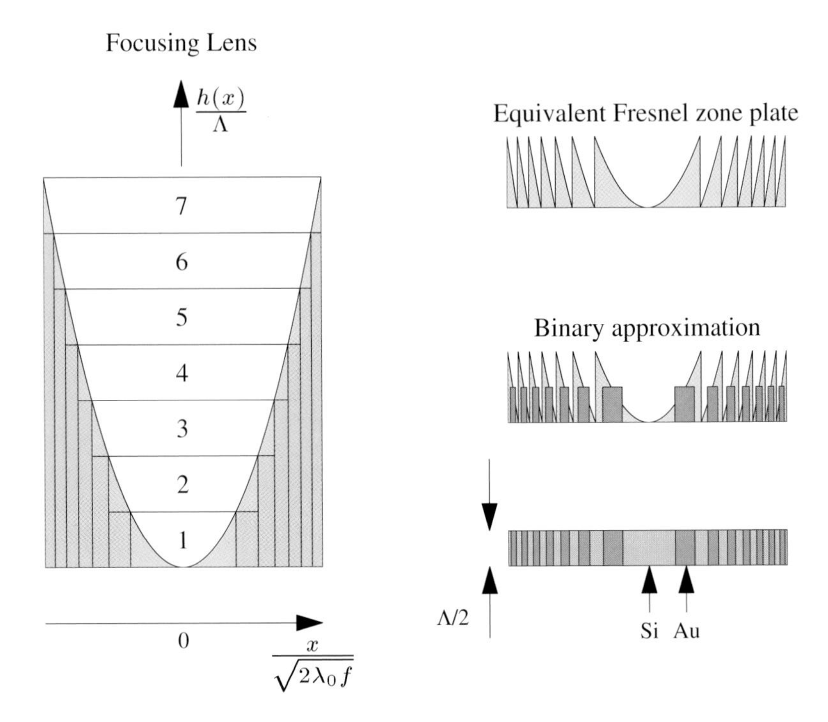

Figure 3.19: Left: The focusing lens. The horizontal lines indicate the levels where the central ray at $x = 0$ is in phase with the rays inside the material. Right, top: The equivalent Fresnel zone plate. Right, bottom: the binary approximation to the Fresnel zone plate.

A Fresnel zone plate has a limited width, or receiving aperture, as there is a technical limit to how narrow one can make a zone of depth Λ. This means in practice that there is an outermost zone number N. To obtain an expression for the width and diameter of this outermost zone it is convenient to first use scaled variables. Let the scaled height be denoted by ν, where $\nu = h(x)/\Lambda$, and the scaled abscissa be denoted by ξ, with $\xi = x/\sqrt{2\lambda_0 f}$. The canonical relation connecting these two scaled variables is, according to Eq. (3.64), $\nu = \xi^2$. The position of the outermost zone N is therefore $\xi_N = \sqrt{N}$, as is clear from the left panel in Fig. 3.19. The scaled *width* of the outermost zone is given by

$$\begin{aligned}\Delta\xi_N &= \xi_N - \xi_{N-1} = \sqrt{N} - \sqrt{N-1} \\ &= \sqrt{N} - \sqrt{N}\sqrt{1 - \frac{1}{N}} \approx \frac{1}{2\sqrt{N}}\end{aligned}$$

while the scaled *diameter* of the lens is $2\xi_N{=}2\sqrt{N}{=}1/\Delta\xi_N$. The equivalent relations for the un-scaled outermost zone width and lens diameter are obtained by multiplying the scaled variables by $\sqrt{2\lambda_0 f}$, i.e.

$$\text{Outermost zone width} = \Delta\xi_N\sqrt{2\lambda_0 f} = \frac{1}{2\sqrt{N}}\sqrt{2\lambda_0 f}$$

and

$$\text{Lens diameter} = \frac{1}{\Delta\xi_N}\sqrt{2\lambda_0 f} = 2\sqrt{N}\sqrt{2\lambda_0 f}$$

The binary approximation

It is technically difficult to produce the Fresnel zones exactly as consecutive pieces of a parabola. As can be seen from the figure, the zones beyond a certain number are almost triangular in shape, but even that turns out to be difficult to manufacture. A zone structure that can be made with a high aspect ratio is the binary zone structure shown in the bottom right part of Fig. 3.19: Stripes of varying width can all be carved to the same depth by means of electron beam lithography followed by etching. A binary approximation to the ideal parabolic zone structure is shown in the right, lower part of Fig. 3.19. The depth of each indentation stripe is $\Lambda/2$. If this zone structure is made in a Silicon wafer, the indentations may afterwards be filled with a material with higher electron density, e.g. gold, and in the formula for Λ one should then insert the difference in electron density of Au and Si as ρ. The binary Fresnel zone plate does not perform nearly as effectively as the parabolic Fresnel zone lens – the efficiency is typically only around 35%.

3.10.2 Curved mirrors

One important application of X-ray reflectivity is the X-ray mirror. The mirror surface is often coated with a heavy material, like gold or platinum, in order to obtain a relatively large electron density. This produces a comparatively large critical angle for total reflection, thereby reducing the required length of the mirror. Such mirrors can be used to filter out the higher-order contamination from a beam monochromatized by Bragg reflection from a single crystal. This is rather obvious from the top part of Fig. 3.6⋆: with a glancing angle $\alpha \leq \alpha_c$ for the fundamental wavevector k of the beam, one obtains close to 100% reflectivity, but the ν'th order wavevector transfer will be $\nu(2\text{k}\sin\alpha)$ and thus larger than Q_c by approximately a factor ν, so the reflectivity is reduced by approximately $(2\nu)^4$. In addition to serving as a low bandpass filter, a mirror may also be curved, and in this way one obtains a focusing optical element for X-rays.

In the ideal mirror device all rays from one particular point will be reflected by the mirror and focused into another point. We distinguish between two cases: in *tangential* focusing (also known *meridional* focusing) all the rays are in one and the same plane, spanned by the incident and reflected central rays; whereas in *sagittal* focusing it is the focusing of rays with a component perpendicular to this plane that is considered.

We shall first consider tangential focusing which is the simpler case since it only requires planar geometry. The ideal mirror curvature is an ellipse. An ellipse can be considered as the projection of a circle, where the projection angle, u, determines the ratio between the minor axis, b, and the major axis, a:

$$\frac{b}{a} = \cos u$$

A ray emitted from the centre of the circle is reflected back into itself. In an ellipse, the circle centre splits into two focal points, F_1 and F_2. A ray emitted

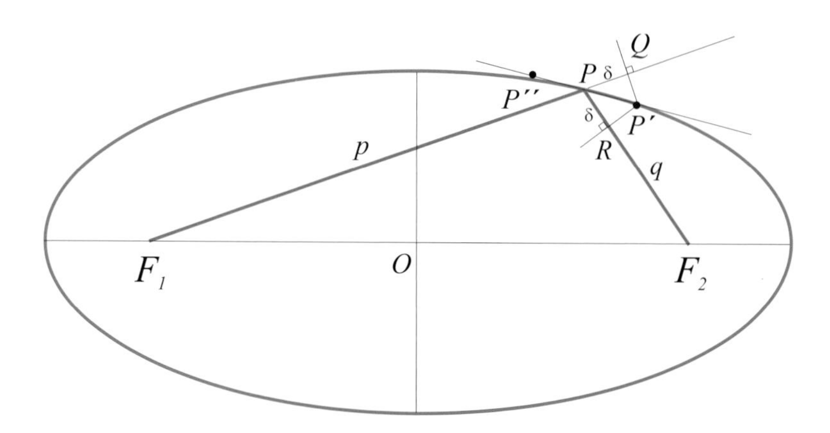

Figure 3.20: Elementary proof that the tangent in a point of an ellipse bisects the angle of the rays from the two foci to the point. The angle RPP' is equal to the angle QPP' and hence a ray from F_1 will be focused at F_2.

in any arbitrary direction from one focal point will be reflected into the other focal point.[3]

For simplicity, we shall in the following restrict the point P to be the symmetric midpoint between F_1 and F_2, providing 1:1 focusing. As shown in Fig. 3.21, the constant sum of distances to the two focal points equals the major diameter $2a$ since $F_1A + F_2A = 2a$, and therefore $F_1B = F_2B = a$. From the general equation of optics:

$$\frac{1}{p} + \frac{1}{q} = \frac{1}{f}$$

where $p(q)$ is the distance from source (image) to the optical element having the focal length f, it follows that for $p = q = a$ one finds $f = a/2$.

Next we consider sagittal focusing. Imagine that the considered ellipse is rotated around the major axis forming an ellipsoid as shown in the bottom part of the figure. In an ellipsoid any ray from one focal point is focused on to the other focal point. In particular the sagittal rays from F_1 to anywhere on a line perpendicular to the plane of central rays through B will be focused in F_2 as indicated in the bottom part of the Fig. 3.21.

In the following we discuss the best approximation to tangential and sagittal cuts through the ellipsoid by circles. The reason is that it is much simpler and cheaper to produce a cylindrical or toroidal surface than it is to produce a true

[3] Although this may be well known we shall give an elementary proof. Consider in Fig. 3.20 a point P on the ellipse at distance $p(q)$ from the focal point $F_1(F_2)$. The basic property of an ellipse is that $p + q$ is constant. Let us find a neighbouring point to P – neighbouring means that the point is on the tangent to the point P. Let point Q be a small amount δ further away from F_1 than P. Consider then all points where the distance from F_1 is increased by this small amount. They must be on the line through Q, perpendicular to F_1P. Similarly, all points with the distance to F_2 diminished by the same amount are on the line through R, perpendicular to PF_2. The intersection of these two lines is therefore the neighbouring point P' since the sum of distances to F_1 and F_2 is maintained constant. A ray from F_1 has an incident angle on the tangent which is $P'PQ$, but since triangles $P'PQ$ and $P'PR$ are congruent this incidence angle is also equal to angle $P'PR$, so the reflected ray will go through F_2.

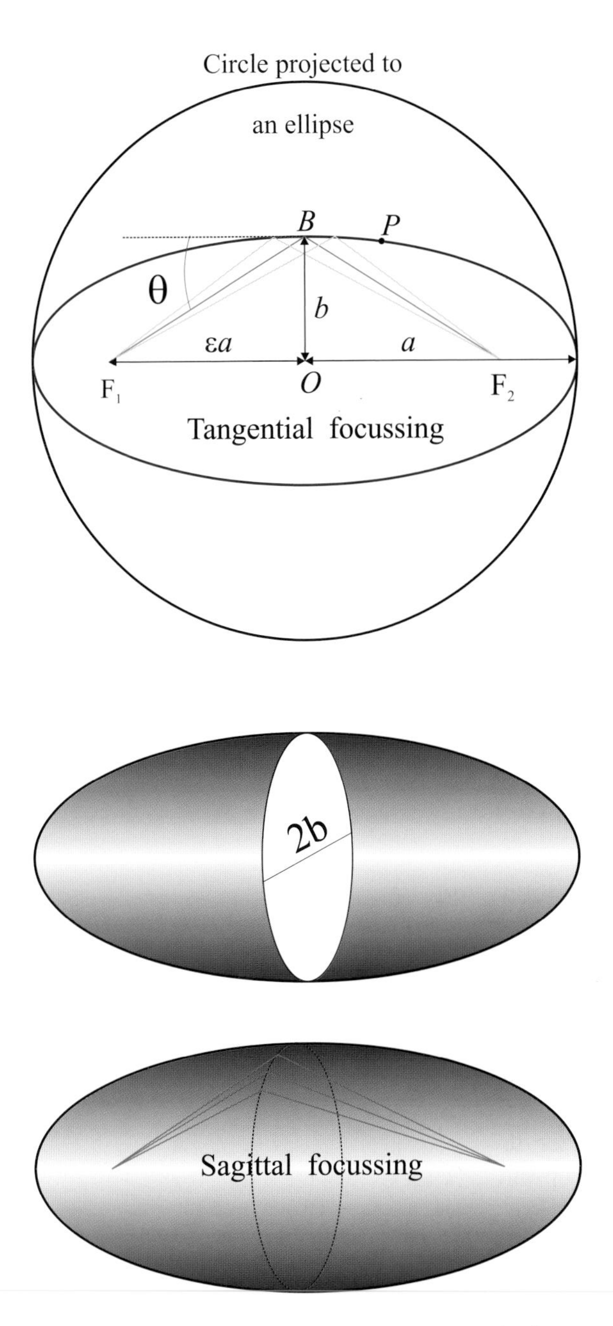

Figure 3.21: By considering an ellipse as a projection of a circle (top), which is the same as a contraction of the vertical axis by the factor (b/a), one realizes that the radius of curvature at B must be the radius of the circle, a, divided by (b/a). Considering the ellipsoid (middle and bottom) it is clear that the best sagittal radius at B must be b.

ellipsoidal surface. The issue is to determine the radius of curvature, ρ, of the best approximating circle.

The glancing angle, θ, at B equals the angle OF_1B, i.e.

$$\sin\theta = \frac{OB}{F_1B} = \frac{b}{a}$$

Furthermore we have just seen that $a = 2f$. The best sagittal circle is clearly the one with radius b as is evident from the middle part of the figure:

$$\rho_{\text{sagittal}} = b = 2f\sin\theta$$

The best tangential circle approximation of the ellipse at B requires a circular radius of

$$\rho_{\text{tangential}} = a\frac{a}{b} = \frac{2f}{\sin\theta}$$

This is clearly correct when $b = a$, and by considering the ellipse as a projection of a circle with radius a it is obvious that the radius at B becomes larger in the ratio of $a : b$.

As stated at the beginning of this section it is desirable to make the angle of incidence of the X-ray as large as possible, so as to reduce the overall length of the mirror, and hence its cost. One way to achieve this is to use a multilayer (as described in Section 3.6) as a mirror. The angle of incidence is then determined by the position of the first principal diffraction maxima, which may be chosen to be many times the critical angle of the materials that constitute the multilayer.

Further reading

X-ray reflectivity studies of liquid surfaces, J. Als-Nielsen, Handbook on Synchrotron Radiation (Eds. G.S. Brown and D.E. Moncton) **3**, 471 (1991)

X-ray and Neutron Reflectivity: Principles and Applications, Eds. J. Daillant and A. Gibaud (Springer-Verlag, 1999)

Focus on Liquid Interfaces, Synchrotron Radiation News, **12** No. 2 (1999),

4. Kinematical diffraction

One of the main uses of X-rays is in the determination of the structure of materials using diffraction. In this chapter we introduce many of the important concepts underlying this subject and derive the essential equations. Our approach is to build on what has been learned already about the interaction of X-rays with a single electron, the Thomson scattering, by starting with simple systems and then gradually adding to the complexity so that we finally arrive at a description of the diffraction from the material in question. Once the formal theory has been developed it is explained how it may be applied in particular situations to solve the structure of a range of materials of contemporary interest, ranging from a bulk powder sample to the structure of a surface. The theme uniting these examples is that the scattering may in a certain sense be considered to be weak. This allows multiple scattering effects to be neglected which leads to considerable simplification in the theory. The weak-scattering limit is also known as the kinematical approximation.

4.1 Two electrons

The most elementary scattering unit that we shall consider is an electron, which is believed to be structureless. Consequently the simplest structure that can be conceived of must be comprised of two electrons. The origin is defined to coincide with one electron, and the second is at a position given by the vector $\mathbf{r}$. Determining the structure of this system therefore amounts to determining $\mathbf{r}$. To do so we imagine that the electrons are illuminated with a monochromatic X-ray beam, and that the *elastically* scattered radiation is observed along a direction $\mathbf{k}'$ as indicated in Fig. 4.1. The incident wave is specified by its wavevector $\mathbf{k}$ and arrives at the electron at $\mathbf{r}$ *after* it has scattered from the electron at the origin. The phase lag for the incident wave, ϕ_{in}, is thus 2π times the ratio of z to the wavelength λ, where z is the projection of $\mathbf{r}$ onto the direction of the incident wave. Thus we can write ϕ_{in}=$\mathbf{k}\cdot\mathbf{r}$. On the other hand, the wave scattered from the electron at $\mathbf{r}$ is *ahead* of the wave scattered from the one at the origin by an amount $|\phi_{\text{out}}|$= $\mathbf{k}'\cdot\mathbf{r}$. It follows that the resulting phase difference is ϕ= $(\mathbf{k}-\mathbf{k}')\cdot\mathbf{r}\equiv\mathbf{Q}\cdot\mathbf{r}$, which defines the wavevector transfer $\mathbf{Q}$. For elastic scattering $|\mathbf{k}|$= $|\mathbf{k}'|$, and from the scattering triangle shown in Fig. 4.1 the magnitude of the scattering vector $\mathbf{Q}$ is related to the scattering angle θ by

$$\boxed{|\mathbf{Q}| = \left(\frac{4\pi}{\lambda}\right)\sin\theta} \qquad (4.1)$$

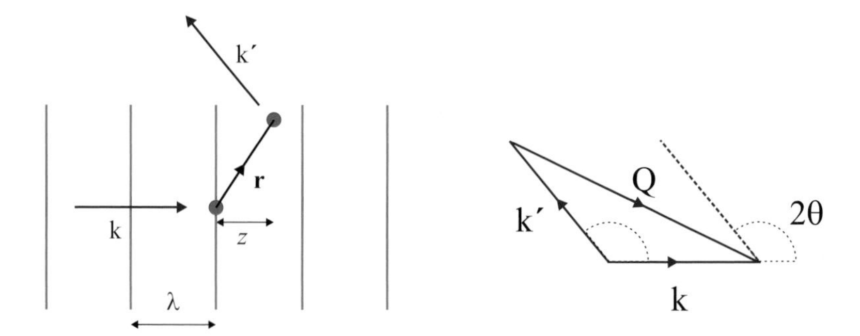

Figure 4.1: The scattering of a monochromatic X-ray beam by a two-electron system. The incident X-ray is labelled by its wavevector $\mathbf{k}$, and has wavefronts represented by the vertical lines. Scattered X-rays are observed in the direction $\mathbf{k}'$. As the scattering is elastic $|\mathbf{k}| = |\mathbf{k}'|$. The phase difference between the incident and scattered X-rays is $\phi = (\mathbf{k} - \mathbf{k}') \cdot \mathbf{r} = \mathbf{Q} \cdot \mathbf{r}$, as shown in the scattering triangle to the right.

The scattering amplitude[1] for the two-electron system may be written as

$$A(\mathbf{Q}) = -r_0(1 + \mathrm{e}^{i\,\mathbf{Q}\cdot\mathbf{r}})$$

and it follows that the intensity is

$$\begin{aligned} I(\mathbf{Q}) &= A(\mathbf{Q})A(\mathbf{Q})^* = r_0^2(1 + \mathrm{e}^{i\,\mathbf{Q}\cdot\mathbf{r}})(1 + \mathrm{e}^{-i\,\mathbf{Q}\cdot\mathbf{r}}) \\ &= 2r_0^2(1 + \cos(\mathbf{Q}\cdot\mathbf{r})) \end{aligned} \tag{4.2}$$

In Fig. 4.2(a) the intensity $I(\mathbf{Q})$ is plotted for the *particular* case where $\mathbf{Q}$ is parallel to $\mathbf{r}$. The natural unit of $\mathbf{Q}$ is Å^{-1} if λ is in Å. However, it is often more convenient to express it in units of 2π divided by the characteristic length scale in the problem, in this case the bond length r. The periodic variation in intensity arises from the interference of waves scattered by the two electrons: it is a maximum when the waves are in phase, and a minimum when they are out of phase. Clearly by measuring $I(\mathbf{Q})$ as a function of $\mathbf{Q}$, that is the diffraction pattern, and fitting $\mathbf{r}$ in Eq. (4.2) to the data, the "structure", i.e. $\mathbf{r}$, of the two-electron system can be determined.

These ideas can be extended to more than two electrons, with the result that the elastic scattering amplitude from any assembly of electrons may be written quite generally as

$$A(\mathbf{Q}) = -r_0 \sum_{\mathbf{r}'_j} \mathrm{e}^{i\,\mathbf{Q}\cdot\mathbf{r}'_j} \tag{4.3}$$

where $\mathbf{r}'_j$ denotes the position of the j'th electron. Of course in the case that the electrons are continuously distributed the sum is replaced by an integral. In this way a model of the diffraction pattern of the sample can be built up gradually by first considering the scattering from all of the electrons in an atom, then all of the atoms in a molecule, etc., until we arrive at a description of the scattering

[1] Here we are assuming that the polarization of the incident beam is perpendicular to the scattering plane spanned by $\mathbf{k}$ and $\mathbf{k}'$ so that the full Thomson acceleration of the electrons is observed at all scattering angles. If this is not the case then the expression for the intensity I must be multiplied by the appropriate polarization factor P, as given in Eq. (1.7).

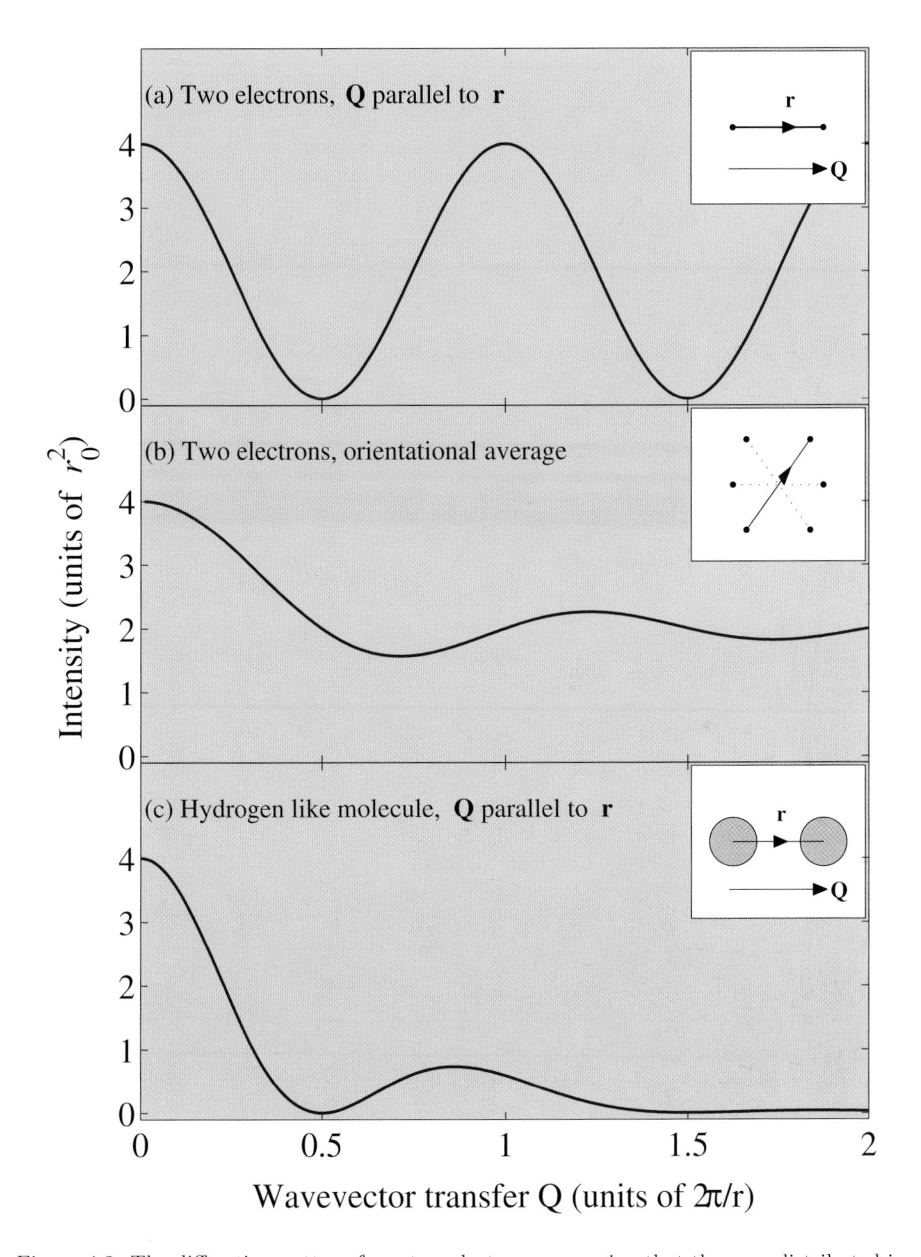

Figure 4.2: The diffraction pattern from two electrons, assuming that they are distributed in various ways. (a) The two electrons are separated by a well-defined vector **r**. The intensity is given by Eq. (4.2), where for the sake of definiteness **r** has been taken parallel to the wavevector transfer **Q**. (b) The electrons are separated by a fixed distance r = |**r**|, but the direction of **r** is randomly oriented in space. In this case the intensity has been calculated from Eqs. (4.5) and (4.6), with $f_1 = f_2 = -r_0$. (c) The electrons are distributed in two charge clouds separated by a distance r = |**r**|, as they would be, for example, in a dumbbell like, diatomic molecule. The wavefunction of each electron has been taken to be the 1s state of the hydrogen atom (see Eq. (4.9)). The effective radius of the electron distribution is specified by the parameter a, and in this example a/r=0.25. Here **Q** has been taken to be parallel to the bond joining the two atoms.

from a crystal. One important and simplifying feature of the diffraction from a crystalline material follows from the fact that a crystal possesses translational symmetry, so that the position vector may be written as $\mathbf{r}'_j = \mathbf{R_n} + \mathbf{r}_j$, where $\mathbf{R_n}$ is a lattice vector specifying the location of a unit cell, and $\mathbf{r}_j$ is the position of the electron within that unit cell. The scattered amplitude then becomes

$$A(\mathbf{Q}) = -r_0 \overbrace{\sum_{\mathbf{R_n}} \mathrm{e}^{i\,\mathbf{Q}\cdot\mathbf{R_n}}}^{\text{lattice sum}} \quad \overbrace{\sum_{\mathbf{r}_j} \mathrm{e}^{i\,\mathbf{Q}\cdot\mathbf{r}_j}}^{\text{unit cell structure factor}} \tag{4.4}$$

The first sum is over the lattice sites of the crystal and is large only when $\mathbf{Q}$ coincides with a reciprocal lattice vector. This is the Laue condition for the observation of diffraction maxima, and is derived later in this chapter. The second sum is the unit cell structure factor which depends on the arrangement of the electrons within the unit cell. The primary goal of crystallography is to determine the unit cell structure factor, and thereby the electron density within the unit cell. In other words to determine how the electrons are distributed in the atoms or molecules that form a crystal structure. However, it is not straightforward to convert the measured diffraction pattern *directly* back to the structure. This is because the measured intensity depends on $|A|^2$, and some special techniques need to be applied to recover information on the phases. This is known as the *phase problem* in crystallography. We shall not discuss these techniques here, although one way to solve the phase problem using resonant scattering is described at the end of Chapter 7. Our aim instead is to illustrate how one can, in a step-by-step procedure, obtain the diffraction pattern from a model of the structure.

It is important to realize that the procedure that we shall follow is valid only if the diffracting volume is small, and hence the scattering is weak. The problem then remains a linear one. In the language of quantum mechanics this means that the sample is only a perturbation on the incident beam, and that the Born approximation is valid. This weak-scattering requirement is also often referred to as the kinematical approximation, which contrasts to the more complicated dynamical one that is treated in Chapter 5. Fortunately the conditions for the validity of the kinematical approximation are met in many applications of X-ray diffraction. In practice this means that the results of an X-ray diffraction experiment are more readily interpreted than those obtained with a strongly interacting probe, such as the electron. Of course there is a price to be paid for this advantage: if the interaction is weak, so is the scattered signal. However, even this drawback is mostly overcome by the availability of intense X-ray beams from today's synchrotron sources.

Orientational averaging

In order to plot the scattered intensity of the two-electron system shown in Fig. 4.2(a) it was necessary to specify the angle between the wavevector transfer, $\mathbf{Q}$, and the position vector, $\mathbf{r}$. For many systems of interest, for example molecules or aggregates in solution, $\mathbf{r}$ is randomly oriented with respect to $\mathbf{Q}$. Here it is explained how the scattering from our simple, prototypical two-electron system is modified by allowing the orientation of $\mathbf{r}$ to vary randomly. The equations derived will enable us to understand the scattering from more realistic systems, for example a gas of molecules, considered later in this chapter.

X-ray scattering is a "fast" probe, in the sense that the time for the transit of the X-ray through the system is short compared to the characteristic time for the motion of the particles comprising the system. Thus in an X-ray experiment a series of snapshots are recorded, which are then averaged. To make the formalism a little more general, we will assume that there are two particles, one at the origin with a scattering amplitude of f_1, and one at a position $\mathbf{r}$ of scattering amplitude f_2, both taken to be real. The instantaneous scattering amplitude from a single snapshot is

$$A(\mathbf{Q}) = f_1 + f_2 \mathrm{e}^{i\,\mathbf{Q}\cdot\mathbf{r}}$$

and the intensity is

$$I(\mathbf{Q}) = f_1^2 + f_2^2 + f_1 f_2 \mathrm{e}^{i\,\mathbf{Q}\cdot\mathbf{r}} + f_1 f_2 \mathrm{e}^{-i\,\mathbf{Q}\cdot\mathbf{r}}$$

If the length of $\mathbf{r}$ remains fixed, but its direction varies randomly with time, then the measured intensity is obtained by performing a spherical or orientational average. This is written as

$$\Big\langle I(\mathbf{Q})\Big\rangle_{\text{orient. av.}} = f_1^2 + f_2^2 + 2\, f_1 f_2 \Big\langle \mathrm{e}^{i\,\mathbf{Q}\cdot\mathbf{r}}\Big\rangle_{\text{orient. av.}} \tag{4.5}$$

Here for simplicity it has been assumed that the scattering particles are spherically symmetric, and it should be clear that $\langle \mathrm{e}^{i\,\mathbf{Q}\cdot\mathbf{r}}\rangle = \langle \mathrm{e}^{-i\,\mathbf{Q}\cdot\mathbf{r}}\rangle$.

The orientational average of the phase factor is

$$\Big\langle \mathrm{e}^{i\,\mathbf{Q}\cdot\mathbf{r}}\Big\rangle_{\text{orient. av.}} = \frac{\int \mathrm{e}^{i\,\mathrm{Qr}\cos\theta}\,\sin\theta\, d\theta d\varphi}{\int \sin\theta\, d\theta d\varphi}$$

The denominator is equal to 4π, while the numerator is

$$\int \mathrm{e}^{i\,\mathrm{Qr}\cos\theta}\,\sin\theta\, d\theta d\varphi = 2\pi \int_0^{\pi} \mathrm{e}^{i\,\mathrm{Qr}\cos\theta}\,\sin\theta\, d\theta = 2\pi \left(\frac{-1}{i\mathrm{Qr}}\right) \int_{i\mathrm{Qr}}^{-i\mathrm{Qr}} \mathrm{e}^{x}\, dx$$

$$= 4\pi \frac{\sin(\mathrm{Qr})}{\mathrm{Qr}}$$

Thus the orientational average of the phase factor is

$$\boxed{\Big\langle \mathrm{e}^{i\,\mathbf{Q}\cdot\mathbf{r}}\Big\rangle_{\text{orient. av.}} = \frac{\sin(\mathrm{Qr})}{\mathrm{Qr}}} \tag{4.6}$$

It is straight forward to generalize this to a system comprising of N particles, with scattering amplitudes of $f_1 \cdots f_N$. The result is

$$\begin{aligned}\left\langle \left| \sum_{j=1}^{N} f_j\, \mathrm{e}^{i\,\mathbf{Q}\cdot\mathbf{r}_j} \right|^2 \right\rangle_{\text{orient. av.}} &= |f_1|^2 + |f_2|^2 + \cdots |f_N|^2 \\ &+ 2f_1 f_2 \frac{\sin(\mathrm{Qr}_{12})}{\mathrm{Qr}_{12}} + 2f_1 f_3 \frac{\sin(\mathrm{Qr}_{13})}{\mathrm{Qr}_{13}} + \cdots + 2f_1 f_N \frac{\sin(\mathrm{Qr}_{1N})}{\mathrm{Qr}_{1N}} \\ &+ 2f_2 f_3 \frac{\sin(\mathrm{Qr}_{23})}{\mathrm{Qr}_{23}} + \cdots + 2f_2 f_N \frac{\sin(\mathrm{Qr}_{2N})}{\mathrm{Qr}_{2N}} \\ &\cdots + 2f_{N-1} f_N \frac{\sin(\mathrm{Qr}_{N-1,N})}{\mathrm{Qr}_{N-1,N}}\end{aligned} \tag{4.7}$$

where $r_{12} = |\mathbf{r}_1 - \mathbf{r}_2|$, etc. This formalism was first derived in 1915 by Debye [Debye, 1915].

In Fig. 4.2(b) the spherically averaged intensity for the two-electron system is plotted as a function of Q. The effect of the averaging is seen to wash out the oscillations in the diffraction pattern at high Q. Mostly, however, we are interested in scattering of electrons bound in atoms, where they may no longer be regarded as point like, but are instead described by a distribution. The fact that atomic electrons have a finite spatial extent also leads to a damping of the diffraction pattern at high Q, as illustrated for a hydrogen-like molecule in Fig. 4.2(c). This is the subject of the next section.

4.2 Scattering from an atom

As our first system of real interest we shall consider the X-ray scattering from an isolated, stationary atom. To begin with the electrons are described as a classical charge distribution, and the elastic scattering is calculated. This introduces the concept of an atomic form factor, which is nothing other than the atomic scattering amplitude. It is then explained, by way of a simple example, how the atomic form factor may be evaluated from a quantum mechanical description of the electrons.

Elastic scattering and the atomic form factor

Classically the atomic electrons are viewed as a charge cloud surrounding the nucleus with a number density $\rho(\mathbf{r})$. The charge in a volume element $d\mathbf{r}$, at a position $\mathbf{r}$ is then $-e\rho(\mathbf{r})d\mathbf{r}$, where the integral of $\rho(\mathbf{r})$ is equal to the total number of electrons Z in the atom. To evaluate the scattering amplitude we must weight the contribution in $d\mathbf{r}$ by the phase factor $\mathrm{e}^{i\,\mathbf{Q}\cdot\mathbf{r}}$, and then integrate over $d\mathbf{r}$, which leads to

$$\boxed{\begin{aligned} f^0(\mathbf{Q}) &= \int \rho(\mathbf{r})\,\mathrm{e}^{i\,\mathbf{Q}\cdot\mathbf{r}}d\mathbf{r} \\ &= \begin{cases} Z & \text{for } \mathbf{Q}\rightarrow 0 \\ 0 & \text{for } \mathbf{Q}\rightarrow \infty \end{cases} \end{aligned}} \tag{4.8}$$

where $f^0(\mathbf{Q})$ is the *atomic form factor* in units of the Thomson scattering length, $-r_0$. The limiting behaviour of $f^0(\mathbf{Q})$ for $\mathbf{Q}\rightarrow 0$ is obvious, as the phase factor then approaches unity and the total number of electrons is the integral of their number density. In the other limit we need to consider how the phases of the waves from the different electrons combine when the wavelength of the radiation becomes much smaller than the atom. The phase factor from any one electron can be represented as a point on the unit circle in the complex plane, since $\mathrm{e}^{i\,\mathbf{Q}\cdot\mathbf{r}} = \cos(\mathbf{Q}\cdot\mathbf{r}) + i\sin(\mathbf{Q}\cdot\mathbf{r})$. Now, in limit of large $\mathbf{Q}$ the phase will be much larger than 2π and the phase factors for the different electrons will fluctuate rapidly around on the unit circle. Therefore the integral, even when weighted by the smoothly varying distribution $\rho(\mathbf{r})$, will tend to zero. In other words, when the wavelength of the radiation becomes small compared to the atom there is a destructive interference of the waves scattered from the different electrons in the atom.

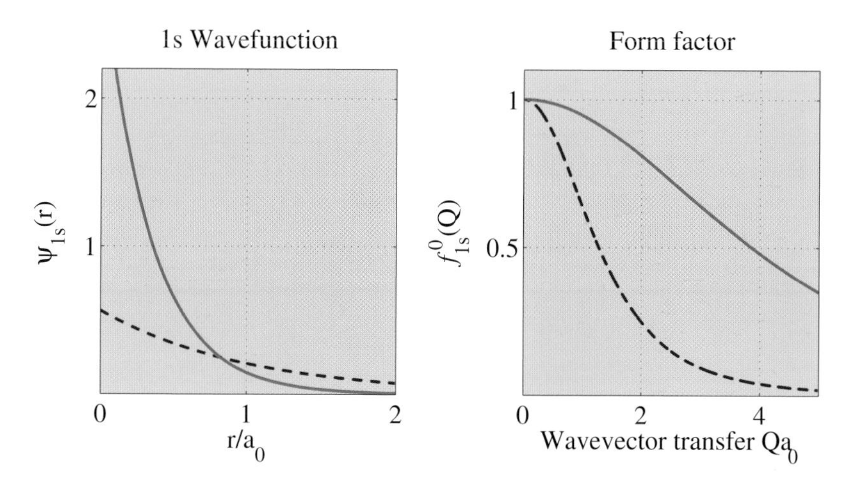

Figure 4.3: The wavefunction and form factor of the 1s state for Z=1 (dashed lines) and Z=3 (solid lines).

In a quantum mechanical description an atomic electron with principal quantum number n is described by its wavefunction $\psi_{\rm n}(\mathbf{r})$. Here we shall take a simple example of the contribution made by the K-shell electrons to the atomic form factor. The wavefunction of a K electron is similar to that of the ground state of the hydrogen atom, and is given by

$$\psi_{1\rm s}(\rm r) = \frac{1}{\sqrt{\pi a^3}}\, e^{-r/a} \tag{4.9}$$

where

$$a = \frac{a_0}{Z - z_{\rm s}}$$

and $a_0 = \hbar^2/me^2$ is the Bohr radius. The effective radius a of the 1s electron is reduced compared with a_0 by the nuclear charge Z. This is itself partly screened by the other 1s electron, and typically $z_{\rm s} \approx 0.3$. The density of a 1s electron is $|\psi_{1\rm s}|^2$ so that the form factor is

$$f^0_{1\rm s}(\mathbf{Q}) = \frac{1}{\pi a^3}\int e^{-2r/a}\, e^{i\,\mathbf{Q}\cdot\mathbf{r}} d\mathbf{r}$$

To evaluate this integral we use spherical polar coordinates $(\rm r, \theta, \phi)$ and note that the integrand is independent of the azimuthal angle ϕ so that the volume element becomes $d\mathbf{r} = 2\pi \rm r^2 \sin\theta\, d\theta d\rm r$. Writing $\mathbf{Q}\cdot\mathbf{r} = \rm Q\, r\cos\theta$ the integral over θ is evaluated in the following way:

$$\begin{aligned} f^0_{1\rm s}(\mathbf{Q}) &= \frac{1}{\pi a^3}\int_0^\infty 2\pi \rm r^2\, e^{-2r/a} \int_{\theta=0}^{\pi} e^{i\,Qr\cos\theta} \sin\theta\, d\theta d\rm r \\ &= \frac{1}{\pi a^3}\int_0^\infty 2\pi \rm r^2\, e^{-2r/a} \frac{1}{i\,\rm Qr}\left[e^{i\,Qr} - e^{-i\,Qr}\right] d\rm r \\ &= \frac{1}{\pi a^3}\int_0^\infty 2\pi \rm r^2\, e^{-2r/a} \frac{2\sin(Qr)}{Qr} d\rm r \end{aligned}$$

The next step is to write sin(Qr) as the imaginary part of a complex exponential, $\sin(\mathrm{Qr})=Im\left\{\mathrm{e}^{i\,\mathrm{Qr}}\right\}$. The form factor then becomes

$$\begin{aligned} f_{1s}^0(\mathbf{Q}) &= \frac{4}{a^3}\frac{1}{\mathrm{Q}}\,Im\left\{\int_0^\infty \frac{\mathrm{r}^2}{\mathrm{r}}\,\mathrm{e}^{-2\mathrm{r}/a}\,\mathrm{e}^{i\,\mathrm{Qr}}\,d\mathrm{r}\right\} \\ &= \frac{4}{a^3}\frac{1}{\mathrm{Q}}\,Im\left\{\int_0^\infty \mathrm{r}\,\mathrm{e}^{-r(2/a-i\,\mathrm{Q})}\,d\mathrm{r}\right\} \end{aligned}$$

which may be integrated by parts to yield the final result

$$f_{1s}^0(\mathbf{Q}) = \frac{1}{[1+(\mathrm{Q}a/2)^2]^2} \tag{4.10}$$

The wavefunction and form factor for two different values of the nuclear charge Z are plotted in Fig. 4.3. The wavefunction has been plotted against r in units of a_0, and the form factor plotted against Q in units of $1/a_0$. As Z is increased the wavefunction becomes more localized around the nucleus, and the form factor correspondingly more extended in Q. Because of this relationship between real space spanned by r and the space spanned by Q, the latter space is known as reciprocal space. Fig. 4.3 serves to illustrate the relationship between a description of objects in the two spaces: objects that are extended in real space, are localized in reciprocal space and *vice versa*. This should be obvious to those familiar with the properties of Fourier transforms, as it is evident from Eq. (4.8) that the atomic form factor is the Fourier transform of the electronic charge distribution.

Considerable effort has been devoted over the years to calculate the form factors of all free atoms (and most of the important ions) from the best available atomic wavefunctions. These are tabulated in the International Tables of Crystallography for different values[2] of $\sin\theta/\lambda = \mathrm{Q}/4\pi$. For computational convenience the calculated form factors have been fitted by the analytical approximation

$$f^0\left(\frac{\mathrm{Q}}{4\pi}\right) = \sum_{j=1}^{4} a_j\,\mathrm{e}^{-b_j \sin^2\theta/\lambda^2} + c = \sum_{j=1}^{4} a_j\,\mathrm{e}^{-b_j\,(\mathrm{Q}/4\pi)^2} + c \tag{4.11}$$

where a_j, b_j and c are fitting parameters. In Table 4.1 we tabulate their values for several of the elements that will be of interest to us later.

The total scattering length f of an atom is the sum of the energy independent part, f^0, and the dispersion correction factors $f' + if''$ that arise from the fact that electrons are bound in an atom. These dispersion corrections were introduced in Chapter 1 and are discussed further in Chapter 7.

4.3 Scattering from a molecule

The next level of complexity we can imagine is to consider the scattering from a group of atoms organized into a molecule. Let the atoms be labelled by j,

[2]Crystallographers tend to prefer to refer to the wavevector transfer as the scattering vector, and to define it without the leading factor of 4π in Eq. (4.1).

	a_1	b_1	a_2	b_2	a_3	b_3	a_4	b_4	c
C	2.3100	20.8439	1.0200	10.2075	1.5886	0.5687	0.8650	51.6512	0.2156
O	3.0485	13.2771	2.2868	5.7011	1.5463	0.3239	0.8670	32.9089	0.2508
F	3.5392	10.2825	2.6412	4.2944	1.5170	0.2615	1.0243	26.1476	0.2776
Si	6.2915	2.4386	3.0353	32.333	1.9891	0.6785	1.5410	81.6937	1.1407
Cu	13.338	3.5828	7.1676	0.2470	5.6158	11.3966	1.6735	64.820	1.5910
Ge	16.0816	2.8509	6.3747	0.2516	3.7068	11.4468	3.683	54.7625	2.1313
Mo	3.7025	0.2772	17.236	1.0958	12.8876	11.004	3.7429	61.6584	4.3875

Table 4.1: Coefficients of the analytical approximation (Eq. (4.11)) to the atomic form factor f^0 for a selection of elements. (Source: International Tables of Crystallography.)

so that we can write the scattering amplitude (again in units of $-r_0$) of the molecule as

$$F^{\rm mol}(\mathbf{Q}) = \sum_{\mathbf{r}_j} f_j(\mathbf{Q})\, \mathrm{e}^{i\,\mathbf{Q}\cdot\mathbf{r}_j} \tag{4.12}$$

To take a specific example, let us consider the molecule CF_4. The four fluorine atoms are tetrahedrally coordinated (at points A,B,C,D) around the central carbon atom (at O) as shown in Fig. 4.4.

The line from D to O intersects the plane spanned by A, B and C at the point O′. Assume that the dimensions are such that OA=OB=OC=OD=1. It is easy to show that $\mathrm{OO}' = z = \frac{1}{3}$ and that angle u between any of the lines from the centre to any of the four apices is given by $\cos u = -\frac{1}{3}$. The proof is as follows. The scalar product of the vectors **OA** and **OD** is

$$\mathbf{OA}\cdot\mathbf{OD} = 1\cdot 1\cdot \cos u = -z$$

but by symmetry we also have

$$\begin{aligned} -z = \mathbf{OA}\cdot\mathbf{OD} &= \mathbf{OA}\cdot\mathbf{OB} = (\mathbf{OO}' + \mathbf{O}'\mathbf{A})\cdot(\mathbf{OO}' + \mathbf{O}'\mathbf{B}) \\ &= z^2 + \mathbf{O}'\mathbf{A}\cdot\mathbf{O}'\mathbf{B} \\ &= z^2 + (\mathrm{O}'\mathrm{A})^2\cos(120^o). \end{aligned}$$

From the right-angle triangle OO′A one immediately finds that $(\mathrm{O}'\mathrm{A})^2 = 1 - z^2$, so that

$$-z = z^2 + (1-z^2)\cos(120^o) = z^2 - \frac{1}{2}(1-z^2) \tag{4.13}$$

from which it follows that $z=\frac{1}{3}$ and $u = \mathrm{acos}(-z) = 109.5^o$.

The molecular form factor of CF_4 is readily evaluated when the scattering vector **Q** is either parallel (+), or anti parallel (−), to a C-F bond:

$$F^{\rm mol}_{\pm}(\mathbf{Q}) = f^{\rm C}(\mathrm{Q}) + f^{\rm F}(\mathrm{Q})(3\times \mathrm{e}^{\mp i\,QR/3} + 1\times \mathrm{e}^{\pm i\,QR}) \tag{4.14}$$

where R is the C-F bond length (1.38 Å). In Fig. 4.5 $|F^{\rm mol}_{\pm}|^2$ is plotted as a function of **Q**, where the values of the form factors $f^{\rm C}(\mathrm{Q})$ and $f^{\rm F}(\mathrm{Q})$ have been

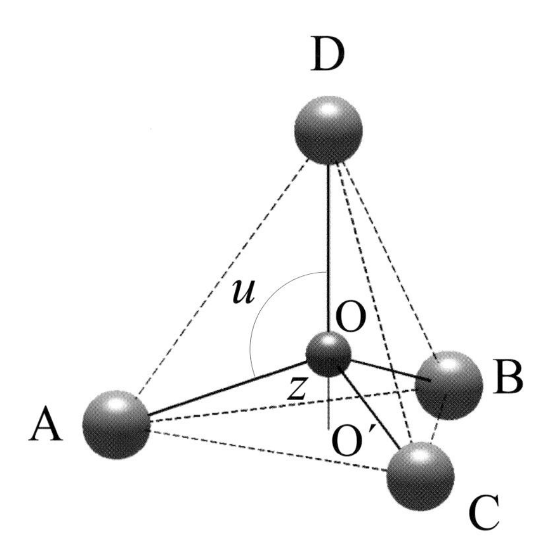

Figure 4.4: The CF_4 molecule. The C-F bond length is 1.38 Å, and the geometry is such that the ratio of the length of OD to OO′ is one third.

calculated from Eq. (4.11) using the coefficients in Table 4.1. The oscillations in the magnitude of $|F_{\pm}^{\rm mol}|^2$ are characteristic features of the scattering from a molecule. They arise from the fact that there are distinct length scales, in this case the C-F (1.38Å) and F-F ($1.38\sqrt{8/3}$) bond lengths, and indeed the first peak in $|F_{\pm}^{\rm mol}|^2$ near $Q = 2\pi/(1.38\sqrt{8/3})$ can be identified with the latter. Clearly, $|F_+|^2 = |F_-|^2$, so that one should not expect too much deviation from the qualitative behaviour shown in Fig. 4.5 when **Q** is not parallel to a C-F bond. That this is indeed the case is evident in the same figure where we have plotted the square of the spherical average of $F^{\rm mol}$. This has been calculated from Eq. (4.7) by noting that there are 4 C-F bonds of length R, and 6 F-F bonds of length $\sqrt{8/3}R$, so that altogether

$$\left|F^{\rm mol}\right|^2 = \left|f^{\rm C}\right|^2 + 4\left|f^{\rm F}\right|^2 + 8\, f^{\rm C} f^{\rm F}\frac{\sin(QR)}{QR} + 12\left|f^{\rm F}\right|^2\frac{\sin(Q\sqrt{8/3}R)}{Q\sqrt{8/3}R} \tag{4.15}$$

(For comparison we also show the squared atomic form factor of molybdenum, the element that has the same number of electrons as CF_4, i.e. Z =6+4×9=42.)

4.4 Scattering from a crystal lattice

A crystalline material is characterized by the fact that it may be constructed by regularly repeating a basic structural unit, known as the unit cell. The points at which the unit cell are located form a *lattice* which may exist in one, two, three and, in certain mathematical models, even higher dimensions. Thus a crystal is constructed by first specifying the lattice, and then associating a collection of atoms (or molecules etc) known as a *basis* with each point in the lattice. The classification of lattices in two and three dimensions is described in introductory texts on solid state physics and crystallography and will not be repeated here. Instead we limit ourselves to reminding the reader of a few important facts that

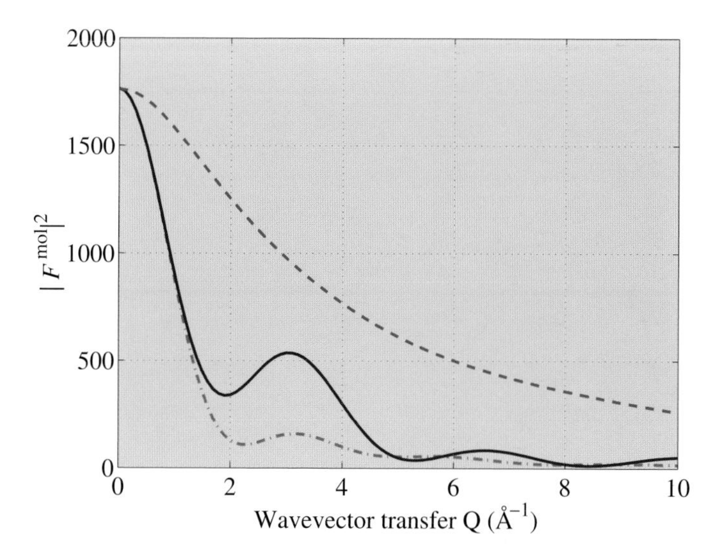

Figure 4.5: The calculated molecular structure factor squared of the CF_4 molecule. The solid line is from Eq. (4.14), while the dashed-dotted line is calculated from a spherical average of the structure factor (Eq. (4.15)). The upper most line (dashed) is the square of the form factor of atomic molybdenum which has the same number of electrons as the CF_4 molecule.

are made use of later.

Lattices and unit cells

A two-dimensional lattice is specified by a set of vectors $\mathbf{R_n}$ with

$$\mathbf{R_n} = n_1\,\mathbf{a}_1 + n_2\,\mathbf{a}_2 \tag{4.16}$$

where $\mathbf{a}_1$ and $\mathbf{a}_2$ are the lattice vectors, and n_1 and n_2 are integers. The vectors $\mathbf{a}_1$ and $\mathbf{a}_2$ define the unit cell, as illustrated in Fig. 4.6(a) for the case of a 2D rectangular lattice. It is important to note that the choice of lattice vectors (including their origin) is to a large extent arbitrary. For example, in the case of our 2D rectangular lattice we could equally well have chosen $\mathbf{a}_2' = 2\mathbf{a}_2$ as shown in Fig. 4.6(b). For any lattice, however, we can always choose the lattice vectors such that the resulting area of the unit cell (or volume in three dimensions) is a minimum. This is known as the *primitive unit cell*, and is defined by the primitive lattice vectors. It follows that the primitive unit cell contains just a single lattice point. That this is the case can be seen by translating the origin by a small amount. When this is done to the unit cell drawn in Fig. 4.6(a) it is clear that it is primitive, whereas the unit cell of Fig. 4.6(b) is non-primitive. From these comments it may seem desirable to always work with a primitive unit cell, as that would seem to offer the best hope of minimizing any possible ambiguities. However, in many situations it turns out to be more convenient to work with a non-primitive unit cell, usually because it is easier to visualize the structure, and the unit cell that is mostly widely used for a given structure is known as the *conventional* unit cell. As an example, in Fig. 4.6(c) we show a cell that is indeed primitive, but one that does not readily reflect the rectangular nature of the lattice.

These considerations of course also apply in three dimensions, and in 3D the lattice is given by a set of vectors of the form

$$\boxed{\mathbf{R_n} = n_1\,\mathbf{a}_1 + n_2\,\mathbf{a}_2 + n_3\,\mathbf{a}_3} \tag{4.17}$$

A given lattice specified by the above has characteristic symmetries, which not only include translations but also rotations. For example, the lattice shown in Fig. 4.6 has a two-fold rotation axis through the origin and perpendicular to the plane of the paper. This enables lattices to be classified into types, and in 1845 Bravais showed that in 2D there are 5 distinct types of lattice consistent with Eq. (4.16) (of which the rectangular lattice is but one), and in 3D there are 14.

To complete the description of a crystal structure we need to associate a *basis* worth of atoms (or molecules) with each and every lattice site. The construction of a two-dimensional crystal from a "lattice+basis" is indicated schematically in Fig. 4.6(d). When the possible symmetries of the basis are combined with those of the lattice it turns out that all crystal structures can be classified into one of 32 possible point groups and one of 230 possible symmetry groups, as described in standard books on crystallography.

Lattices that exist in the real space occupied by the crystal are known as *direct lattices* to distinguish them from ones that may be defined in other spaces.

Lattice planes and Miller indices

X-ray diffraction from a crystalline material is concerned with the scattering from atoms that lie within families of planes in the crystal, and it is obviously desirable to have some way to specify a given family of planes. The Miller indices turn out to be the most convenient way to achieve this. For a given family of planes, the Miller indices (h, k, l) are defined such that the plane closest to the origin (but not including the origin) has intercepts (a_1/h,a_2/k,a_3/l) on the axes ($\mathbf{a}_1$,$\mathbf{a}_2$,$\mathbf{a}_3$). (We note that by convention a negative intercept is represented by writing a bar over the relevant Miller index.)

In Fig. 4.7 we indicate the (10) and (21) planes for the 2D rectangular lattice. This example serves to illustrate two important features of planes specified by their Miller indices. The first is that the density of lattice points in a given family of planes is the same, and that all lattice points are contained within each family. The second is that, again for a given family, the planes are equally spaced, so that it is possible to define a lattice spacing d_{hkl}. For example, it may be shown that the d spacings of a cubic lattice are given by

$$d_{hkl} = \frac{\mathrm{a}}{\sqrt{h^2 + k^2 + l^2}} \tag{4.18}$$

where a is the lattice parameter.

4.4.1 The Laue condition and reciprocal space

Having introduced a way of describing the structure of a crystal we can now proceed to calculate the scattering amplitude. A given atom in the crystal may be thought of as belonging to a basis which is itself associated with a particular unit cell. The position of the atom in the crystal may then be written as

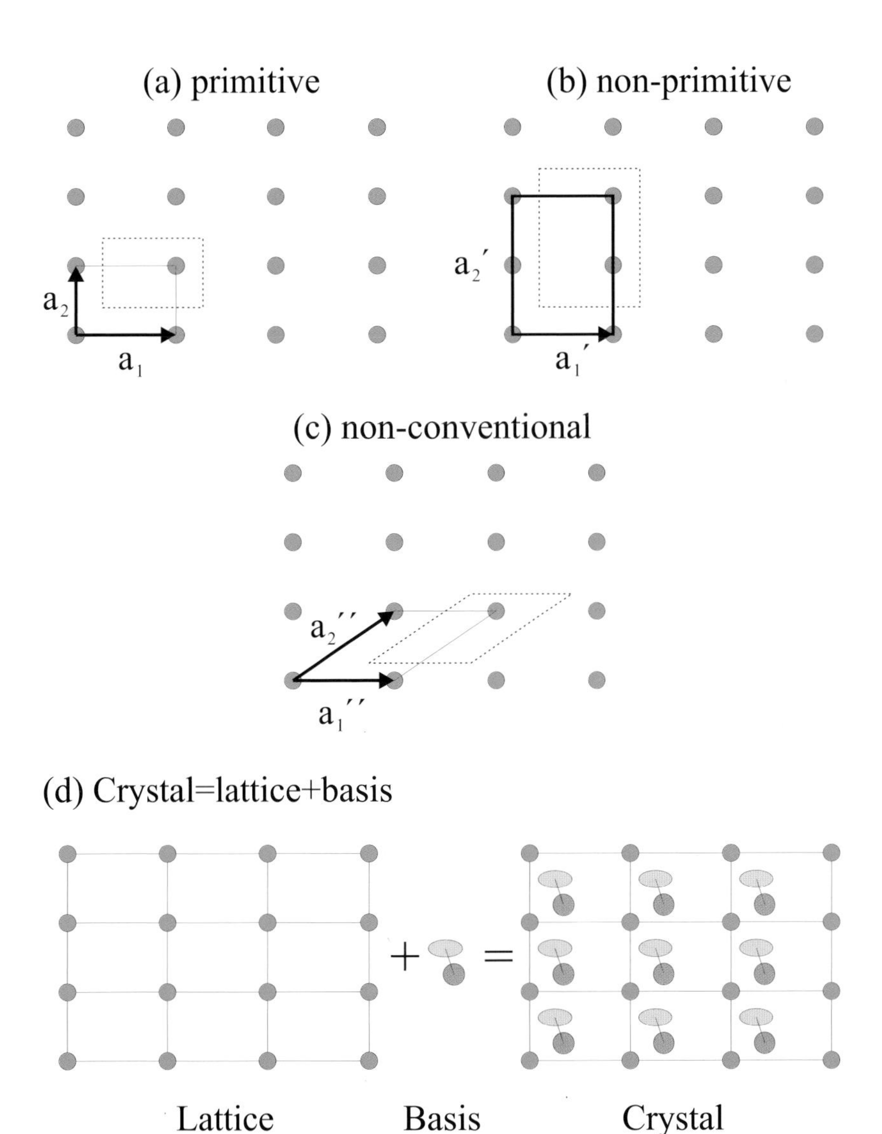

Figure 4.6: Possible unit cells of the 2D rectangular lattice. (a) a primitive unit cell defined by $\mathbf{a}_1$ and $\mathbf{a}_2$. If we translate the origin and produce a new unit cell (indicated by the dotted line) then it is apparent that the unit cell contains one lattice point and is hence primitive. This is also the conventional cell of the 2D rectangular lattice. (b) non-primitive unit cell defined by $\mathbf{a}_1'$ and $\mathbf{a}_2'$, with $\mathbf{a}_2' = 2\mathbf{a}_2$. The unit cell produced by a shift of the origin is seen to contain two lattice points. (c) primitive, unconventional unit cell defined by $\mathbf{a}_1''$ and $\mathbf{a}_2''$. (d)The construction of a two-dimensional crystal structure from a "lattice+basis".

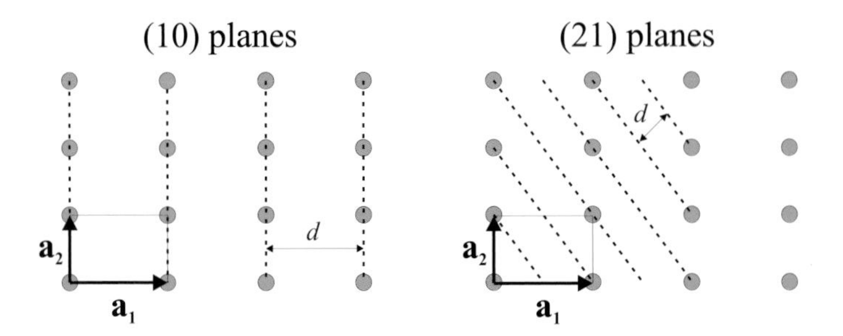

Figure 4.7: Lattice planes and Miller indices for the 2D rectangular lattice: (a) the (10) planes; (b) the (21) planes. In both cases the d spacing of the planes is indicated.

$\mathbf{R_n}+\mathbf{r}_j$, where $\mathbf{R_n}$ specifies the origin of the unit cell and $\mathbf{r}_j$ is the position of the atom relative to that origin. It is immediately apparent from Eq. (4.3) that the scattering amplitude for the crystal factorizes into two terms so that it may be written as

$$F^{\text{crystal}}(\mathbf{Q}) = \overbrace{\sum_{\mathbf{r}_j} F_j^{\text{mol}}(\mathbf{Q})\, \mathrm{e}^{i\,\mathbf{Q}\cdot\mathbf{r}_j}}^{\text{unit cell structure factor}} \overbrace{\sum_{\mathbf{R_n}} \mathrm{e}^{i\,\mathbf{Q}\cdot\mathbf{R_n}}}^{\text{lattice sum}} \tag{4.19}$$

The first term is the scattering amplitude from the basis of molecules (or atoms) contained within the unit cell and is known as the *unit cell structure factor*:

$$F^{\text{u.c.}}(\mathbf{Q}) = \sum_{\mathbf{r}_j} F_j^{\text{mol}}(\mathbf{Q})\, \mathrm{e}^{i\,\mathbf{Q}\cdot\mathbf{r}_j} \tag{4.20}$$

Here $\mathbf{r}_j$ is the position of the j'th molecule in the unit cell, and the term molecule is used in a general sense that may equally apply to a single atom. The second term in Eq. (4.19) is the *lattice sum.*

Although conceptually the lattice sum is yet another step in building up the total scattering amplitude from a crystalline material, it is in practice quite different from all of the summations we have considered so far. The reason is that the number of terms in the lattice sum is enormous. A small crystallite may be of order of 1 micron on each side, which is of order 10^4 times the length of a basis vector, so that the number of terms is of order 10^{12} or more. Each of the terms is a complex number, $\mathrm{e}^{i\,\phi_n}$, located somewhere on the unit circle. The sum of phase factors is of order unity, except when all phases are 2π or a multiple thereof, in which case the sum will be equal to the huge number of terms. The problem is then to solve

$$\mathbf{Q}\cdot\mathbf{R_n} = 2\pi \times \text{ integer} \tag{4.21}$$

To find a solution, suppose that we now construct a lattice in the wavevector space (which has dimensions of reciprocal length) spanned by basis vectors $(\mathbf{a}_1^*, \mathbf{a}_2^*, \mathbf{a}_3^*)$ which fulfill

$$\mathbf{a}_i \cdot \mathbf{a}_j^* = 2\pi\, \delta_{ij} \tag{4.22}$$

where δ_{ij} is the Kronecker delta, defined so that $\delta_{ij} = 1$ if $i = j$ and is zero otherwise. The points on this *reciprocal lattice* are specified by vectors of the

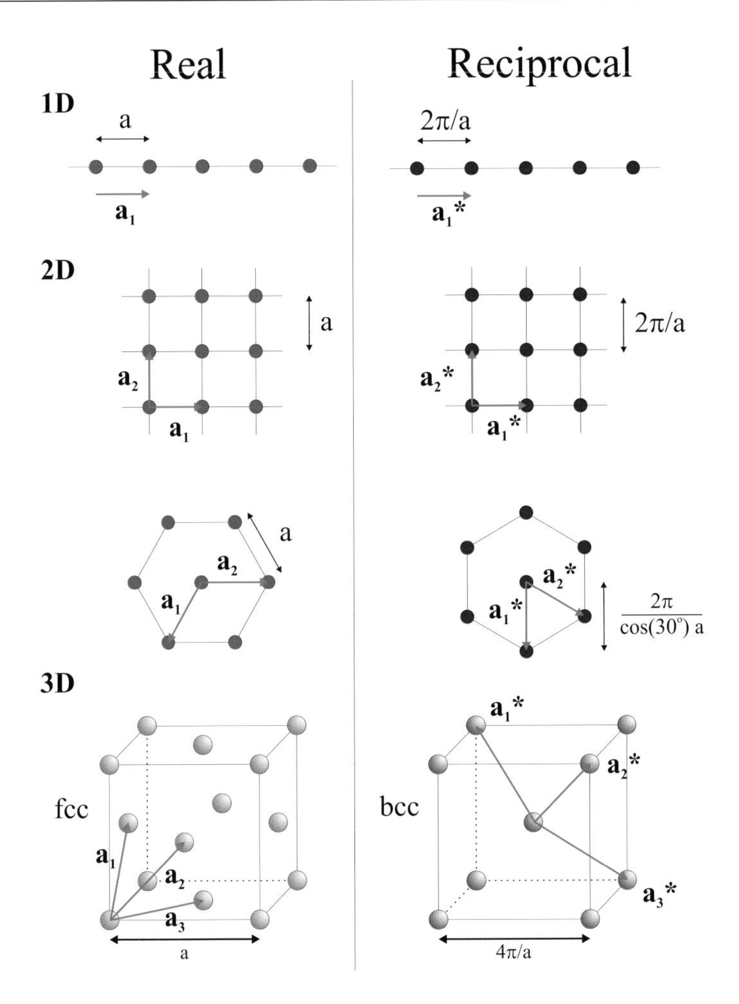

Figure 4.8: Example of the construction of reciprocal lattices in one, two, and three dimensions.

type

$$\boxed{\mathbf{G} = h\,\mathbf{a}_1^* + k\,\mathbf{a}_2^* + l\,\mathbf{a}_3^*} \tag{4.23}$$

where h, k, l are all integers. It is now apparent that the reciprocal lattice vectors $\mathbf{G}$ satisfy Eq. (4.21) since the scalar product of $\mathbf{G}$ and $\mathbf{R_n}$ is

$$\mathbf{G} \cdot \mathbf{R_n} = 2\pi(hn_1 + kn_2 + ln_3) \tag{4.24}$$

and as all of the variables in the parenthesis are integers, the sum of their product is also an integer. In other words, only if $\mathbf{Q}$ coincides with a reciprocal lattice vector will the scattered amplitude from a crystallite be non-vanishing. This is the Laue condition for the observation of X-ray diffraction:

$$\boxed{\mathbf{Q} = \mathbf{G}} \tag{4.25}$$

It is worth emphasizing that the Laue condition is a vector equation, requiring that each component of the wavevector transfer equals the corresponding component of the reciprocal lattice vector. Only when this condition is fulfilled will all of the phases of the scattered waves add up coherently to produce an intense signal. The Laue condition provides a mathematically elegant, but powerful way to visualize diffraction, as we shall see[3]. In order to calculate intensities it is of course necessary to explicitly calculate the lattice sum, and we shall return to this in Section 4.4.4.

Reciprocal lattices

One remaining problem is to find an algorithm to generate the basis vectors of the reciprocal lattice. In one dimension the construction of the reciprocal lattice is obvious from Eq. (4.22), and is given in the top part of Fig. 4.8. In two and three dimensions the situation is a little more complex, and it may be shown that the reciprocal lattice basis vectors are

$$\mathbf{a}_1^* = \frac{2\pi}{v_c}\,\mathbf{a}_2 \times \mathbf{a}_3 \qquad \mathbf{a}_2^* = \frac{2\pi}{v_c}\,\mathbf{a}_3 \times \mathbf{a}_1 \qquad \mathbf{a}_3^* = \frac{2\pi}{v_c}\,\mathbf{a}_1 \times \mathbf{a}_2 \tag{4.26}$$

where $v_c = \mathbf{a}_1 \cdot (\mathbf{a}_2 \times \mathbf{a}_3)$ is the volume of the unit cell. This may be verified by direct substitution into Eq. (4.22).

In two dimensions $\mathbf{a}_3$ is chosen to be a unit vector normal to the 2D plane spanned by $\mathbf{a}_1$ and $\mathbf{a}_2$. For the 2D square lattice it is readily apparent that the reciprocal lattice is also square with a lattice spacing of $2\pi/\mathrm{a}$. If the axes are not orthogonal, as is the case for the 2D hexagonal lattice, then the basis vectors in real and reciprocal space are not necessarily parallel, as indicated in the middle section of Fig. 4.8.

As a final example we show in the bottom part of Fig. 4.8 a three dimensional example. The primitive basis vectors for the face centred cubic lattice are

$$\mathbf{a}_1 = \frac{a}{2}(\hat{\mathbf{y}} + \hat{\mathbf{z}}) \quad , \quad \mathbf{a}_2 = \frac{a}{2}(\hat{\mathbf{z}} + \hat{\mathbf{x}}) \quad , \quad \mathbf{a}_3 = \frac{a}{2}(\hat{\mathbf{x}} + \hat{\mathbf{y}})$$

where we have chosen a set of Cartesian axes parallel to the cube edges. The volume of the unit cell is $v_c{=}\mathbf{a}_1 \cdot (\mathbf{a}_2 \times \mathbf{a}_3)$, and the basis of the reciprocal lattice are

$$\mathbf{a}_1^* = \frac{4\pi}{a}\left(\frac{\hat{\mathbf{y}}}{2} + \frac{\hat{\mathbf{z}}}{2} - \frac{\hat{\mathbf{x}}}{2}\right) \quad , \quad \mathbf{a}_2^* = \frac{4\pi}{a}\left(\frac{\hat{\mathbf{z}}}{2} + \frac{\hat{\mathbf{x}}}{2} - \frac{\hat{\mathbf{y}}}{2}\right) \quad , \quad \mathbf{a}_3^* = \frac{4\pi}{a}\left(\frac{\hat{\mathbf{x}}}{2} + \frac{\hat{\mathbf{y}}}{2} - \frac{\hat{\mathbf{z}}}{2}\right)$$

These are in fact the primitive basis vectors of a body centred cubic lattice with a cube edge of $4\pi/\mathrm{a}$.

Equivalence of the Laue and Bragg conditions

It may be shown that the Laue condition is exactly equivalent to Bragg's Law. In Fig. 4.9(a) the proof of this equivalence is indicated for the specific case of a two-dimensional square lattice. The left hand part of the figure shows the construction normally used to derive Bragg's Law. X-rays are specularly

[3] We have defined the wavevector transfer as $\mathbf{Q} = \mathbf{k} - \mathbf{k}'$, with the implication that $\mathbf{Q}$ points into the origin of reciprocal space. The Laue condition should then read $\mathbf{Q} = -\mathbf{G}$, but this change of sign does not affect any of the discussion.

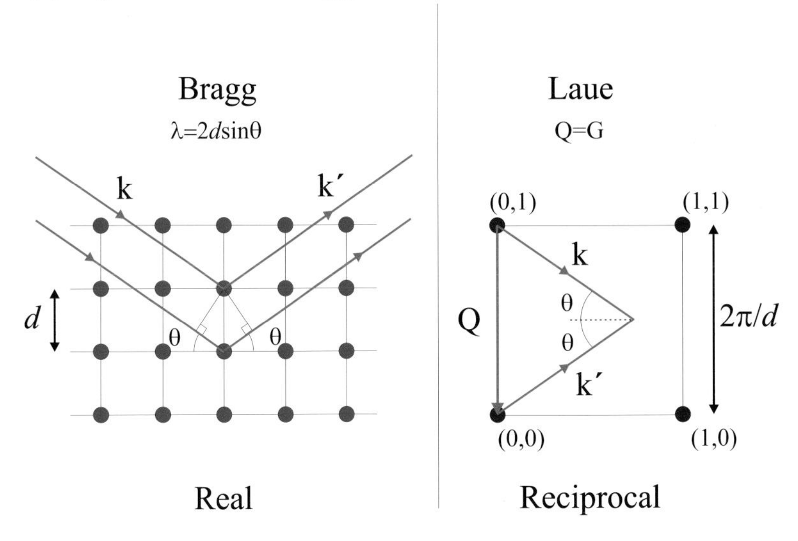

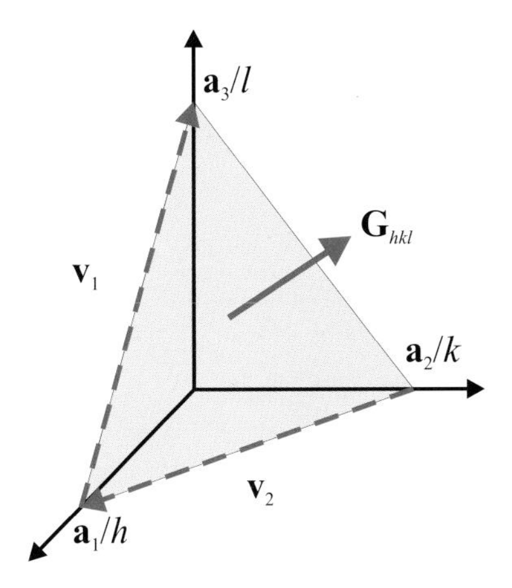

Figure 4.9: (a) The equivalence of Bragg's Law and the Laue condition for the particular case of the 2D square lattice. (b) Construction to prove that the reciprocal lattice vector $\mathbf{G}_{hkl}$ is perpendicular to the (h, k, l) planes, and has a magnitude equal to $2\pi/d_{hkl}$.

reflected from atomic planes with a spacing of d, and the requirement that the path length difference is an integer multiple of the wavelength leads to the well-known statement of Bragg's law: $\lambda = 2d\sin\theta$. The same scattering event is drawn in reciprocal space in the right hand panel. The Laue condition requires that $\mathbf{Q} = \mathbf{G}$. The reciprocal lattice in this case is also square with a lattice spacing of $2\pi/d$, and in the figure we have chosen $\mathbf{Q} = \frac{2\pi}{d}(0,1)$. From the geometry $\mathrm{Q} = 2\mathrm{k}\sin\theta$, since $|\mathbf{k}| = |\mathbf{k}'|$, and we thus have $2\pi/d = 2\mathrm{k}\sin\theta$ which can be rearranged to yield Bragg's Law.

The general proof of the equivalence of Laue's and Bragg formulations follows from the intimate relationship between *points* in reciprocal space and *planes* in the direct lattice. We shall now show that for each point of the reciprocal lattice given by Eq. (4.23) there exists a set of planes in the direct lattice such that

1. $\mathbf{G}_{hkl}$ is perpendicular to the planes with Miller indices (h, k, l).
2. $|\mathbf{G}_{hkl}| = \frac{2\pi}{d_{hkl}}$, where d_{hkl} is the lattice spacing of the (h, k, l) planes.

Consider the plane with Miller indices (h, k, l) shown in Fig. 4.9(b). Two vectors in this plane are given by

$$\mathbf{v}_1 = \frac{\mathbf{a}_3}{l} - \frac{\mathbf{a}_1}{h} \qquad \mathbf{v}_2 = \frac{\mathbf{a}_1}{h} - \frac{\mathbf{a}_2}{k} \tag{4.27}$$

Hence any point in this plane is specified by $\mathbf{v} = \epsilon_1\mathbf{v}_1 + \epsilon_2\mathbf{v}_2$, where ϵ_1 and ϵ_2 are parameters. From Eq. (4.22) the scalar product of $\mathbf{G}$ and $\mathbf{v}$ is

$$\begin{aligned}\mathbf{G}\cdot\mathbf{v} &= (h\mathbf{a}_1^* + k\mathbf{a}_2^* + l\mathbf{a}_3^*)\cdot\left((\epsilon_2 - \epsilon_1)\frac{\mathbf{a}_1}{h} - \epsilon_2\frac{\mathbf{a}_2}{k} + \epsilon_1\frac{\mathbf{a}_3}{l}\right)\\ &= 2\pi(\epsilon_2 - \epsilon_1 - \epsilon_2 + \epsilon_1) = 0\end{aligned}$$

This establishes the first assertion. The plane spacing d is the distance from the origin to the plane, and is found by taking the scalar product of $\hat{\mathbf{G}} = \mathbf{G}/|\mathbf{G}|$, the unit vector along $\mathbf{G}$, and any vector connecting the origin to the plane, $\mathbf{a}_1/h$ say. The d spacing is thus

$$\boxed{d = \frac{\mathbf{a}_1}{h}\cdot\frac{\mathbf{G}}{|\mathbf{G}|} = \frac{2\pi}{|\mathbf{G}|}} \tag{4.28}$$

as required.

To complete the general proof of the equivalence we note that the Laue condition may be re-written in the form, $\mathbf{k} = \mathbf{G} + \mathbf{k}'$. Taking the square of both sides and using the fact that the scattering is elastic ($|\mathbf{k}| = |\mathbf{k}'|$) yields the result

$$\mathrm{G}^2 = 2\mathbf{G}\cdot\mathbf{k} \tag{4.29}$$

where we have also utilized the fact that, if $\mathbf{G}$ is a reciprocal lattice vector, then so is $-\mathbf{G}$. From the scattering triangle (Fig. 4.9(a)) it is apparent that $\mathbf{G}\cdot\mathbf{k} = \mathrm{Gk}\sin\theta$, and since we have already shown above that $\mathrm{G} = 2\pi/d$, Eq. (4.29) can be rearranged as $\lambda = 2d\sin\theta$, thus completing the proof.

The relationship between $|\mathbf{G}|$ and d is an extremely useful one, as once $\mathbf{G}$ is known for the Bragg reflection of interest, d can be calculated. For example, for the simple cubic lattice $\mathbf{G} = \frac{2\pi}{\mathrm{a}}(h, k, l)$, from which it follows that $|\mathbf{G}| = \frac{2\pi}{\mathrm{a}}\sqrt{h^2 + k^2 + l^2}$, and hence $d = \mathrm{a}/\sqrt{h^2 + k^2 + l^2}$ as stated in Eq. (4.18).

4.4.2 The Ewald sphere

A useful way to visualize diffraction events in reciprocal space is provided by the Ewald sphere, or in two dimensions the Ewald circle, construction. First consider the case where a monochromatic beam is incident on a sample. In Fig. 4.10(a) part of a 2D reciprocal lattice is shown. The Laue condition requires that the wavevector transfer $\mathbf{Q}$ is equal to a reciprocal lattice vector $\mathbf{G} = h\mathbf{a}_1^* + k\mathbf{a}_2^*$. In Fig. 4.10(b) the incident X-ray beam is labelled by $\mathbf{k}$ and originates at A and terminates at the origin O. A circle is now drawn centred at A with a length of k, and hence passes through the origin. As shown in Fig. 4.10(c), if any reciprocal lattice points fall on the circle, then the Laue condition is fulfilled, and a diffraction peak observed if the detector is set in the direction of $\mathbf{k}'$. The figure shows an example where we have chosen the point $h = 1$ and k=2 to lie on the circle. Rotating the crystal (equivalent to rotating the Ewald circle about the origin O) brings other reciprocal lattice points onto the Ewald circle. These ideas can be generalized to three dimensions and give rise to the concept of the Ewald sphere.

In certain settings it could occur that more than one reciprocal lattice point falls on the Ewald circle at the same time, giving rise to the simultaneous observation of several reflections (Fig. 4.10(d)). This is known as multiple scattering.

A beam that is not completely monochromatic may be represented by allowing the Ewald circle to have a finite width. Obviously, in the limit that the incident beam is "white" all reflections will be observed within the circles of radius equal to the maximum and minimum $\mathbf{k}$ vector in the beam, as illustrated in Fig. 4.11(a). The discovery of X-ray diffraction by Knipping, Friderich and von Laue was performed in this way. They used the bremsstrahlung spectrum from an X-ray tube to record the diffraction pattern from a single crystal of ZnS. Diffraction data taken with a white beam are now known as Laue patterns. This method is particularly suited to the study of complex structures, such as proteins, where it is necessary to record the intensity of perhaps thousands of Bragg reflections. It may also be desirable to follow the kinetics of a chemical or biological process by monitoring the changes to the structure that occur as the process proceeds. Using modern X-ray sources it is possible to collect a complete Laue pattern from the radiation produced by a single bunch of electrons in the storage ring (Fig. 4.11(b)). This allows the kinetics to be studied on a time scale of 100 ps.

4.4.3 The unit cell structure factor

We now turn to the problem of how to calculate the unit cell structure factor defined in Eq. (4.20). It is important realize that a choice has to be made over which unit cell to use, as this affects the definition of the basis of atoms within the unit cell. The first example is the face centred cubic (fcc) structure shown in bottom part of Fig. 4.8. Here the conventional cubic unit cell is chosen, as it reflects in a more obvious way the symmetry of the problem. With this choice of unit cell the lattice is simple cubic with a lattice spacing of a, and the basis consists of four atoms at

$$\mathbf{r}_1 = 0 \quad , \quad \mathbf{r}_2 = \frac{1}{2}(\mathbf{a}_1 + \mathbf{a}_2) \quad , \quad \mathbf{r}_3 = \frac{1}{2}(\mathbf{a}_2 + \mathbf{a}_3) \quad , \quad \mathbf{r}_4 = \frac{1}{2}(\mathbf{a}_3 + \mathbf{a}_1)$$

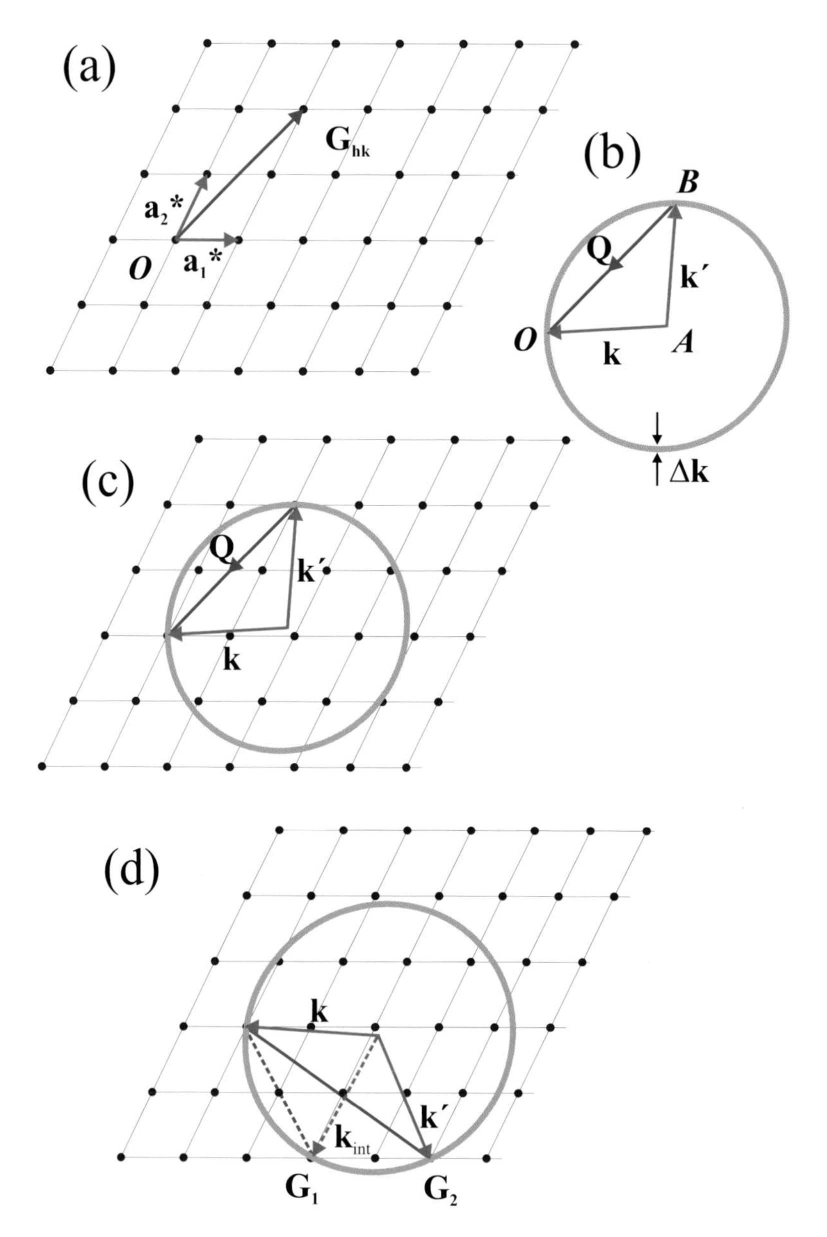

Figure 4.10: The Ewald circle in two dimensions. (a) In 3D the reciprocal lattice is generated by integer coordinates (h, k, l) of the reciprocal basis vectors $(\mathbf{a}_1^*, \mathbf{a}_2^*, \mathbf{a}_3^*)$. For simplicity a 2D lattice at points given by $\mathbf{G} = h\mathbf{a}_1^* + k\mathbf{a}_2^*$ is shown. (b) The scattering triangle. Monochromatic incident radiation specified by $\mathbf{k} = \mathbf{AO}$ can be scattered to any wavevector $\mathbf{k}' = \mathbf{AB}$ terminating on the sphere of radius k. The bandwidth of the incident radiation Δk is indicated by the thickness of the circle. The scattering vector is defined as the vector $\mathbf{Q} = \mathbf{BO}$. (c) The Ewald circle (or Ewald sphere in 3D) is a superposition of (a) and (b) with $\mathbf{k}$ terminating on the origin of the reciprocal lattice. (d) Multiple scattering occurs if two or more reciprocal lattice points fall on the Ewald sphere. The rotation of the crystal and detector are set to record the $\mathbf{G}_2$ reflection, but as $\mathbf{G}_1$ is on the circle, the incident wave will also be scattered to $\mathbf{k}_{\mathrm{int}}$. *Inside* the crystal $\mathbf{k}_{\mathrm{int}}$ is scattered to $\mathbf{k}'$ by the reflection $\mathbf{G}_2 - \mathbf{G}_1$, and intensity may appear in the direction of $\mathbf{k}'$, even if the unit cell structure factor for this reflection vanishes.

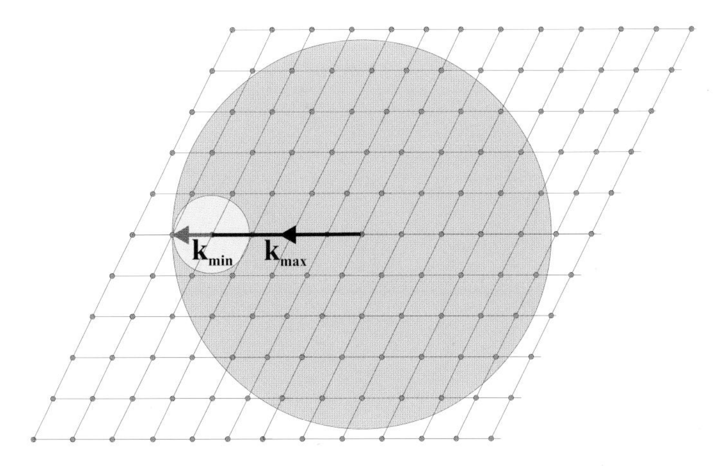

Figure 4.11: The Ewald sphere construction for a white beam containing all wavevectors from k_{min} to k_{max}. All the reciprocal lattice points in the shaded area will Bragg reflect simultaneously. Knipping, Friderich and von Laue's discovery of the diffraction of X-rays from a single crystal of ZnS was performed in this way using the bremsstrahlung spectrum from an X-ray tube, and with a photographic film as detector. The exposure time was several hours. With today's third generation synchrotron sources one can register of order a 1000 reflections on an area detector within the duration of a single pulse from the electron bunch, i.e. about 100 ps.

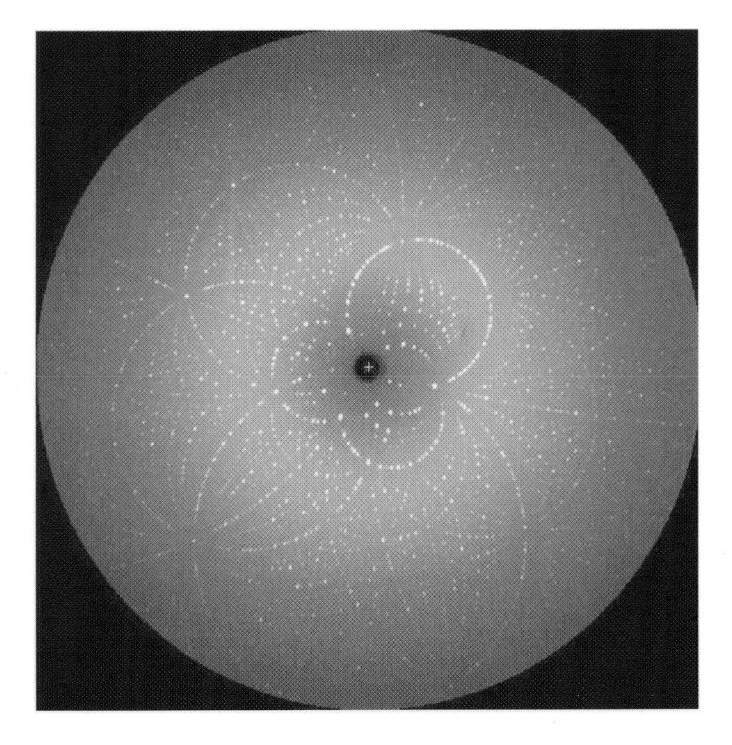

Figure 4.12: Pulsed Laue diffraction pattern from the photo-active yellow protein. The diffraction pattern was collected by averaging over 10 exposures, each of 100 ps duration. This image contains about 3700 usable reflections from which the structure could be obtained. (Data courtesy of Michael Wulff, European Radiation Facility, and Benjamin Perman, University of Chicago.)

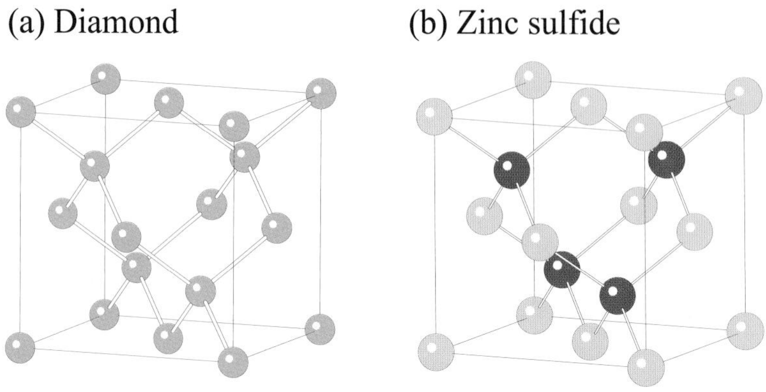

(a) Diamond (b) Zinc sulfide

Figure 4.13: The diamond lattice (a) can be formed from two inter-penetrating fcc lattices displaced by $(\frac{1}{4}\frac{1}{4}\frac{1}{4})$ with respect to each other. For the zinc sulfide (also known as zinc blende) structure (b) the two lattices are occupied by different types of atom.

Here $\mathbf{a}_1$, $\mathbf{a}_2$ and $\mathbf{a}_3$ are parallel to the cube edges. The choice of unit cell implies that the reciprocal lattice is also simple cubic with a lattice spacing of $2\pi/\mathrm{a}$, and as a result a reciprocal lattice vector is of the form $\mathbf{G}=(\frac{2\pi}{\mathrm{a}})\,(h,k,l)$.

For simplicity assume that all of the atoms in the unit cell are identical. The atomic scattering factor can then be taken outside the summation in Eq. (4.20), and the problem then is to sum the phase factors. The unit cell structure factor is evaluated as

$$\begin{aligned} F_{hkl}^{fcc} &= f(\mathbf{Q}) \sum_{\mathbf{r}_j} \mathrm{e}^{i\,\mathbf{Q}\cdot\mathbf{r}_j} \\ &= f(\mathbf{Q})(1 + \mathrm{e}^{i\,\pi(h+k)} + \mathrm{e}^{i\,\pi(k+l)} + \mathrm{e}^{i\,\pi(l+h)}) \\ &= f(\mathbf{Q}) \times \begin{cases} 4 & \text{if } h,k,l \text{ are all even or all odd} \\ 0 & \text{otherwise} \end{cases} \end{aligned}$$

The (1,0,0) reflection, which is the shortest reciprocal lattice vector, has a vanishing structure factor, since h is odd, but k and l are even: the reflection is said to be forbidden. The shortest reciprocal lattice vector for an allowed reflection is the (1,1,1) reflection – all indices are odd. The next one is the (2,0,0) reflection – here all indices are even.

Next, consider the diamond lattice shown in Fig. 4.13(a) which is composed of two fcc lattices, displaced $\frac{1}{4}$ of a cube diagonal relative to each other. This is the structure of Si and Ge, and of course of carbon in its diamond form. The structure factor is

$$F_{hkl}^{\mathrm{diamond}} = F_{hkl}^{fcc}\,(1 + \mathrm{e}^{i\,2\pi(h/4+k/4+l/4)})$$

By inspection, the (1,1,1) reflection has a structure factor of $4(1-i)$, the (2,0,0) reflection is forbidden, the (4,0,0) reflection has a structure factor of 8, the (2,2,2) reflection is forbidden, etc. However, as shown schematically in Fig. 4.10(d) one can accidentally observe intensity with the spectrometer set for a forbidden reflection such as (2,2,2) through a multiple scattering event. For example, the forbidden (2,2,2) reflection can be thought of as the sum of two allowed reflections, (3,1,1)+ ($\bar{1}$,1,1), and if both of these reciprocal lattice points happen to fall on the Ewald sphere, then scattered intensity is observed.

One important variant of the diamond structure is when the two *fcc* lattices are occupied by different types of atom. This is known as the zinc sulfide (or zinc blende) structure, see Fig. 4.13(b), and is the structure adopted by many semiconducting materials such as GaAs, InSb, CdTe, etc. In the former Ga atoms occupy one *fcc* lattice, and As atoms the other. In that case the different phase factors must be weighted by their respective atomic scattering factors. The structure factor for the (2,0,0) reflection, which is forbidden in the diamond structure, is

$$F_{200}^{\mathrm{GaAs}} = 4(f^{\mathrm{Ga}}(2,0,0) - f^{\mathrm{As}}(2,0,0)) \tag{4.30}$$

4.4.4 Lattice sums in 1, 2 and 3 dimensions

The next ingredient that needs to be considered before the intensity of a given Bragg reflection can be calculated is the lattice sum defined in Eq. (4.19) as

$$S_N(\mathbf{Q}) = \sum_{\mathbf{R_n}} \mathrm{e}^{i\,\mathbf{Q}\cdot\mathbf{R_n}} \tag{4.31}$$

Here the sum is evaluated in one, two and three dimensions. The reader is reminded that the subscript **n** refers to the fact that the lattice vector $\mathbf{R_n}$ is specified by a set of integers that reflects the dimensionality of the lattice. In three dimensions we require a set of integers (n_1, n_2, n_3), etc. As our main aim is to arrive at an expression for the intensity we will also evaluate the squared lattice sum $|S_N(\mathbf{Q})|^2$.

One dimension

In one dimension the lattice points are specified by $\mathrm{R}_n = n\,\mathrm{a}$ where n is an integer and a is the lattice parameter. For a finite 1D lattice with N unit cells the sum may be written as

$$S_N(\mathrm{Q}) = \sum_{n=0}^{N-1} \mathrm{e}^{i\,\mathrm{Q}n\mathrm{a}}$$

Evaluation of this geometric series has already been discussed on page 51. In order to study its behaviour when the Laue condition is almost fulfilled, a small parameter ξ is introduced which is defined by

$$\mathrm{Q} = (h+\xi)\mathrm{a}^*$$

where $\mathrm{a}^* = 2\pi/\mathrm{a}$ is a reciprocal lattice basis vector, and h is an integer. The lattice sum may then be written as

$$S_N(\xi) = \mathrm{e}^{i\pi\xi(N-1)} \frac{\sin(N\pi\xi)}{\sin(\pi\xi)}$$

and it follows immediately that its absolute value is

$$|S_N(\xi)| = \left[\frac{\sin(N\pi\xi)}{\sin(\pi\xi)}\right] \to N$$

where we have indicated the limit for $\xi \to 0$ and for large N.

The width of the absolute value of the lattice sum may be estimated by setting $\xi = 1/2N$ whence

$$|S_N(\xi = \frac{1}{2N})| \approx \left(\frac{2}{\pi}\right) N \approx \frac{1}{2} \text{(Peak Height)}$$

It is evident that the peak height is equal to N, and the full width at half maximum is approximately $1/N$, so that the area and hence the integrated intensity is approximately equal to unity. In fact it is possible to show that the area is exactly unity. As N becomes large the width narrows and the lattice sum can then be written as

$$|S_N(\xi)| \to \delta(\xi)$$

where δ is the Dirac delta function. The reader is reminded of the definition and properties of the Dirac delta function in the box on page 147. The lattice sum may be written in the equivalent form

$$|S_N(\mathrm{Q})| \to \mathrm{a}^* \, \delta(\mathrm{Q} - \mathrm{G}_h) \tag{4.32}$$

where $\mathrm{G}_h \equiv h\mathrm{a}^*$, the 1D reciprocal lattice vector. The factor of a^* arises when converting from the delta function $\delta(\xi)$ to the delta function $\delta(\mathrm{Q} - \mathrm{G}_h)$, since $\delta(\mathrm{Q} - \mathrm{G}_h) = \delta(\xi \mathrm{a}^*) = \delta(\xi)/\mathrm{a}^*$.

In a diffraction experiment it is the squared lattice sum that is of interest. Using similar arguments to those given above it is straight forward to establish that the squared lattice sum can also be written in terms of a delta function:

$$|S_N(\mathrm{Q})|^2 \to N\,\mathrm{a}^* \, \delta(\mathrm{Q} - \mathrm{G}_h) \tag{4.33}$$

The squared sum is plotted in the box on page 51.

Two and three dimensions

A two-dimensional lattice is shown in Fig. 4.6. The unit cell is spanned by the two basis vectors $\mathbf{a}_1$ and $\mathbf{a}_2$. A special case is when the macroscopic crystal has the shape of a parallelepiped, so that the number of unit cells along the $\mathbf{a}_1$ direction is always N_1, independent of the row number $1, 2, \cdots, N_2$. Following the same method outlined for the 1D case above it is obvious that

$$|S_N(\xi_1, \xi_2)|^2 \to N_1 N_2 \, \delta(\xi_1) \, \delta(\xi_2)$$

for large (N_1, N_2). Again use is made of the Dirac delta function to write this in the form

$$|S_N(\mathbf{Q})|^2 \to (N_1 \mathrm{a}_1^*)(N_2 \mathrm{a}_2^*) \, \delta(\mathbf{Q} - \mathbf{G}) = N A^* \, \delta(\mathbf{Q} - \mathbf{G}) \tag{4.34}$$

where $\mathbf{G} = h\,\mathbf{a}_1^* + k\,\mathbf{a}_2^*$, A^* is the area of the unit cell in reciprocal space, and $N = N_1 N_2$ is the number of unit cells. In the general case, one cannot evaluate the sum analytically and afterwards square it to look at the limiting behaviour for large numbers of unit cells. However, the delta function character will be maintained for any crystal shape as long as the number of unit cells in both directions is large.

Generalization to three dimensions is straightforward: for a parallelepiped the summation can be carried out analytically, but for a general shape it cannot. When the number of unit cells in all three dimensions is large, then independently of the actual crystal shape

$$|S_N(\mathbf{Q})|^2 \to N\, v_c^*\, \delta(\mathbf{Q}-\mathbf{G}) \tag{4.35}$$

where $\mathbf{G} = h\,\mathbf{a}_1^* + k\,\mathbf{a}_2^* + l\,\mathbf{a}_3^*$, N is the total number of unit cells, and v_c^* is the volume of the unit cell in reciprocal space.

4.4.5 Quasiperiodic lattices

The defining property of a crystal, in contrast to a liquid or gas, is that it displays long-range order at the atomic level. Up to this point long-range order has been interpreted as meaning that the crystal structure is periodic. This allows it to be described in terms of a lattice of unit cells, with lattice vectors given by Eq. (4.17). The requirement of periodicity restricts the set of transformations which leave the properties of the lattice invariant. If we consider possible rotations, then a periodic lattice may be invariant under a n-fold rotation only if n is equal to 2, 3, 4, or 6. In two dimensions, for example, the fact that it is not possible to have a 5-fold axis corresponds to the well-known problem of trying to tile a 2D surface with pentagons, which cannot be achieved without leaving holes.

It is no exaggeration to state that the whole edifice of crystallography was built on the assumption of periodicity, and it therefore came as a considerable shock when a new class of materials was discovered by Shechtman and co-workers in 1982 which displayed a sharp diffraction pattern with a 10-fold axis of rotation. This discovery was met with a large degree of disbelief, and indeed it took over two years for the results to be accepted for publication in a scientific journal [Shechtman et al., 1984]. The paradox posed by these materials was that they had a 10-fold axis of rotation, which is forbidden for periodic materials, while at the same time they produced sharp Bragg peaks, which can only occur if the system has long-range order at the atomic level. The materials are now known as quasicrystals, and the solution to the paradox is that they have long-range *quasiperiodic* order. The discovery of quasicrystals has led to a profound redefinition of what constitutes a crystal, as will be explained here.

Before proceeding it is worth pointing out that even prior to the discovery of quasicrystals it had been known for a long time that certain crystalline materials are not periodic. In these materials the position of the atoms is modulated with a wavelength that is not a rational fraction of a lattice parameter. Such materials are said to be *incommensurate*, or *modulated.* This is illustrated in Fig. 4.14(a) for the case of a 1D lattice. The positions of the atoms are given by

$$x_n = \mathrm{a}n + u\,\cos(\mathrm{qa}n) \tag{4.36}$$

where a is the lattice parameter, n is a positive integer, u is the amplitude of the displacement, and $\mathrm{q}=2\pi/\lambda_m$ is its wavevector. In the case of an incommensurate material the modulation wavelength is given by $\lambda_m = c\,\mathrm{a}$, where c is an irrational number. If, as is sometimes found, the wavelength is expressible as a rational fraction, then the material is said to posses a *commensurate* modulation. Examples of commensurate and incommensurate modulations are shown in Fig. 4.14(a).

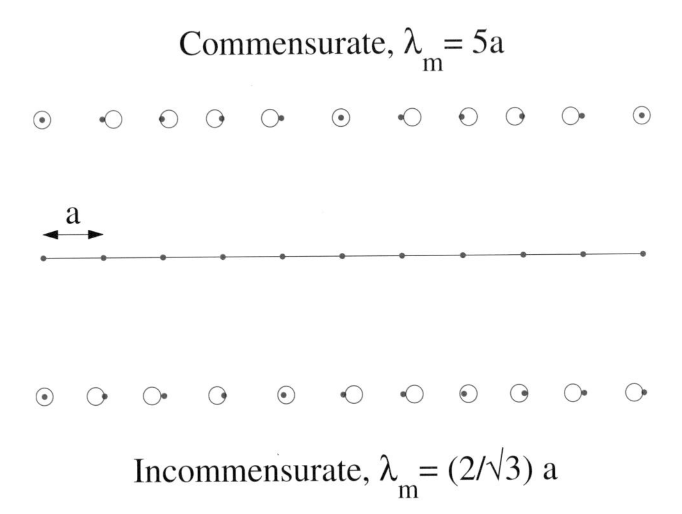

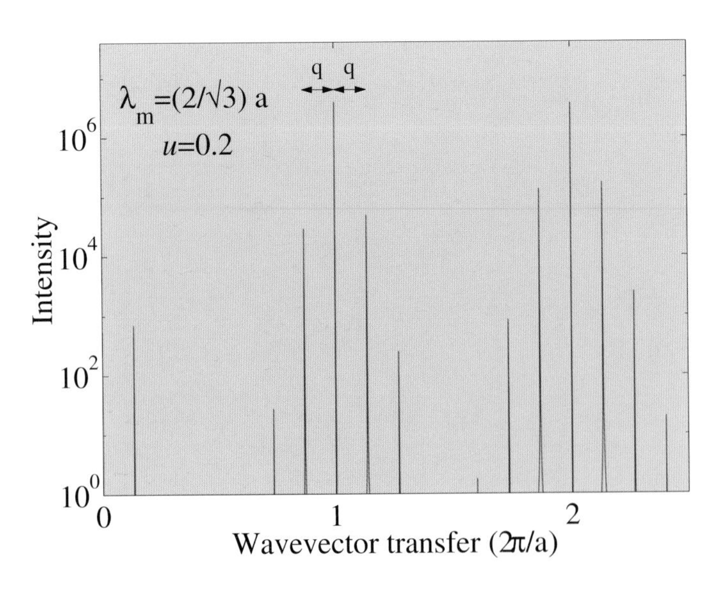

Figure 4.14: Scattering from a 1D incommensurate chain. The top panel shows the position of atoms in a 1D chain with a commensurate, $\lambda_m = 5$a, and an incommensurate, $\lambda_m = (2/\sqrt{3})$a modulation wavevector. The bottom panel is the calculated scattered intensity for a chain of N=2000 atoms, with positions given by Eq. (4.36), a modulation wavelength of $\lambda_m = (2/\sqrt{3})$a, and a displacement amplitude of u=0.2. For simplicity it has been assumed that the atomic scattering length is unity, and independent of wavevector transfer Q.

For incommensurate materials it is still possible to define an average, periodic lattice. The scattering then consists of Bragg peaks from the average lattice, plus additional Bragg peaks known as satellite reflections from the modulation. This is easy to verify by calculating the intensity numerically of the scattering from an incommensurate chain of N atoms with positions given by Eq. (4.36). An example of such a calculation is shown in Fig. 4.14, where for simplicity the atomic scattering length has been set equal to unity. The main Brag peaks occur at integer multiples of the modulus of the reciprocal lattice vector, $(2\pi/\mathrm{a})$, while the satellite peaks are displaced from the main peaks by multiples of the modulation wavevector $\mathrm{q}=2\pi/\lambda_m$.

It is not a difficult exercise to show analytically that a modulation produces regularly spaced satellite peaks. The scattering amplitude for an incommensurate 1D chain is

$$A(\mathrm{Q}) = \sum_{n=0}^{N-1} \mathrm{e}^{i\,\mathrm{Q}x_n} = \sum_{n=0}^{N-1} \mathrm{e}^{i\,\mathrm{Q}(\mathrm{a}n+u\,\cos(\mathrm{qa}n))} = \sum_{n=0}^{N-1} \mathrm{e}^{i\,\mathrm{Qa}n}\mathrm{e}^{i\,\mathrm{Q}u\,\cos(\mathrm{qa}n)}$$

For simplicity the atomic scattering length has again been set equal to unity. The approximation is now made that the displacement u is small. This allows the second phase factor to be expanded, and the amplitude becomes

$$\begin{aligned} A(\mathrm{Q}) &\approx \sum_{n=0}^{N-1} \mathrm{e}^{i\,\mathrm{Qa}n}\,(1 + i\mathrm{Q}u\cos(\mathrm{qa}n)) + \cdots) \\ &= \sum_{n=0}^{N-1} \mathrm{e}^{i\,\mathrm{Qa}n} + i\left(\tfrac{\mathrm{Q}u}{2}\right)\sum_{n=0}^{N-1}\left[\mathrm{e}^{i\,(\mathrm{Q}+\mathrm{q})\mathrm{a}n} + \mathrm{e}^{i\,(\mathrm{Q}-\mathrm{q})\mathrm{a}n}\right] \end{aligned}$$

In the limit that N becomes large, the scattered intensity is given by

$$\begin{aligned} I(\mathrm{Q}) &= N\left(\tfrac{2\pi}{a}\right)\delta(\mathrm{Q}-\mathrm{G}_h) \\ &\quad + N\left(\tfrac{\mathrm{Q}u}{2}\right)^2\left(\tfrac{2\pi}{a}\right)\left[\delta(\mathrm{Q}+\mathrm{q}-\mathrm{G}_h) + \delta(\mathrm{Q}-\mathrm{q}-\mathrm{G}_h)\right] \end{aligned} \tag{4.37}$$

where $\mathrm{G}_h = (2\pi/\mathrm{a})h$ is a reciprocal lattice vector, h the Miller index of the 1D lattice, and the results derived in Section 4.4.4 have been used to replace the squared sums by Dirac delta functions. The first term generates the main Bragg peaks at $\mathrm{Q}=\mathrm{G}_h$, while the second generates satellite reflections at $\mathrm{Q}=\mathrm{G}_h \pm \mathrm{q}$. In the numerical example shown in Fig. 4.14 satellites are also evident at $\pm 2\mathrm{q}$. These do not appear in the analytical calculation of Eq. (4.37), as the expansion of the exponential was truncated after the second term.

To index a given Bragg peak from an incommensurate system it is first necessary to specify the main Bragg peak with which it is associated. In three dimensions this requires the usual three Miller indices (h, k, l). The satellite peaks then require additional indices. In the example of a one dimensional modulation shown in Fig. 4.14 one extra index suffices. It turns out that incommensurate systems regain their periodicity if they are described in an abstract mathematical higher dimensional space. For a 1D modulation in a 3D crystal the dimension would be four, and the actual physical structure is obtained by making a particular three dimensional cut through this 4D space.

Quasicrystals are fundamentally different from incommensurate crystals, as for one thing they lack anything that can be identified with an average, periodic

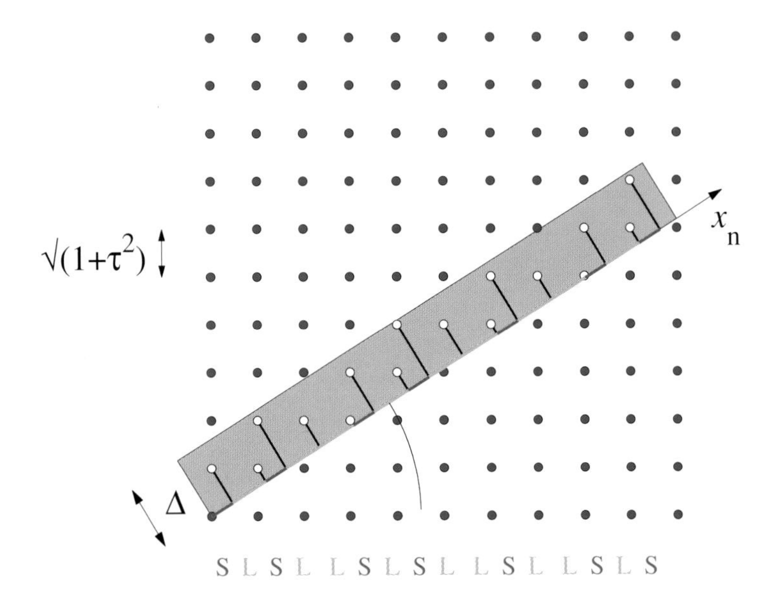

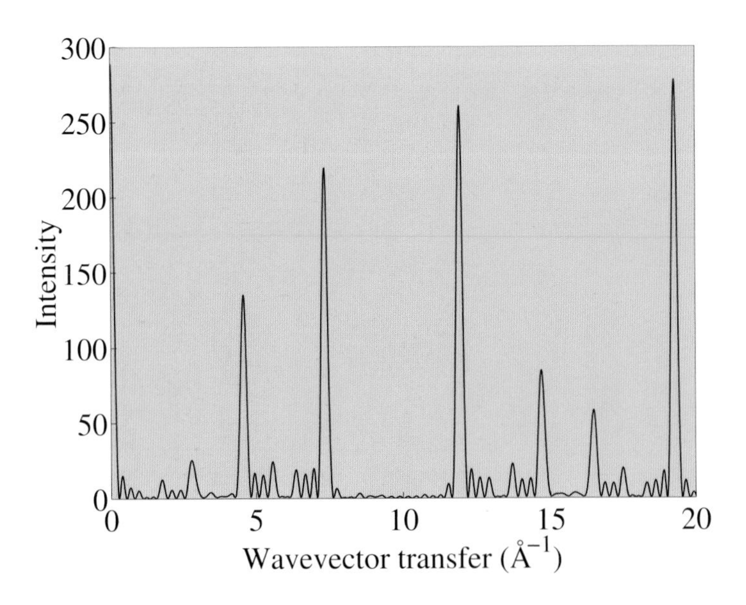

Figure 4.15: ⋆ The Fibonacci chain. The top panel illustrates how the Fibonacci chain may be obtained from a 2D square lattice using the strip-projection method. The slope of the strip is irrational, equal to $1/\tau$, where $\tau=(1+\sqrt{5})/2$ is the golden mean. Lattice points inside the strip are projected down onto the axis x_n, and the chain is then formed from two tiles S and L in the sequence shown. The bottom panel plots the calculated intensity from the Fibonacci chain derived from the lattice shown above. The predominant peaks are not regularly spaced, as would be the case for a periodic lattice. As the lattice size is increased the peaks become sharper, showing that long-range quasiperiodic order produces sharp Bragg peaks.

lattice. To make this clear we shall discuss the properties of the Fibonacci chain. This is an example of a quasiperiodic system, and is often used as a 1D model of a quasicrystal. There are several ways to derive the Fibonacci chain. One employs what is known as a substitution rule to generate a chain composed of two types of objects, or tiles, here labelled S for short and L for long. The chain starts with S, and then the substitution rule S→L and L→LS is used to inflate it. The chain develops in successive inflation steps as: S, L, LS, LSL, LSLLS, LSLLSLSL, etc. In the limit of an infinitely long chain it can be shown that the letter L appears with a frequency of $1/\tau$, where $\tau=(1+\sqrt{5})/2$ is the golden mean. The fact that this frequency is an irrational number is proof of the fact that the chain is not periodic. It does, however, posses long-range order. To see this consider Fig. 4.15⋆, which illustrates how the Fibonacci chain may be derived using the so-called strip projection method. A strip of thickness $\Delta=1+\tau$ is drawn with a slope of $1/\tau$ on a 2D lattice. All lattice points that fall inside the strip are projected down onto the 1D axis x_n. The position of the points along the x_n axis are 0, 1, $1+\tau$, $2+\tau$, $2+2\tau$, $2+3\tau$, etc., which is equivalent to the Fibonacci chain if S=1 and L=τ. As the 1D chain was derived from a periodic lattice with long-range order in 2D, the chain itself must have long-range order. In general, a strip drawn at an irrational slope will generate a quasiperiodic lattice; the choice of the slope $1/\tau$ was made to obtain the Fibonacci chain.

The scattered intensity calculated for a Fibonacci chain derived from a 10x10 2D lattice is shown in Fig. 4.15⋆. As the size of the lattice is increased the predominant peaks become sharper, but their positions remain fixed. It is therefore apparent that a quasiperiodic lattice, which lacks anything that may be identified with an average, periodic lattice, still produces a sharp diffraction pattern. Following the discovery of quasicrystals, the International Union of Crystallographers in 1991 decided to change the definition of a crystal to include the statement

> "...by crystal we mean any solid having an essentially discrete diffraction diagram....."

The definition thus shifted emphasis from a crystal thought of as a periodic structure in real space, to one that produces sharp diffraction peaks in reciprocal space.

Quasicrystals have many fascinating properties, and the reader is referred to the book by Janot for further information [Janot, 1992].

4.4.6 Crystal truncation rods

In Section 4.4.4 it was shown that in the case of an infinite three dimensional crystal the lattice sum produces a delta function. Scattering events are then restricted by the Laue condition such that $\mathbf{Q} = \mathbf{G}$, and as this is a vector equation it applies to all three components of $\mathbf{Q}$. For a *finite* size crystal this condition is relaxed and the scattering then extends over a volume in reciprocal space inversely proportional to the size of the crystal. This is illustrated in Fig. 4.16(a). We now imagine that the crystal is cleaved, so as to produce a flat surface. The scattering will no longer be isotropic and streaks of scattering appear in the direction parallel to the surface normal as shown in in Fig. 4.16(b). These are the crystal truncation rods (CTR) [Andrews and Cowley, 1985, Robinson, 1986].

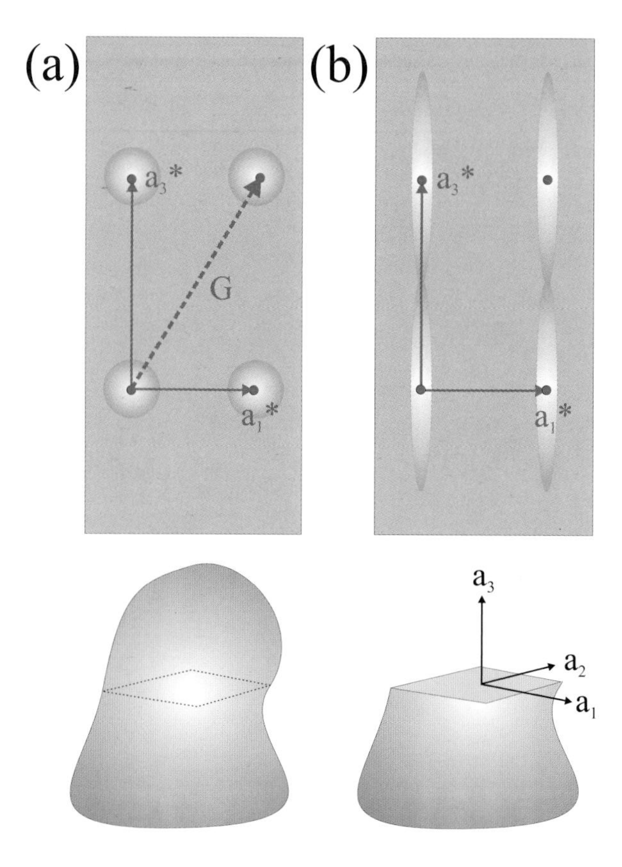

Figure 4.16: (a) Top: a map of reciprocal space in the plane spanned by $\mathbf{a}_1^*$ and $\mathbf{a}_3^*$ for the crystal shown in the bottom part of the figure. (b) Same as (a) except that the crystal has been cleaved to produce a surface perpendicular to the $\mathbf{a}_3$ axis. This produces streaks of scattering – known as crystal truncation rods – through all Bragg peaks in a direction perpendicular to the surface.

Insight into the origin of the CTR's follows once it is realized that the act of cleaving the crystal may be represented mathematically as a multiplication of the original density of the crystal, $\rho(z)$, by a step function, $h(z)$. (Here the coordinate system is chosen with z perpendicular to the surface.) The scattering *amplitude* is proportional to the Fourier transform of the product of the density $\rho(z)$ and $h(z)$. From the Convolution Theorem (see page 85) this is equivalent to the convolution of the Fourier transforms of $\rho(z)$ and $h(z)$. The Fourier transforms of $\rho(z)$ and $h(z)$ are a delta function and i/q_z respectively, as described in Appendix E. Away from a Bragg peak the scattering amplitude is thus proportional to $1/\mathrm{q}_z$ and the intensity to $1/\mathrm{q}_z^2$. The effect of the surface is therefore to produce streaks of scattering, known as crystal truncation rods, in the direction normal to the surface.

To develop an expression for the intensity distribution of the CTR we need only consider the lattice sum in the direction of the surface normal, $\mathbf{a}_3$; the sum over the other two directions leads to the usual product of delta functions $\delta(\mathrm{Q}_x - ha_1^*)\ \delta(\mathrm{Q}_y - ka_2^*)$. If $A(\mathbf{Q})$ is the scattering amplitude from a layer of atoms (here for simplicity assumed to be the same for all layers), then the

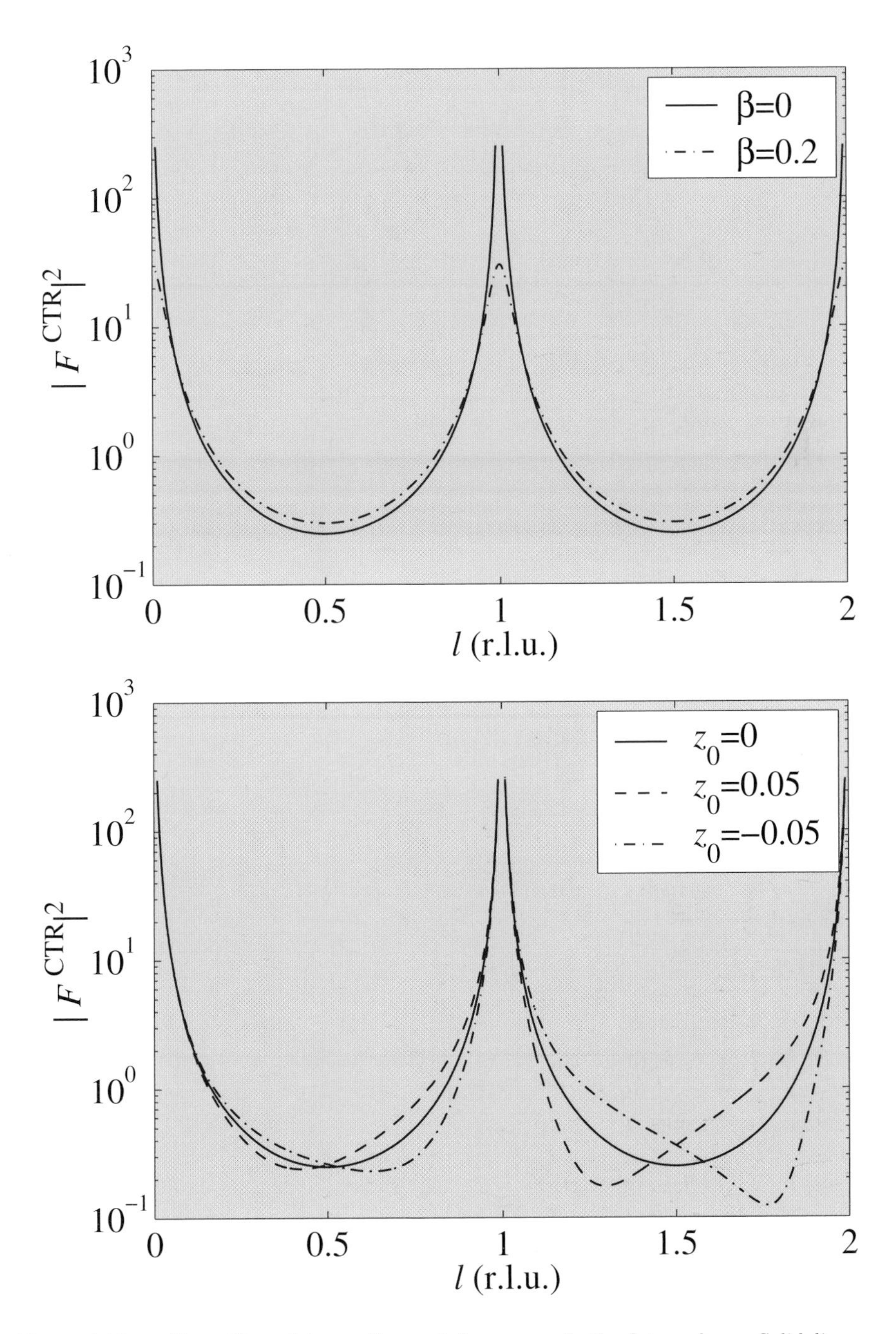

Figure 4.17: ⋆ Top: Crystal truncation rod from a perfectly flat surface. Solid line, no absorption ($\beta = 0$); dashed-dotted line with absorption. Typically β is of order 10^{-5}, as may be seen from the values of the absorption coefficient μ listed in Table 3.1. Here β has been chosen to be 0.2, an unrealistically high value, but one that serves to illustrate the effects of absorption. For simplicity $A(\mathbf{Q})$ has been chosen to be unity. Bottom: Crystal truncation rod ($\beta = 0$) from a flat surface with an overlayer. The relative displacement of the overlayer from the bulk lattice spacing is given by z_0. The effect of the displacement of the layer is seen to become more pronounced at higher wavevector transfers.

scattering amplitude from an infinite stack of such layers is

$$F^{\rm CTR} = A(\mathbf{Q}) \sum_{j=0}^{\infty} e^{i\,Q_z a_3\,j} = \frac{A(\mathbf{Q})}{1 - e^{i\,Q_z a_3}} = \frac{A(\mathbf{Q})}{1 - e^{i\,2\pi l}} \tag{4.38}$$

where the wavevector transfer along the surface normal is $Q_z = 2\pi l/a_3$. The intensity distribution along the crystal truncation rod is

$$\begin{aligned} I^{\rm CTR} = |F^{\rm CTR}|^2 &= \frac{|A(\mathbf{Q})|^2}{(1 - e^{i\,2\pi l})(1 - e^{-i\,2\pi l})} \\ &= \frac{|A(\mathbf{Q})|^2}{4\sin^2(\pi l)} \end{aligned} \tag{4.39}$$

This is plotted in the top panel of Fig. 4.17⋆. The expression for $I^{\rm CTR}$ is clearly only valid away from the Bragg peaks when l is not an integer, otherwise $\sin(\pi l)$ is zero and the intensity diverges. To examine how the intensity falls off close to the Bragg peak the wavevector transfer is written as $Q_z = q_z + 2\pi/a_3$, where q_z is the deviation in wavevector from the Laue condition. This is then inserted into argument of the sine in the denominator of Eq. (4.39). Since q_z is small the sine can be expanded to give the result that $I^{\rm CTR}$ is proportional to $1/q_z^2$, as expected from the introductory remarks made at the beginning of this section.

Absorption is included by re-writing Eq. (4.38) as

$$F^{\rm CTR} = A(\mathbf{Q}) \sum_{j=0}^{\infty} e^{i\,Q_z a_3\,j} e^{-\beta j} = \frac{A(\mathbf{Q})}{1 - e^{i\,Q_z a_3} e^{-\beta}} \tag{4.40}$$

where $\beta = a_3\mu/\sin\theta$ is the absorption parameter per layer. The main effect of absorption is in the vicinity of the Bragg peak, as represented by the dashed-dotted line in Fig. 4.17⋆.

The intensity distribution along the CTR depends on the exact way in which the surface is terminated, and measurements of CTR's have become a very useful probe of the structure of the surface and near-surface region of single crystals. This can be illustrated by imagining that the top most layer, having $j = -1$ in Eq. (4.38), of the crystal has a lattice spacing that is different from the bulk value. The total scattering amplitude is then

$$\begin{aligned} F^{\rm total} &= F^{\rm CTR} + F^{\rm top\ layer} \\ &= \frac{A(\mathbf{Q})}{1 - e^{i\,2\pi l}} + A(\mathbf{Q})\, e^{-i\,2\pi(1+z_0)l} \end{aligned} \tag{4.41}$$

where z_0 is the relative displacement of the top layer away from the bulk lattice spacing of a_3. For $z_0 = 0$ the same intensity distribution along the rod is found. If z_0 is non-zero then the interference between the scattering from the top layer and the rest of the crystal leads to characteristic features in the CTR, as shown in the bottom panel of Fig. 4.17⋆. In Section 4.6.3 it is explained through the example of oxygen deposited on the copper (110) surface exactly how determination of the CTR helps in solving the surface structure.

In deriving Eq. (4.38) we assumed for simplicity a specular scattering geometry, so that the angle of incidence is equal to the angle of reflection. The argument presented here can be generalized to show that CTR's arise from all Bragg peaks, with the direction of the rods being parallel to the surface normal.

4.4.7 Lattice vibrations, the Debye-Waller factor and TDS

The lattices considered so far have been assumed to be perfectly rigid. Atoms arranged on a lattice in a crystal vibrate, and here we explore the effect of these vibrations on the scattered intensity. The vibrations are due to two distinct causes. The first is purely quantum mechanical in origin and arises from the uncertainty principle. These vibrations are independent of temperature, and occur even at the absolute zero of temperature. For this reason they are known as the zero-point fluctuations. At finite temperatures elastic waves (or phonons) are thermally excited in the crystal, thereby increasing the amplitude of the vibrations.

To start with we shall consider the scattering from a simple crystal structure in which there is one type of atom located at each lattice point. From Eq. (4.19) the scattering amplitude is then

$$F^{\text{crystal}} = \sum_{\text{n}} f(\mathbf{Q})\, \text{e}^{i\,\mathbf{Q}\cdot\mathbf{R}_{\text{n}}} \tag{4.42}$$

The effects of vibrations are allowed for by writing the *instantaneous* position of an atom as $\mathbf{R}_{\text{n}} + \mathbf{u}_{\text{n}}$, where $\mathbf{R}_{\text{n}}$ is the time-averaged mean position, and $\mathbf{u}_{\text{n}}$ is the displacement. By definition, $\langle \mathbf{u}_{\text{n}} \rangle = 0$, the angle brackets $\langle \cdots \rangle$ indicating a temporal average. The scattered intensity is calculated by taking the product of the scattering amplitude and its complex conjugate, and then evaluating the time average. The intensity is thus

$$\begin{aligned} I &= \left\langle \sum_{\text{m}} f(\mathbf{Q})\, \text{e}^{i\,\mathbf{Q}\cdot(\mathbf{R}_{\text{m}}+\mathbf{u}_{\text{m}})} \sum_{\text{n}} f^*(\mathbf{Q})\, \text{e}^{-i\,\mathbf{Q}\cdot(\mathbf{R}_{\text{n}}+\mathbf{u}_{\text{n}})} \right\rangle \\ &= \sum_{\text{m}} \sum_{\text{n}} f(\mathbf{Q}) f^*(\mathbf{Q})\, \text{e}^{i\,\mathbf{Q}\cdot(\mathbf{R}_{\text{m}}-\mathbf{R}_{\text{n}})} \left\langle \text{e}^{i\,\mathbf{Q}\cdot(\mathbf{u}_{\text{m}}-\mathbf{u}_{\text{n}})} \right\rangle \end{aligned} \tag{4.43}$$

For convenience, the last term on the right hand side is rewritten as

$$\left\langle \text{e}^{i\,\mathbf{Q}\cdot(\mathbf{u}_{\text{m}}-\mathbf{u}_{\text{n}})} \right\rangle = \left\langle \text{e}^{i\,\text{Q}(\text{u}_{\text{Qm}}-\text{u}_{\text{Qn}})} \right\rangle \tag{4.44}$$

where u_{Qn} is the component of the displacement parallel to the wavevector transfer $\mathbf{Q}$ for the n'th atom. This expression can be further simplified by using the Baker-Hausdorff theorem, which was introduced in the context of the scattering from rough surfaces in Section 3.8 on page 84. (The proof of the Baker-Hausdorff theorem is given in Appendix D.) This theorem states that if x is described by a Gaussian distribution then

$$\left\langle \text{e}^{i\,x} \right\rangle = \text{e}^{-\frac{1}{2}\langle x^2 \rangle} \tag{4.45}$$

Using this result the temporal average becomes

$$\begin{aligned} \left\langle \text{e}^{i\,\text{Q}(\text{u}_{\text{Qm}}-\text{u}_{\text{Qn}})} \right\rangle &= \text{e}^{-\frac{1}{2}\langle \text{Q}^2(\text{u}_{\text{Qm}}-\text{u}_{\text{Qn}})^2 \rangle} \\ &= \text{e}^{-\frac{1}{2}\text{Q}^2\langle (\text{u}_{\text{Qm}}-\text{u}_{\text{Qn}})^2 \rangle} \\ &= \text{e}^{-\frac{1}{2}\text{Q}^2\langle \text{u}_{\text{Qm}}^2 \rangle} \text{e}^{-\frac{1}{2}\text{Q}^2\langle \text{u}_{\text{Qn}}^2 \rangle} \text{e}^{\text{Q}^2\langle \text{u}_{\text{Qm}}\text{u}_{\text{Qn}} \rangle} \end{aligned} \tag{4.46}$$

To proceed we write the last term as

$$\text{e}^{\text{Q}^2\langle \text{u}_{\text{Qm}}\text{u}_{\text{Qn}} \rangle} = 1 + \left\{ \text{e}^{\text{Q}^2\langle \text{u}_{\text{Qm}}\text{u}_{\text{Qn}} \rangle} - 1 \right\} \tag{4.47}$$

This allows the scattered intensity to be separated into two terms:

$$I = \sum_m \sum_n f(\mathbf{Q})\,\mathrm{e}^{-\frac{1}{2}\mathrm{Q}^2\langle \mathrm{u}_{\mathrm{Qm}}^2\rangle}\mathrm{e}^{i\,\mathbf{Q}\cdot\mathbf{R}_\mathrm{m}}\,f^*(\mathbf{Q})\,\mathrm{e}^{-\frac{1}{2}\mathrm{Q}^2\langle \mathrm{u}_{\mathrm{Qn}}^2\rangle}\mathrm{e}^{-i\,\mathbf{Q}\cdot\mathbf{R}_\mathrm{n}}$$
$$+\sum_m \sum_n f(\mathbf{Q})\,\mathrm{e}^{i\,\mathbf{Q}\cdot\mathbf{R}_\mathrm{m}}\,f^*(\mathbf{Q})\,\mathrm{e}^{-i\,\mathbf{Q}\cdot\mathbf{R}_\mathrm{n}}\left\{\mathrm{e}^{\mathrm{Q}^2\langle \mathrm{u}_{\mathrm{Qm}}\mathrm{u}_{\mathrm{Qn}}\rangle}-1\right\} \quad (4.48)$$

The first term is recognizable as the elastic scattering from a lattice except that the atomic form factor is replaced by

$$f^{\mathrm{atom}} = f(\mathbf{Q})\,\mathrm{e}^{-\frac{1}{2}\mathrm{Q}^2\langle \mathrm{u}_{\mathrm{Q}}^2\rangle} \equiv f(\mathbf{Q})\,\mathrm{e}^{-M} \quad (4.49)$$

where the exponential term in known as the Debye-Waller factor. As the first term contains contributions for large values of $|\mathbf{R}_\mathrm{m} - \mathbf{R}_\mathrm{n}|$ it still gives rise to a delta function in the scattering. This shows that the elastic Bragg scattering is reduced in intensity by atomic vibrations, but its width is not increased. The contribution from the second term in Eq. (4.48) has a distinctly different character. It has an intensity that actually increases as the the mean-squared displacement increases, and has a width determined by the correlations, $\langle \mathrm{u}_{\mathrm{Qm}}\mathrm{u}_{\mathrm{Qn}}\rangle$, between the displacements of different atoms. These turn out to be correlated significantly only over short distances, so that the lattice sum extends only over a few lattice sites, and the scattering has an appreciable width, much greater than the width of a Bragg peak. For these reasons this contribution is known as *thermal diffuse scattering*, or TDS for short. In crystallographic experiments TDS gives rise to a background signal which sometimes needs to be subtracted from the data. Alternatively, the study of TDS may also be of interest in its own right, as it provides information on the low-energy elastic waves in lattices. In this case the diffuse nature of TDS requires that the scattering is mapped out over large volumes of reciprocal space. To compare the experimental results with theory it is necessary to evaluate the second term in Eq. (4.48). This is achieved by performing a calculation of the lattice dynamics, which yields the atomic displacements $\mathbf{u}_\mathrm{n}$, and hence the correlation term $\langle \mathrm{u}_{\mathrm{Qm}}\mathrm{u}_{\mathrm{Qn}}\rangle$.

An example of TDS from Si is shown in Fig. 4.18 [Holt et al., 1999]. The data (left panels) were recorded in a transmission geometry with a (111) (top) and a (100) (bottom) axis parallel to the incident beam. The data are plotted on a logarithmic scale so that the weak diffuse scattering which peaks along high-symmetry directions connecting reciprocal lattice points is enhanced. In the right panels are shown the corresponding images obtained by fitting a model of the lattice dynamics to the data. Good agreement was found for the phonon dispersion curves derived from this model and earlier neutron scattering experiments.

The separation of the total diffracted intensity into a sharp Bragg and a diffuse component in the presence of thermal vibrations is analogous to the separation of the reflectivity from a rough interface into specular and diffuse components, as described in Section 3.8. Indeed this separation is useful whenever there are random atomic displacements, be they static or dynamic, from the lattice sites. Here we have considered the case of dynamic displacements by elastic waves, but static distortions, caused for example by lattice defects, can be studied through the diffuse scattering.

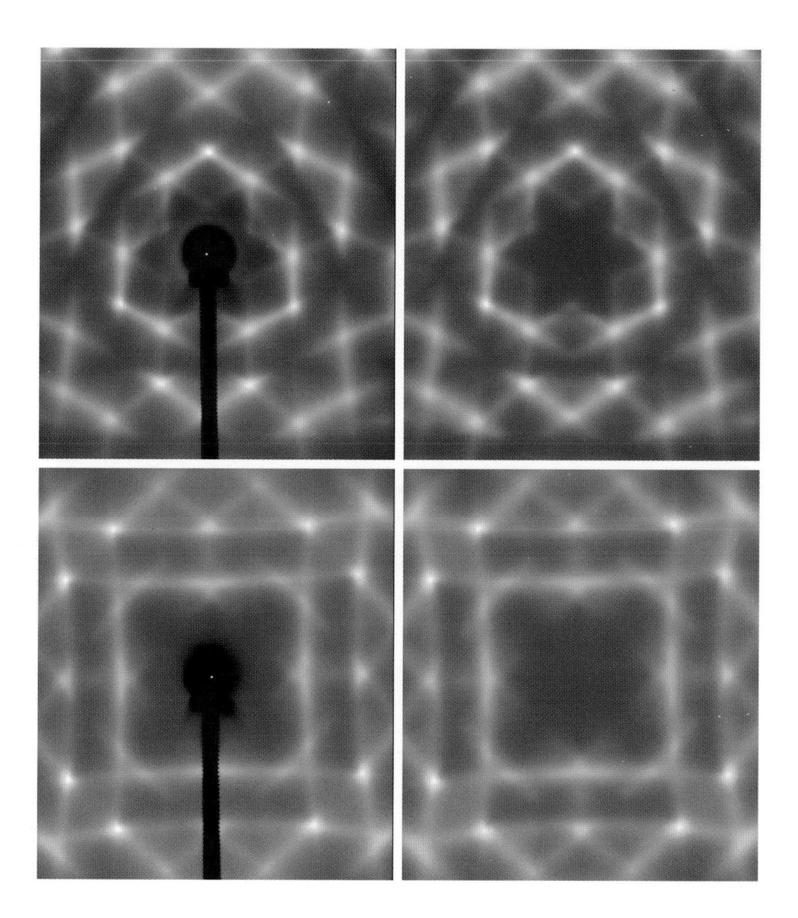

Figure 4.18: Thermal diffuse scattering (TDS) from Si. The data were collected in a transmission geometry (photon energy 28 keV) using an image plate detector. The data were collected on the UNI-CAT beamline at the Advanced Photon Source in an exposure time of ~10 s. The top and bottom left panels show the data taken with a (111) and a (100) axis parallel to the incident beam respectively. The data are plotted on a logarithmic scale. The brighter spots are not Bragg peaks, as the Laue condition is never exactly fulfilled, but are due to the build up of TDS close to the position of where the Bragg peaks would occur. The right panels show the corresponding calculated images based on a simultaneous pixel-by-pixel fit to the data [Holt et al., 1999].

In the rest of this section we consider the properties of the Debye-Waller factor. It is straight forward to generalize the above results to the case of a crystal with several different types of atom in the unit cell. In this case the unit cell structure factor (Eq. (4.20)) becomes

$$F^{\text{u.c.}} = \sum_{\mathbf{r}_j} f_j(\mathbf{Q})\, \mathrm{e}^{-M_j}\, \mathrm{e}^{i\,\mathbf{Q}\cdot\mathbf{r}_j} \tag{4.50}$$

where

$$M_j = \frac{1}{2}\mathrm{Q}^2\langle \mathrm{u}_{\mathrm{Q}j}^2\rangle = \frac{1}{2}\left(\frac{4\pi}{\lambda}\right)^2 \sin^2\theta\,\langle \mathrm{u}_{\mathrm{Q}j}^2\rangle = B_T^j\left(\frac{\sin\theta}{\lambda}\right)^2 \tag{4.51}$$

refers to the j'th atom in the unit cell, and $B_T^j = 8\pi^2 \langle u_{Qj}^2 \rangle$. The reason for writing it in this form is that crystallographers prefer to express the wavevector transfer as $\sin\theta/\lambda$ instead of as $Q = 2k \sin\theta$. If the atom vibrates isotropically then $\langle u^2 \rangle = \langle u_x^2 + u_y^2 + u_z^2 \rangle = 3 \langle u_x^2 \rangle = 3 \langle u_Q^2 \rangle$, so that

$$B_{T,\text{isotropic}} = \frac{8\pi^2}{3} \langle u^2 \rangle \tag{4.52}$$

The effect of the vibrations can be viewed as being equivalent to a smearing of the electron distribution around the point at **R** with a Gaussian distribution of radius σ, with $\langle u^2 \rangle / 6 = \sigma^2/2$. If the vibrations are anisotropic, they can be described by a "vibrational ellipsoid" with three principal axes of different magnitude.

In a compound each type of atom will in general have a different Debye-Waller factor, as it should be obvious that lighter atoms will generally vibrate more than heavier ones. The Debye-Waller factors need not be isotropic, as the bonding will also restrict the vibrations along certain directions. For example, it usually costs less energy to change a bond angle than a bond length, so the vibrations of atoms at the ends of bonds will have a larger amplitude perpendicular to the bond than along it. These subtleties are usually taken into account by including extra fitting parameters in the data analysis. Here, however, we can simplify the discussion by restricting ourselves to consider only one type of atom in cubic symmetry so that the vibrations are isotropic. Then within the harmonic approximation the Debye-Waller factor depends on $\langle u^2 \rangle$ only. The vibrations in a solid are dominated by the low-energy, long-wavelength modes, and they can be characterized by a single effective parameter known as the Debye temperature, Θ. This is usually obtained from bulk properties such as specific heat measurements. The mean-squared amplitude of the vibrations is readily calculated in terms of Θ with the result that

$$B_T = \frac{6\,h^2}{m_A k_B \Theta} \left\{ \frac{\phi(\Theta/T)}{\Theta/T} + \frac{1}{4} \right\} \tag{4.53}$$

with

$$\phi(x) \equiv \frac{1}{x} \int_0^x \frac{\xi}{e^\xi - 1}\, d\xi \tag{4.54}$$

where Θ and T are in degrees Kelvin, and m_A is the atomic mass. The parameter B_T has dimension of length squared, which in practical units of Å^2 is

$$B_T[\text{Å}^2] = \frac{11492\, T[\text{K}]}{A\Theta^2[\text{K}^2]} \phi(\Theta/T) + \frac{2873}{A\Theta[\text{K}]} \tag{4.55}$$

and A is the atomic mass number. The function $\phi(x)$ is shown in the top panel of Fig. 4.19⋆. At or close to absolute zero the first term in Eq. (4.53) is negligible, but B_T remains finite due to the second term of $1/4$. This term arises from the zero-point motion, a purely quantum mechanical effect consistent with the uncertainty principle. With increasing temperature it can be seen from Eq. (4.53) that B_T increases once the temperature becomes comparable to Θ.

To illustrate how the Debye-Waller factor alters the scattering let us take the example of Al, which crystallizes in the face centred cubic structure (Fig. 4.8)

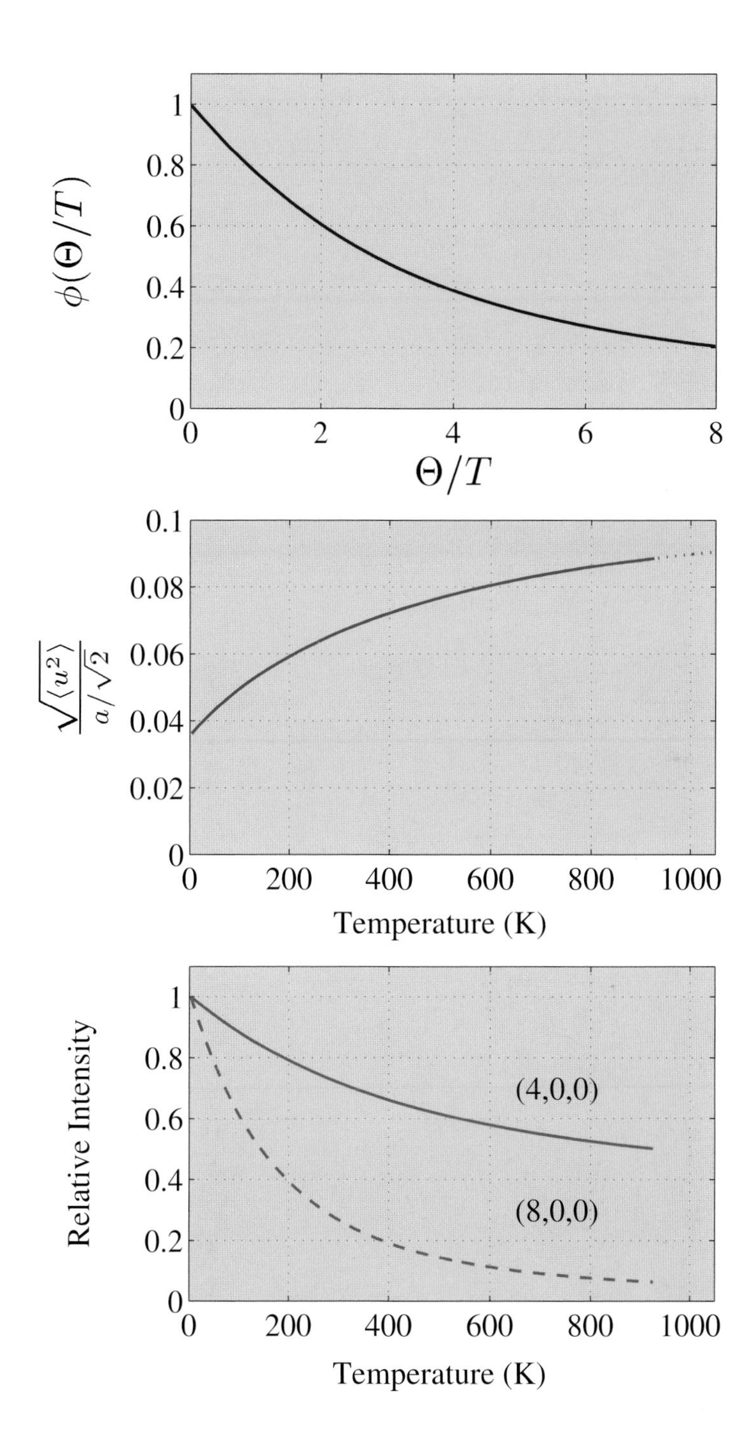

Figure 4.19: ⋆ Top: plot of the value of the integral $\phi(x)$ versus $x = \Theta/T$. Middle: temperature dependence of the rms fluctuation **u** in units of $a/\sqrt{2}$ for Al. Bottom: the relative intensity of the scattered intensity from Al as a function of temperature. The curves were calculated for the (4,0,0) (solid line) and the (8,0,0) (dashed line) Bragg peaks respectively. The melting temperature of Al is 933 K.

	A	Θ (K)	$B_{4.2}$	B_{77} (Å^2)	B_{293}
Diamond	12	2230	0.11	0.11	0.12
Al	27	394	0.25	0.30	0.72
Si	28.1	645	0.17	0.18	0.33
Cu	63.5	343	0.13	0.17	0.47
Ge	72.6	374	0.11	0.13	0.35
Mo	96	450	0.06	0.08	0.18

Table 4.2: The Debye temperature Θ, and the Debye-Waller factor B_T at temperatures of 4.2, 77 and 293 K, for a selection of cubic elements. The Debye-Waller factors have been calculated from the stated Debye temperatures using Eq. (4.55).

with a cube edge of a=4.04 Å, and with Θ=394 K, A=27. It is interesting to compare the rms vibrational amplitude as the temperature is raised towards the melting temperature (933 K). This is shown in Fig. 4.19⋆ (middle panel) where we have normalized $\sqrt{\langle u^2 \rangle}$ by the nearest- neighbour distance $a/\sqrt{2}$. We note in passing that just below the melting temperature $\sqrt{\langle u^2 \rangle}$ divided by the nearest-neighbour distance is approximately 0.1, which is consistent with Lindemann's empirical criterion for the melting of a solid: when the thermal vibrations approach about 10% of the nearest-neighbour distance the solid melts. The lower part of Fig. 4.19⋆ shows that the Q^2 dependence of the Debye-Waller factor has a dramatic effect on the intensity when the scattering vector is increased in length. In Table 4.2 the Debye temperatures are given for a number of cubic elements, along with calculated values of B_T at different temperatures.

4.5 The measured intensity from a crystallite

In this section the integrated intensity of a Bragg reflection from a small crystal is evaluated, as this is the quantity that is readily determined in an experiment. This requires that we specify exactly how the integrated intensity is to be measured. The starting point is to assemble the various expressions that have been developed in the previous sections into a single formula for the intensity. However, instead of referring to the intensity we shall be a little more precise and use instead the *differential cross-section* $(d\sigma/d\Omega)$, which is discussed more fully in Appendix A. For the case considered here where the sample is fully illuminated by the beam the differential cross-section is defined by

$$\left(\frac{d\sigma}{d\Omega}\right) = \frac{\text{Number of X-rays scattered per second into } d\Omega}{(\text{Incident flux})(d\Omega)}$$

where $d\Omega$ is the solid angle. From Eqs. (1.6), (4.19), (4.20) and (4.35) we have that

$$\boxed{\left(\frac{d\sigma}{d\Omega}\right) = r_0^2 P\, |F(\mathbf{Q})|^2 N\, v_c^*\, \delta(\mathbf{Q}-\mathbf{G})} \tag{4.56}$$

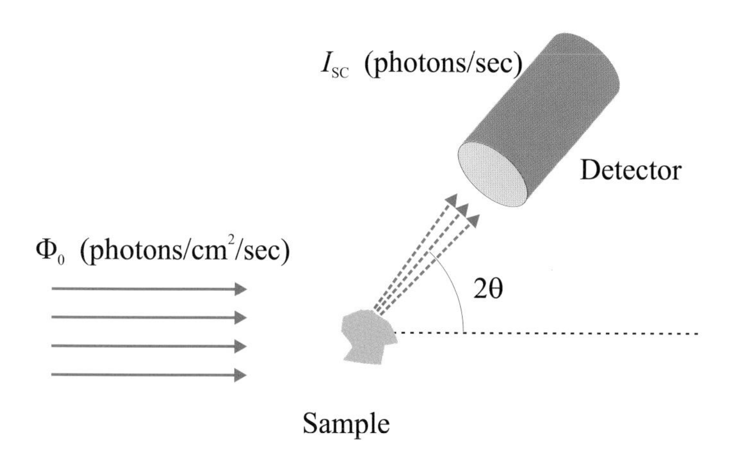

Figure 4.20: Scattering from a small crystal. The incident beam is assumed to be perfectly collimated and monochromatic, and to fully illuminate the crystal. The scattered intensity I_{sc} is proportional to the flux Φ_0 and to the differential cross-section $(d\sigma/d\Omega)$ of the sample.

where the superscript on $F(\mathbf{Q})$, the unit cell structure factor, has been dropped, and P is the polarization factor (Eq. (1.7)).

The experimental arrangement typically used for determining the integrated intensity of a Bragg peak is sketched in Fig. 4.20. The incident beam is assumed to be both perfectly monochromatic and collimated. The scattered beam will then also be perfectly monochromatic, since the scattering is elastic. However, it will not necessarily be perfectly collimated. In Section 4.4.4 it has been shown that the width of a Bragg peak is inversely proportional to N, the number of unit cells, and since N is not infinite, the Bragg peak has a finite width. This means that the Laue condition does not have to be exactly fulfilled for a measurable intensity to be recorded. This is represented in Fig. 4.21 by the existence of an elliptical contour: if $\mathbf{Q}$ falls within this contour then appreciable intensity is obtained, and the scattered beam will have some divergence. Let us assume that the geometry is such that all of the slightly divergent scattered rays hit the detector. Therefore, referring to Fig. 4.21, all of the scattering processes where $\mathbf{k}'$ terminates on the heavy red line will be recorded. However, we are interested in the sum of all the scattering processes where $\mathbf{Q}$ terminates within (or in the vicinity) of the smeared Bragg point contour. This means that the crystal has to be rotated (or rocked) a little with respect to the incident beam, and the measurement repeated, corresponding then to one of the other light red lines in Fig. 4.21, and in this way the integrated intensity is accumulated. (We note in passing, that the varying angle between $\mathbf{k}$ and $\mathbf{G}$ in the scan is equivalent to allowing the incident beam not to be perfectly collimated.)

Thus the integrated intensity is recorded by rotating the crystal so that the angle θ varies. The formula given in Eq. (4.56) applies to a single setting of the instrument, and in order to compare it with the integrated intensity that is measured in an experiment we have to allow for both the integration over $\mathbf{k}'$ and over θ. This gives rise to an additional term known as the Lorentz factor, which is derived in the following section. It is important to appreciate that the Lorentz factor depends on exactly how the intensity is integrated and hence on the details of the experiment.

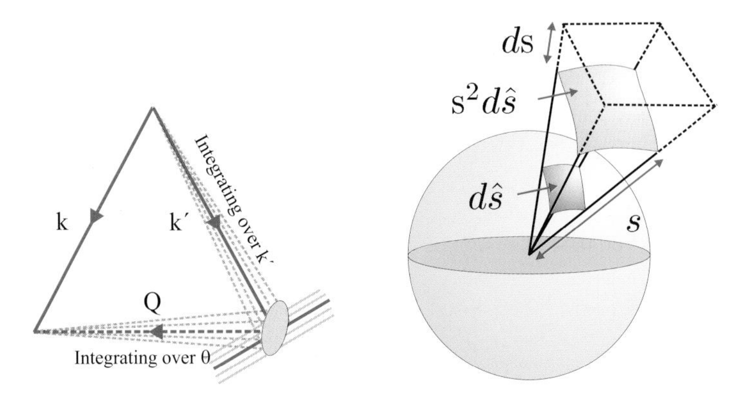

Figure 4.21: Left: The scattering from a small crystallite is represented by the grey ellipse, which reflects the reciprocal of the shape of the crystal. For a given orientation of the crystal the detector accepts all of the scattered wavevectors $\mathbf{k}'$ which fall on the red line. As the crystal is rotated the integration corresponds to the other light red lines. Right: The solid angle element $d\hat{\mathbf{s}}$ and the volume element $d\mathbf{s}$ are related by $d\mathbf{s}{=}s^2\, d\hat{\mathbf{s}}\, ds$.

4.5.1 The Lorentz factor

Integration over k′

A unit vector along $\mathbf{k}'$ is indicated by a hat. The element of solid angle $d\hat{\mathbf{k}}'$ is two-dimensional, and integration over the *directions* of $\mathbf{k}'$ is therefore equivalent to integrating over $d\hat{\mathbf{k}}'$. Instead of $\mathbf{k}'$ we introduce the vector $\mathbf{s} = \mathrm{k}'\hat{\mathbf{s}}$, where $\hat{\mathbf{s}}$ is a unit vector (right panel Fig. 4.21). The problem then is to integrate the delta function in equation Eq. (4.56) over $d\hat{\mathbf{k}}'$, i.e.

$$\int d\hat{\mathbf{k}}'\, \delta(\mathbf{Q}-\mathbf{G}) = \int d\hat{\mathbf{k}}'\, \delta(\mathbf{k}-\mathbf{k}'-\mathbf{G})$$

The integral is rewritten as

$$\int d\hat{\mathbf{k}}'\, \delta(\mathbf{k}-\mathbf{k}'-\mathbf{G}) = \overbrace{\frac{2}{\mathrm{k}'}\int \mathrm{s}^2\delta(\mathrm{s}^2-\mathrm{k}'^2)\, d\mathrm{s}}^{1} \int \delta(\mathbf{k}-\mathbf{s}-\mathbf{G})\, d\hat{\mathbf{s}}$$

since the first integral on the right hand side is unity, as proven in the delta function box on the facing page, and in the second integral $\mathbf{k}'$ has been replaced by $\mathbf{s}$, as they are equal by definition. The point of this trick is to transform the two-dimensional integral into a three-dimensional one. This is made clear if the above is rearranged as

$$\begin{aligned}\int d\hat{\mathbf{k}}'\, \delta(\mathbf{k}-\mathbf{k}'-\mathbf{G}) &= \frac{2}{\mathrm{k}'}\int \delta(\mathrm{s}^2-\mathrm{k}'^2)\,\delta(\mathbf{k}-\mathbf{s}-\mathbf{G})\; \mathrm{s}^2\, d\hat{\mathbf{s}}\, d\mathrm{s} \\ &= \frac{2}{\mathrm{k}'}\int \delta(\mathrm{s}^2-\mathrm{k}'^2)\,\delta(\mathbf{k}-\mathbf{s}-\mathbf{G})\; d^3\mathbf{s}\end{aligned}$$

where $d^3\mathbf{s}{=}\mathrm{s}^2\, d\hat{\mathbf{s}}\, d\mathrm{s}$ is the three-dimensional volume element. To proceed, the second delta function is used to require that $\mathbf{s} = \mathbf{k} - \mathbf{G}$, and this is then substi-

The Dirac δ function

Usually a mathematical function such as $e^x, \sin(x)\cdots$, etc., can be tabulated and plotted. This is not the case for the Dirac δ function. It represents the limiting case for a number of functions with a peak, such as a box, triangle, Gaussian, or a Lorentzian, when the width tends towards zero while the area remains constant.

The Dirac δ function is used in connection with integration. When an arbitrary function, $f(x)$, is multiplied by the δ function and integrated, the result is by definition $f(x=0)$:

$$f(0) = \int f(x)\delta(x)\,dx$$

If the argument of the δ function is not x, but rather a function of x, $t(x)$ say, one can use the following procedure:

$$\begin{aligned}\int f(x)\delta(t(x))\,dx &= \int f(t(x))\delta(t)\,(dt/dx)^{-1}\,dt \\ &= \left[f(t)(dt/dx)^{-1}\right]_{t=0}\end{aligned}$$

Suppose for example that $t(x) = x - a$. Then $dt/dx = 1$ and

$$\int f(x)\delta(x-a)\,dx = f(a)$$

Another linear function is $t(x) = x/a$:

$$\int f(x)\delta(x/a)\,dx = a\,f(0)$$

In connection with the derivation of the Lorentz factor we shall use and evaluate

$$F(\mathrm{k}) \equiv \int x^2\,\delta(x^2 - \mathrm{k}^2)\,dx \qquad \text{for} \quad \mathrm{k} > 0$$

With $t = x^2 - \mathrm{k}^2$ we obtain $dt/dx = 2x$ and therefore

$$F(\mathrm{k}) = \left[\frac{x^2}{2x}\right]_{t=0} = \frac{\mathrm{k}}{2}$$

or in other words

$$1 = \frac{2}{\mathrm{k}}\int x^2\,\delta(x^2 - \mathrm{k}^2)\,dx \qquad \text{for} \quad \mathrm{k} > 0$$

tuted into the first δ function, with the result that the integral becomes

$$\int d\hat{\mathbf{k}}'\;\delta(\mathbf{k}-\mathbf{k}'-\mathbf{G}) = \frac{2}{\mathrm{k}'}\,\delta((\mathbf{k}-\mathbf{G})\cdot(\mathbf{k}-\mathbf{G})-\mathrm{k}'^2)$$
$$= \frac{2}{\mathrm{k}}\,\delta(\mathrm{G}^2 - 2\,\mathrm{k}\,\mathrm{G}\,\sin\theta) \qquad (4.57)$$

In the second equation use has been made of the fact that the scattering is elastic, i.e. $\mathrm{k}' = \mathrm{k}$. When integrated over the directions of $\mathbf{k}'$ the cross-section is

$$\left(\frac{d\sigma}{d\Omega}\right)_{\text{int. over }\mathbf{k}'} = r_0^2 P\,|F(\mathbf{Q})|^2\,N\,v_c^*\,\frac{2}{\mathrm{k}}\,\delta(\mathrm{G}^2 - 2\mathrm{kG}\sin\theta)$$

Integration over θ

Evaluation of the integrated intensity is completed by integrating over the angular variable θ. The remaining delta function is itself a function of θ, and using the results given in the box on the page before the integral is

$$\int \delta(\mathrm{G}^2 - 2\mathrm{kG}\sin\theta)\,d\theta = \int \delta(t(\theta))\,d\theta = \left[\left(\frac{dt}{d\theta}\right)^{-1}\right]_{t=0}$$

The derivative of the argument of the delta function is

$$\frac{d(\mathrm{G}^2 - 2\mathrm{kG}\sin\theta)}{d\theta} = -2\mathrm{kG}\cos\theta$$

with the result that

$$\int \delta(\mathrm{G}^2 - 2\mathrm{kG}\sin\theta)\,d\theta = \left[\frac{-1}{2\mathrm{kG}\cos\theta}\right]_{t=0} = \frac{-1}{2\mathrm{k}^2\sin 2\theta}$$

Therefore the differential scattering cross-section integrated over both the directions of $\mathbf{k}'$ and over θ is

$$\left(\frac{d\sigma}{d\Omega}\right)_{\text{int. over }\mathbf{k}',\theta} = r_0^2 P\,|F(\mathbf{Q})|^2\,N\,v_c^*\,\frac{2}{\mathrm{k}}\,\frac{1}{2\mathrm{k}^2\sin 2\theta}$$
$$= r_0^2 P\,|F(\mathbf{Q})|^2\,N\,\frac{\lambda^3}{v_c}\,\frac{1}{\sin 2\theta}$$

In the second equation the volume v_c of the unit cell in real space has been introduced, rather than the volume v_c^* of the unit cell in reciprocal space, and we have utilized the fact that $2\pi/\mathrm{k} = \lambda$.

The integrated intensity I_{sc} is then found by multiplying the above by the incident flux Φ_0 to yield the final result

$$I_{\mathrm{sc}}\left(\frac{\text{photons}}{\text{sec}}\right) = \Phi_0\left(\frac{\text{photons}}{\text{unit area}\times\text{sec}}\right)\,r_0^2 P\,|F(\mathbf{Q})|^2\,N\,\frac{\lambda^3}{v_c}\,\frac{1}{\sin 2\theta} \qquad (4.58)$$

Each electron in the unit cell has a differential scattering cross section of $r_0^2 P$, where P is a polarization factor. The differential scattering cross section of a

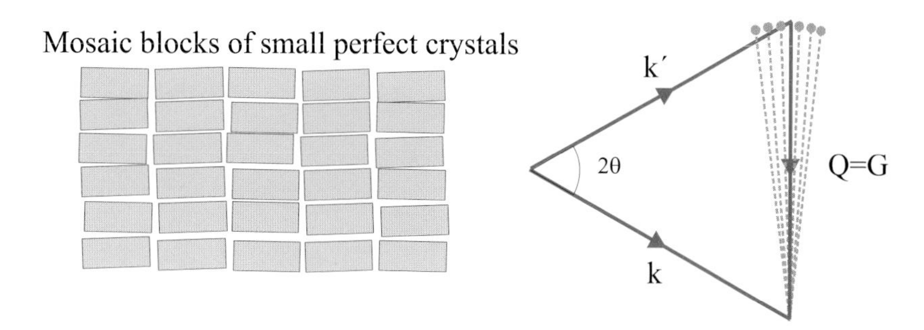

Figure 4.22: Left: Real single crystals are often composed of small, ideal crystal grains, also called mosaic blocks which have a narrow distribution of orientations, the so-call mosaic distribution. Right: For a given (h, k, l) reflection the crystal is rotated, corresponding to rotating $\mathbf{G}$ about the origin. In this way the integrated intensity from each mosaic block is accumulated.

unit cell is $r_0^2\, P\, |F(\mathbf{Q})|^2$ where, in the limit $\mathbf{Q} \to 0$, $F(\mathbf{Q}){=}\sum_j Z_j$, the number of electrons in the unit cell. For general $\mathbf{Q} > 0$ the scattering is diminished due to different optical path lengths as given by $F(\mathbf{Q})$. The total scattering is proportional to the number of unit cells, N. Summation and proper integrations give rise to the last two factors, the last of which is sometimes is called the Lorentz factor.

Extinction

The formula given in Eq. (4.58) applies to a single and idealized "small" but otherwise perfect crystal, with all of the diffracting planes in exact registry. Real macroscopic crystals on the other hand are often imperfect, and may be thought of as being composed of small perfect blocks with a distribution of orientations around some average value. The crystal is then said to be mosaic, as it is considered to be composed of a mosaic of small blocks as shown in the left panel of Fig. 4.22. Typically the mosaic blocks may have orientations distributed over an angular range of between $0.01°$ and $0.1°$.

Each block is "small" in the sense that there is a negligible chance of the diffracted beam being re-scattered before it exits the block, and the kinematical approximation applies. As the block size becomes larger this approximation breaks down, and instead it is necessary to allow for multiple scattering effects. These are discussed in Chapter 5, where it is shown (Eq. (5.34)) that the integrated intensity from a macroscopic perfect crystal takes on a form very different from that given in Eq. (4.58). In fact, for reasons that will be discussed in Chapter 5, the integrated intensity from a macroscopic perfect crystal is in general lower than that of an imperfect one. Thus, if the mosaic blocks are not sufficiently small, then the measured integrated intensity of a given Bragg reflection will be less than predicted by Eq. (4.58). The reflection is then said to be reduced by *primary extinction* effects.

There is a second way that the scattering from a mosaic crystal may be less than given by Eq. (4.58). For a mosaic crystal it may happen that one or several mosaic blocks are shadowed by blocks which have an identical orientation. The beam incident on one of the shadowed blocks will then necessarily be weaker, since some of the beam has already been diffracted into the exit beam by the

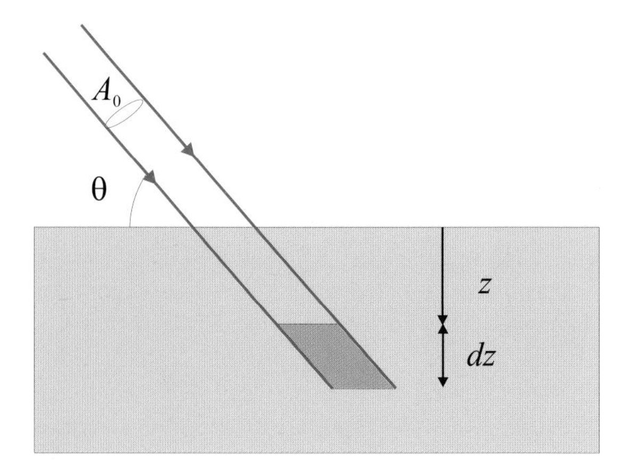

Figure 4.23: Extended face geometry. A beam of X-rays is incident on the flat face of a crystal at an angle θ. The cross-sectional area of the beam is A_0, and the illuminated volume of sample at a depth of z from the surface is $(A_0/\sin\theta)dz$.

higher lying blocks. If this is the case, then *secondary extinction* is said to be present. In the event that both primary and secondary extinction are negligible, the crystal is said to be *ideally imperfect.*

The scattering diagram for an ideally imperfect crystal is shown in Fig. 4.22. Each mosaic block is indicated by a little dot in the fan of reciprocal lattice vectors denoted by $\mathbf{G}$. The spectrometer is first set for a given (hkl) reflection at scattering angle $2\theta_{hkl}$. The crystal is then rotated or scanned through the Laue condition $\mathbf{Q} = \mathbf{G}$, over an angular range that is large enough to capture the entire mosaic fan and the integrated intensity recorded. Formula Eq. (4.58) can then be used to calculate the integrated intensity from the mosaic crystal. All that needs to be done is to replace N, the number of unit cells by $N'N_{\text{mb}}$, the product of the N', the number of unit cells in a single mosaic block and N_{mb}, the number of mosaic blocks. This follows from the fact that there is no definite phase relationship between waves scattered from different blocks and it suffices to add the intensities. This should be contrasted with the case of a single block where amplitudes are first added, and then the result multiplied by its complex conjugate to give the scattered intensity.

The idea that an imperfect crystal is composed of mosaic blocks is an over-simplification. Although it is true that dislocations and other defects produce a broadening of Bragg peaks, the micro-structure of an imperfect crystal rarely resembles that shown in the left panel of Fig. 4.22. It follows that the division of extinction effects into primary and secondary causes is not always valid. Indeed in many situations the use of a mosaic model to describe an imperfect crystal is an expediency, and is justified only in that it allows mathematical treatments of extinction effects, which can then be used to correct data. One way to avoid extinction effects altogether is to make a very fine powder of the sample, and powder diffraction is considered further in Section 4.6.1.

Absorption effects: extended face geometry

Equation (4.58) has been derived by neglecting absorption effects. It should be clear that in general these depend on the shape of the sample, and in practice various approximations are applied to correct the measured intensities. (The correction of integrated intensities for absorption effects is described in the International Tables of Crystallography.) One particularly useful and simple geometry for which an analytical solution can be found is the case of the diffraction from a crystal with an extended, flat face, as shown in Fig. 4.23, where it is assumed that the crystal is large enough to intercept the whole beam. If N is the total number of unit cells that are illuminated by the incident beam, then for a mosaic crystal we write $N = N' \times N_{\text{mb}}$, where N' is the number of unit cells in a mosaic block, and N_{mb} is the number of mosaic blocks. The number of mosaic blocks illuminated by a beam of cross-sectional area A_0 in the interval from z to $z + dz$ is

$$N_{\text{mb}} = \frac{A_0\, dz}{\sin\theta} \times \frac{1}{V'} \tag{4.59}$$

where V' is the volume of a mosaic block, and z is the depth of the beam in the crystal. At a depth z from the surface absorption reduces the intensity by $\text{e}^{-2\mu z/\sin\theta}$, where the factor of 2 allows for the path length of the incident and exit beams through the crystal. The integrated intensity is found from

$$\begin{aligned} I_{\text{sc}} &= \frac{\Phi_0 r_0^2 P\, |F(\mathbf{Q})|^2\, \lambda^3}{v_c \sin 2\theta} N' \int_0^\infty \text{e}^{-2\mu z/\sin\theta} \frac{A_0\, dz}{V' \sin\theta} \\ &= \frac{\Phi_0 r_0^2 P\, |F(\mathbf{Q})|^2\, \lambda^3}{v_c \sin 2\theta} \frac{A_0 N'}{V'} \left[\frac{-1}{2\mu} \text{e}^{-2\mu z/\sin\theta} \right]_0^\infty \\ &= \left(\frac{1}{2\mu}\right) \frac{\Phi_0 A_0 r_0^2 P\, |F(\mathbf{Q})|^2\, \lambda^3}{v_c^2 \sin 2\theta} \end{aligned} \tag{4.60}$$

The fact the scattered intensity is now proportional to the intensity of the incident beam, i.e. the product of the flux Φ_0 and cross-sectional area of the incident beam A_0, arises from the fact that in deriving the above it has been assumed that the face of the crystal is extended enough that it intercepts the entire beam (see also Appendix A for a discussion of the definition of the differential cross-section). The above result for the integrated intensity from an extended face mosaic crystal with absorption effects included is compared in Chapter 5 with the case of the scattering from a perfect crystal with an extended face.

4.5.2 The Lorentz factor in 2D

In this section we consider briefly the two dimensional (2D) equivalent of the Lorentz factor derived in the last section for a 3D crystal. The differential cross-section in 2D has the form

$$\left(\frac{d\sigma}{d\Omega}\right)^{\text{2D}} = r_0^2\, P\, |F_{hk}|^2\, N\, A^*\, \delta(\text{Q}_x - ha_1^*)\delta(\text{Q}_y - ka_2^*) \tag{4.61}$$

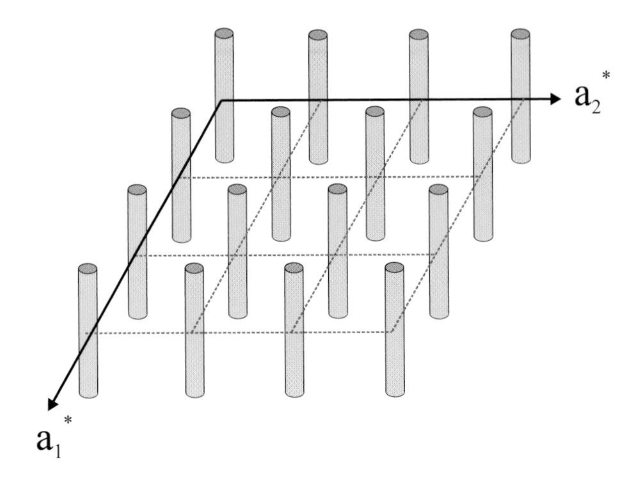

Figure 4.24: Scattering from a two dimensional crystal consists of rods of constant intensity passing through the points of the 2D reciprocal lattice given by $\mathbf{G}_{hk} = ha_1^* + ka_2^*$.

where N is the number of unit cells in the 2D lattice, A^* is the area of the unit cell in reciprocal space (see Eqs. (4.19), (4.20) and (4.34)). The Dirac delta functions restrict the scattering within the 2D plane to points forming the 2D reciprocal lattice given by $\mathbf{G}_{hk} = ha_1^* + ka_2^*$. The fact that there is no restriction on the scattering in the direction perpendicular to the 2D means that the scattering consists of rods of constant intensity passing through the points $\mathbf{G}_{hk}$, as illustrated in Fig. 4.24.

In the 3D case the delta function in Eq. (4.57) is three dimensional, and in order to derive the scattered intensity, Eq. (4.58), it was necessary to perform three integrations: one over the angle θ, and a two-dimensional integral over the element of solid angle. It is the latter integration which is different in the 2D case. Here it is only the in-plane part of the solid angle that involves an integration over a delta function; the out-of-plane part of the solid angle must be considered separately.

Let us first carry out the in-plane integrations, and in order to keep the notation simple let all vectors in the equations below be confined tacitly to the 2D plane. Utilizing the identity

$$\int x\,\delta(x^2 - \mathrm{k}^2)\,dx = \left[\frac{x}{2x}\right]_{x=\mathrm{k}} = \frac{1}{2}$$

we obtain

$$\begin{aligned}
\int d\hat{\mathbf{k}'}\,\delta(\mathbf{k} - \mathbf{k}' - \mathbf{G}) &= \overbrace{2\int \mathrm{s}\,\delta(\mathrm{s}^2 - \mathrm{k}^2)\,ds}^{1} \int \delta(\mathbf{k} - \mathrm{k}\hat{\mathbf{s}} - \mathbf{G})\,d\hat{\mathbf{s}} \\
&= 2\int \delta(\mathrm{s}^2 - \mathrm{k}^2)\,\delta(\mathbf{k} - \mathbf{s} - \mathbf{G})\;d^2\mathbf{s} \\
&= 2\,\delta((\mathbf{k} - \mathbf{G})\cdot(\mathbf{k} - \mathbf{G}) - \mathrm{k}^2) \\
&= 2\,\delta(\mathrm{G}^2 - 2\,\mathrm{k}\,\mathrm{G}\,\sin\theta)
\end{aligned}$$

where $d^2\mathbf{s}$=s $d\hat{\mathbf{s}}\,ds$ is the two dimensional volume element. The integral over θ

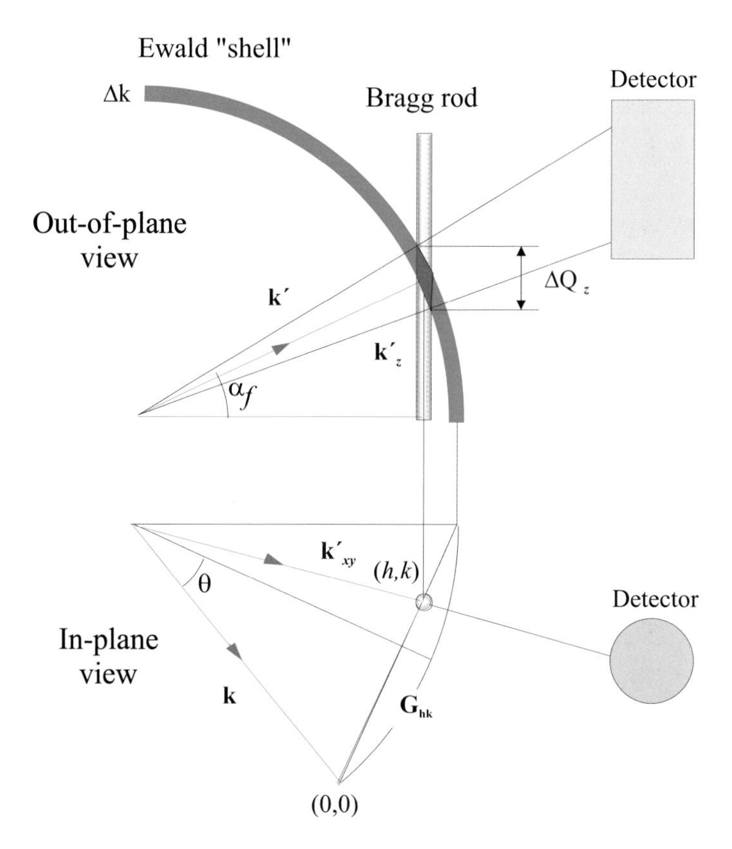

Figure 4.25: Reciprocal space diagram for a 2D system showing the intersection of part of the Ewald sphere with the Bragg rod. The in-plane projection of reciprocal space is shown in the bottom part, and the out-of-plane projection in the top.

is performed in a similar way to the 3D case and yields

$$\int 2\delta(\mathrm{G}^2 - 2\mathrm{kG}\sin\theta)\, d\theta = \frac{-1}{\mathrm{k}^2 \sin 2\theta}$$

leading to

$$\left(\frac{d\sigma}{d\Omega}\right)_{\text{int. over } \mathbf{k}'_{xy},\, \theta} = r_0^2 P \left|F_{hk}\right|^2 N \, \frac{\lambda^2}{A} \, \frac{1}{\sin 2\theta}$$

Here A is the unit cell area in direct space, and N is the number of illuminated unit cells. The scattered wavevector has been written with the subscript xy to emphasize that it is the 2D part that has been considered so far.

Having performed the integration over the in-plane part of the scattered wavevector $\mathbf{k}'_{xy}$ it is now necessary to integrate over the out-of-plane part of the solid angle. This is illustrated in Fig. 4.25, which shows the in-plane and out-of-plane projections of reciprocal space. In contrast to the 3D case the *projected* scattering triangle is no longer isosceles as $|\mathbf{k}'_{xy}| \leq |\mathbf{k}|$ since $\sqrt{(\mathbf{k}'_{xy})^2 + (\mathbf{k}'_z)^2} = |\mathbf{k}|$, where $\mathbf{k}'_{xy}$ and $\mathbf{k}'_z$ are the in-plane and out-of-plane components of the scattered wavevector respectively. The out-of-plane projection is shown in the top of the figure. As the scattering is elastic the projection of $\mathbf{k}'$ must lie within the circular band of the Ewald sphere. The finite thickness of the band represents the bandwidth Δk of the incident beam. Only scattered wavevectors with an out-of-plane component within the interval ΔQ_z are allowed. The corresponding component of the solid angle element is $\Delta\Omega_z = \Delta Q_z/\text{k}$ and the scattered intensity is therefore

$$I_{\text{sc}}^{2D}\left(\frac{\text{photons}}{\text{sec}}\right) = \Phi_0\left(\frac{\text{photons}}{\text{unit area}\times\text{sec}}\right) r_0^2 P\left|F_{hk}\right|^2 N\,\frac{\lambda^2}{A}\frac{1}{\sin 2\theta}\left(\frac{\Delta Q_z}{\text{k}}\right) \tag{4.62}$$

4.6 Applications of kinematical diffraction

The objective of an X-ray diffraction experiment is to determine the structure, which for a crystalline material means determining the unit cell and basis. The formulae derived so far in this chapter apply mostly to the diffraction from a single crystal. Under favourable circumstances materials may indeed be available as single crystals. The three-dimensional structure of the material is then solved by measuring as many Bragg peaks as possible as a function of the Miller indices (h, k, l). In rough terms, the size and symmetry of the unit cell are found from the position of the Bragg peaks, while the nature of the basis and the position of the atoms (or molecules) within it determine the Bragg peak intensities. Sophisticated techniques have been developed for going from the measured intensities to the final structure, and for unit cells containing a modest number of atoms the whole process of structure solution using single crystals is now routine and highly automated.

Many important materials cannot be obtained in single-crystal form, and may instead be in the form of powders, or fibres. Alternatively, it may be that it is the two-dimensional structure of the crystal surface, and not the three-dimensional structure of the bulk that is of interest. Understanding the diffraction pattern in these situations requires further concepts to be developed. In this section these concepts are introduced, and examples are used to illustrate how they apply in practice.

4.6.1 Powder diffraction

A good crystalline powder consists of many thousands of tiny crystallites oriented at random. Let us focus our interest on a particular reciprocal lattice vector $\mathbf{G}_{hkl}$ specified by its Miller indices (h, k, l). In the ideal powder sample the directions of the $\mathbf{G}_{hkl}$ vectors are isotropically distributed over the sphere indicated in Fig. 4.26. Some of the grains have the correct orientation, relative to the incident wavevector $\mathbf{k}$, for Bragg scattering – in the figure they are represented by the circle, which is a cut through the sphere of the plane perpendicular to $\mathbf{k}$. The scattered wavevectors $\mathbf{k}'$ are thus distributed evenly on a cone with $\mathbf{k}$ as the axis and an apex half angle of 2θ. This cone is called the Debye-Scherrer cone after the two physicists who first correctly interpreted X-ray scattering from a powder.

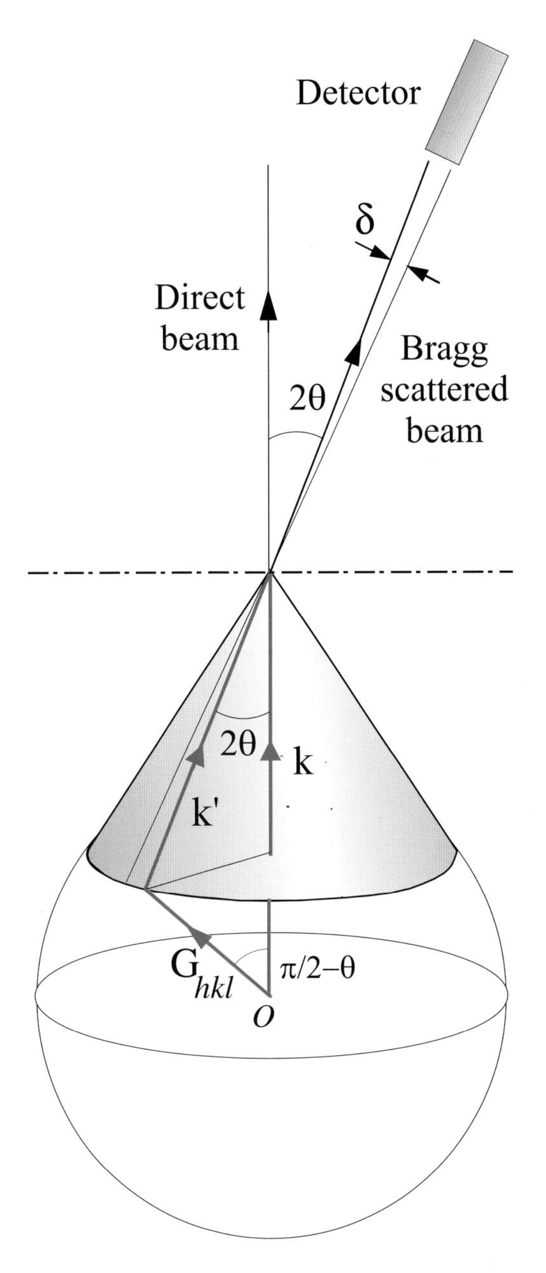

Figure 4.26: In an ideal powder there is an isotropic distribution of crystal grain orientations, as indicated by the sphere which represents the terminal points of the reciprocal lattice vector **G** from all of the grains. For fixed incident wavevector **k** all of the **G** vectors terminating on the circle will Bragg reflect, so that the scattered wavevectors $\mathbf{k}'$ span a cone, the so-called Debye-Scherrer cone. The angular acceptance of the detector is δ.

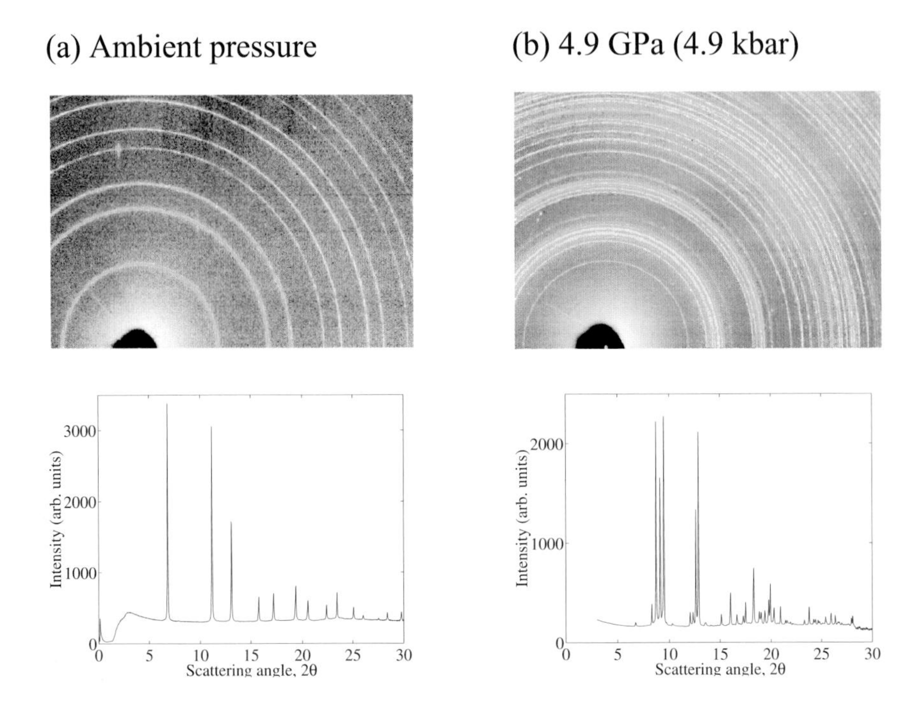

Figure 4.27: Powder diffraction patterns from InSb at (a) ambient pressure, and (b) at a pressure of 4.9 GPa. The patterns recorded on an image plate detector are shown in the top row, and display rings where the detector intercepts the Debye-Scherrer cones. The data were recorded with an incident wavelength of $\lambda = 0.447$ Å. In the bottom row the radially averaged patterns as a function of 2θ are displayed. The results show that InSb undergoes a phase transition from the zinc sulfide structure to a phase with an orthorhombic structure at pressures above 4.9 GPa. (Data courtesy of Malcolm McMahon, University of Edinburgh.)

At first sight it might appear to be an impossible task to solve a full three-dimensional crystal structure from a powder diffraction pattern, which is a two-dimensional projection. However, several methods for achieving this have been developed. Probably the most commonly one used is Rietveld refinement [Rietveld, 1969]. This seeks to use the entire diffraction profile, and not just the integrated intensities of the powder lines, to constrain or refine the parameters in the structural model. The Rietveld refinement which gives the method its name is in fact the last step in a process. The first step is known as indexing, and involves finding the size and symmetry of the unit cell so that the powder lines can be labelled with the appropriate values of (h, k, l). The second step is to extract the measured intensities and to convert them into structure factors. The third is to use the measured structure factors to build a structural model. Finally, the structural model is refined using the entire diffraction profile.

Here we restrict our discussion to showing how the measured intensity in a powder diffraction experiment is related to the structure factor. For a certain fixed (h, k, l) reflection the number of powder grains oriented to reflect is proportional to the circumference of the base-circle of the Debye-Scherrer cone shown in Fig. 4.26. The circumference is given by $\mathrm{G}_{hkl} \sin(\frac{\pi}{2} - \theta) = \mathrm{G}_{hkl} \cos\theta$. However, permutations of (h, k, l) may have the same sphere of $\mathbf{G}_{hkl}$ vectors, and this is taken into account by introducing the *multiplicity* of a reflection m_{hkl}. For example, for a cubic lattice the multiplicity of the $(h, 0, 0)$ reflections is 6 as the $(\pm h, 0, 0)$, $(0, \pm h, 0)$ and $(0, 0, \pm h)$ reflections will all Bragg reflect to the same 2θ. So for a given $\mathbf{G}_{hkl}$ the intensity must be proportional to $m_{hkl} \cos\theta$. At a *different* $\mathbf{G}_{hkl}$, and hence a different value of 2θ, the detector will see a different fraction of the base circle. Independent of $\mathbf{G}_{hkl}$, the circumference of the circle is $2\pi \mathrm{k} \sin 2\theta$. Thus the fraction seen by the detector is $\mathrm{k}\delta/(2\pi \mathrm{k} \sin 2\theta)$, which is proportional to $1/\sin 2\theta$. Finally, for a single crystallite the observed intensity will be proportional to the Lorentz factor we have already derived, $1/\sin 2\theta$. Altogether then the observed intensity will be proportional to

$$L_{\text{powder}} = m_{hkl} \cos\theta \frac{1}{\sin 2\theta} \frac{1}{\sin 2\theta} = \frac{m_{hkl}}{2 \sin\theta \sin 2\theta}$$

The structure factors squared can be derived on a relative scale from the observed diffraction intensities. Suppose, for example, that it is required to determine the ratio of squared structure factors for the (1,1,1) and (2,0,0) reflections from an *fcc* crystal. In addition to the combined Lorentz factor given above, it is also necessary to allow for the polarization factor P which depends on the scattering angle, so that the ratio of the intensities is

$$\frac{I_{111}}{I_{200}} = \frac{|F_{111}|^2}{|F_{200}|^2} \frac{L_{\text{powder}}(\theta_{111})}{L_{\text{powder}}(\theta_{200})} \frac{P(\cos 2\theta_{111})}{P(\cos 2\theta_{200})} \tag{4.63}$$

The assumption of an isotropic distribution of orientations of the crystallites is not fulfilled trivially in practice. When this condition is not fulfilled the powder is said to posses preferred orientations. Grains in a metal ingot, for example, may be highly textured in orientation due to mechanical rolling. The texture is in fact of importance for the mechanical properties of the metal, and it can be determined by suitable rotation of the sample. In other cases a powdered sample is prepared by crushing the material into a powder and loading it into a glass capillary tube. An isotropic distribution is then ensured by rotating the capillary tube around its axis during exposure. Powder diffraction is particularly useful for studying the structure of materials under extreme conditions, such as the study of phase transitions as a function of applied pressure.

In Fig. 4.27 data are shown from a study of the semiconducting material InSb as a function of pressure. The data were obtained by loading a small quantity of powdered InSb into a diamond anvil pressure cell, and then recording the powder diffraction patterns with an image plate detector. The data shown in Fig. 4.27(a) were recorded at ambient pressure where InSb adopts the zinc sulphide structure (see Fig. 4.13(b)) with a lattice parameter of 6.48 Å. The X-ray wavelength was 0.447 Å, and Debye-Scherrer cones were observed at scattering angles of $2\theta =$ 6.81°, 11.15°, 13.06°, 15.78°, 17.12°, $\cdots$, corresponding to the (1,1,1), (2,2,0), (3,1,1), (4,0,0), (3,3,1), $\cdots$, Bragg peaks respectively. When pressure is applied to the system a phase transition occurs, as shown in Fig. 4.27(b) for an applied pressure of 4.9 GPa, and the crystal structure transforms to become orthorhombic.

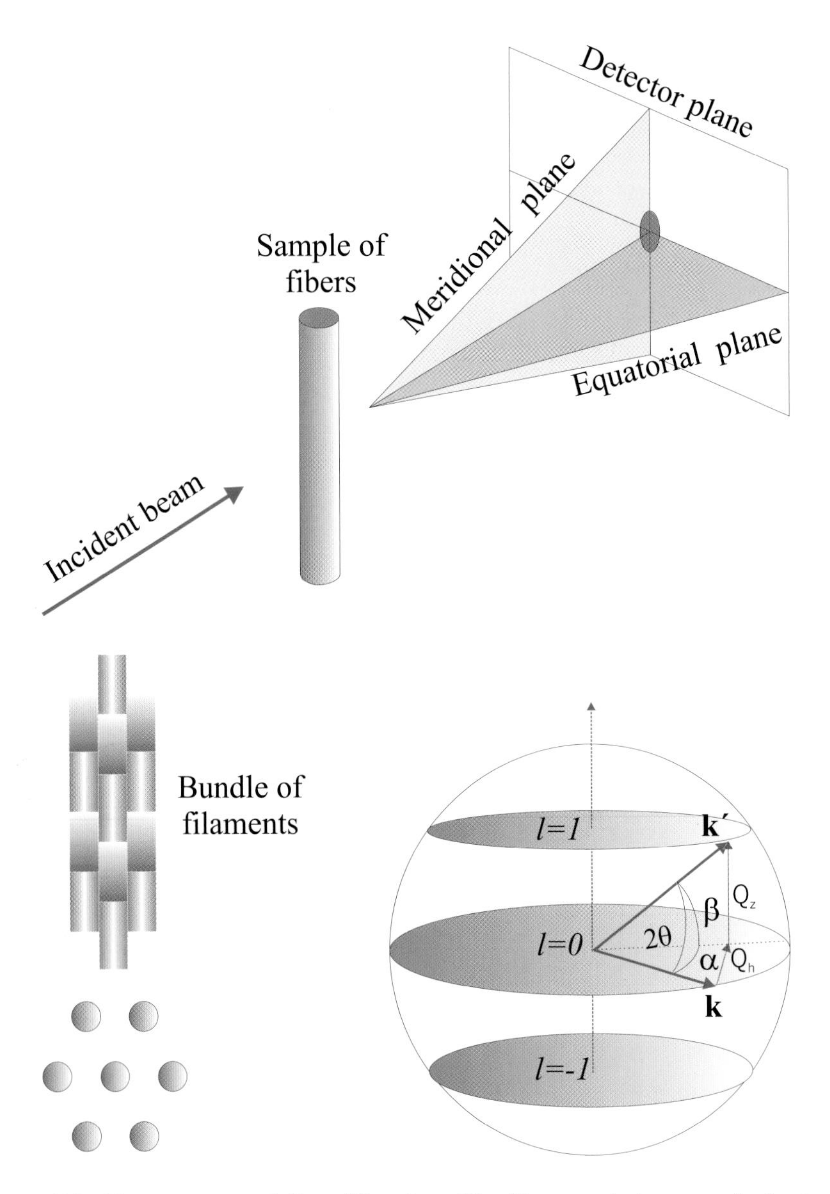

Figure 4.28: Top: geometry of fibre diffraction. The fibre sample is perpendicular to the incident monochromatic beam. Bottom left: the fibre sample comprises a large number of filaments, randomly orientated in the azimuthal angle. Bottom right: periodicity along the fibre implies that Bragg reflections are restricted to layers.

4.6.2 Diffraction from a fibre

From a crystallographic point of view a fibre can be considered as an extreme limit of anisotropy in a powder: all the crystallites have one of their crystallographic axes, here denoted the c-axis, aligned along the fibre axis, whereas the azimuthal orientation of the $a-b$ plane is random. Fibres occur frequently in nature, for example, in muscles and in collagen, and artificial fibres are widely used in industry. The structure of fibres, as revealed by diffraction studies, is therefore of general interest and is the subject of this section.

In fibre diffraction geometry, the monochromatic incident beam with wavevector $\mathbf{k}$ is perpendicular to the vertical fibre axis, as shown in Fig. 4.28. Bragg reflections occur in the horizontal (equatorial) plane due to the fact that each individual fibre in the sample is formed from a bundle of thinner long filaments, which pack in a two-dimensional lattice perpendicular to the fibre axis. Each filament may also exhibit periodicity along the axis, thus giving rise to Bragg reflections in the vertical (meridional) plane.

The lower right part of Fig. 4.28 shows reciprocal space. The scattering vector $\mathbf{Q}$ is decomposed into the vertical component $\mathbf{Q}_z$ and the horizontal component $\mathbf{Q}_h$. For Bragg scattering $\mathbf{Q}$ must terminate in layers at $\mathbf{Q}_z = lc^*$ with $l = 0, \pm1, \pm2, \cdots$, etc, and $c^* = 2\pi/c$, where c is the period along the filament axis. Furthermore, if the filaments making up the fibre are arranged in a 2D-lattice, Bragg reflections will occur in the different l-layers outside the meridional axis. The l=0 layer is called the equatorial layer. For a certain l-layer, the angle β is constant. The scattered wavevector $\mathbf{k}'$ must terminate on a circle in the l-layer since the scattering is elastic; the scattering angle 2θ varies with the azimuthal angle α according to $\cos 2\theta = \cos\alpha\,\cos\beta$, so for certain values of α the Bragg condition $\lambda = 2d_{hkl}\sin\theta$ will be fulfilled. In the bundle of many fibres making up the sample there will always be one with the correct (a^*, b^*) orientation for Bragg reflection. Independent of the fibre structure, symmetry implies that the Bragg spots occur symmetrically around the meridional plane at $\pm\alpha$.

Example: Helices in biology and the structure of DNA

The primary structure of a protein is a polypeptide backbone, depicted in Fig. 4.29, onto which is attached a sequence of amino acids. Around 1950 Linus Pauling formulated a seminal idea on the structure of proteins which has had far reaching consequences [Pauling et al., 1951]. In Pauling's laboratory they had been studying the building blocks of polypeptide chains. As a result of this work Pauling became convinced that a protein was formed by structural units which could be considered to be rigid and planar, or at least approximately so. This is illustrated in Fig. 4.29 where the shaded parallelograms indicate that the carbonyl and amide groups are planar. It follows that when the polypeptide chain folds to form a protein the main degrees of structural freedom are the rotation angles around the links between these rigid structural units. One of Pauling's most important insights was that the formation of a hydrogen bond between the carbonyl C=O of one unit and the amino N-H group four units further along would cause the chain to curl up into a helix. This structure was christened the α-helix, and has 3.7 residues for one period of rotation. Subsequent experiments established that Pauling was indeed correct, and it

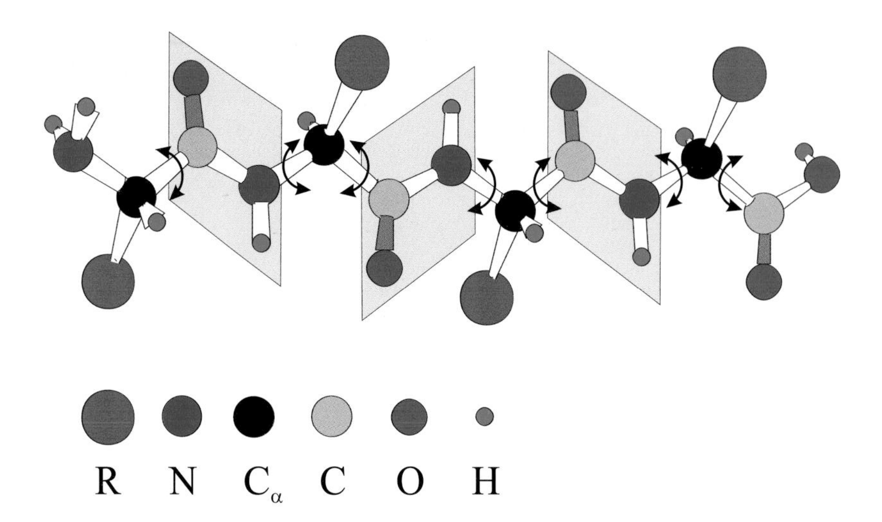

Figure 4.29: A polypeptide chain is composed of planar moieties of carbonyl and amide groups. These planar moieties can be rotated around either the N-C_α or the C_α-C bonds. Hydrogen bonding between the N-H and C=O groups causes the chain to fold into a helical structure known as the α-helix after Pauling. Here R stands for an amino acid residue.

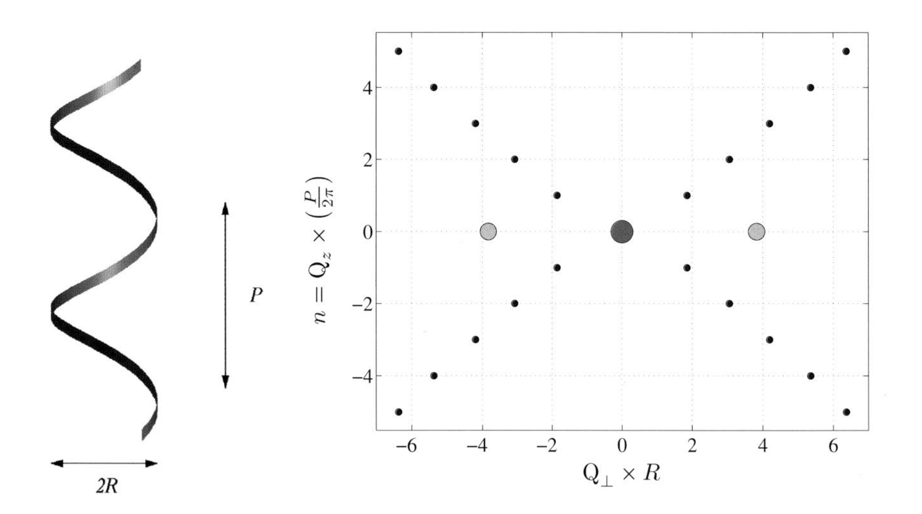

Figure 4.30: Scattering from a single, infinitely long helix of radius R and period P. The structure factor squared has principle maxima arising from the peaks in the Bessel functions, and form a cross in reciprocal space as indicated by the blue circles. The grey circles on the equatorial axis are the secondary maxima from the zeroth-order Bessel function. Here Q_z and $Q_\perp$ are the components of the scattering vector parallel and perpendicular to the axis of the helix respectively.

is now known that α-helices are an important structural component of many proteins.

Inspired by Pauling's ideas, Cochran et al. calculated the generic diffraction pattern from a helix [Cochran et al., 1952]. As the scattering from helices has assumed such significance in structural biology an outline of this calculation is given here. The starting point is to imagine that a uniform and continuous distribution of material lies on a infinitely long helical string of period P. The problem in calculating the diffraction pattern is to add up the phase factor for each differential element along the helix. As the material is uniformly distributed the scattering amplitude is found by evaluating the integral

$$A(\mathbf{Q}) \propto \int \mathrm{e}^{i\,\mathbf{Q}\cdot\mathbf{r}}\, dz$$

where z is taken to be along the axis of the helix. For a helix with a period of P and a radius of R, any point $\mathbf{r}$ on the helix is given by

$$\mathbf{r} = \left(R\cos(\tfrac{2\pi\, z}{P}),\ R\sin(\tfrac{2\pi\, z}{P}),\ z\right)$$

As the helix is periodic, the integral decomposes into a sum over all periods (or lattice sites) multiplied by the structure factor of a single period. The scattering amplitude then becomes

$$\begin{aligned} A(\mathbf{Q}) &\propto \sum_{m=0}^{\infty} \mathrm{e}^{i\,\mathrm{Q}_z mP} \int_{z=0}^{z=P} \mathrm{e}^{i\,\mathbf{Q}\cdot\mathbf{r}}\, dz \\ &\propto \int_{z=0}^{z=P} \delta(\mathrm{Q}_z - \tfrac{2\pi n}{P})\, \mathrm{e}^{i\,\mathbf{Q}\cdot\mathbf{r}}\, dz \end{aligned} \tag{4.64}$$

Here n is an integer, and use has been made of Eq. (4.32), which allows the sum over lattice sites to be written as a delta function.

To evaluate the phase $\mathbf{Q}\cdot\mathbf{r}$ it is convenient to use cylindrical coordinates and express the scattering vector as

$$\begin{aligned} \mathbf{Q} &= (\mathrm{Q}_\perp \cos(\Psi),\ \mathrm{Q}_\perp \sin(\Psi),\ \mathrm{Q}_z) \\ &= \left(\mathrm{Q}_\perp \cos(\Psi),\ \mathrm{Q}_\perp \sin(\Psi),\ \tfrac{2\pi n}{P}\right) \end{aligned}$$

where Q_z is the axial component, $\mathrm{Q}_\perp$ is the radial component and Ψ is the azimuthal angle. The scattering amplitude from a helix then assumes the form

$$\boxed{A_1(\mathrm{Q}_\perp, \Psi, \mathrm{Q}_z) \propto \mathrm{e}^{i\,n\Psi}\, J_n(\mathrm{Q}_\perp R)}$$

where $J_n(\mathrm{Q}_\perp R)$ is the n'th order Bessel function of the first kind, and $\mathrm{Q}_\perp R$ is a dimensionless argument. The mathematics leading to this expression are explained in the box on the following page. The subscript "1" is used as a reminder of the fact that the expression refers to the scattering from a single helix. The scattered intensity given by the above equation is plotted in Fig. 4.30.

Perhaps the most celebrated helical structure in biology is the *double* helix of DNA (deoxyribose nucleic acid). The structure of DNA was first solved by Watson and Crick [Watson and Crick, 1953], who mainly used stereo-chemical arguments to build a model which helped them deduce the correct structure. They were assisted greatly in their work by the X-ray diffraction experiments

Structure factor of a helix and the Bessel function $J_n(\xi)$

From Eq. (4.64) the scattering amplitude from a helix of period P and radius R is

$$A(\mathbf{Q}) \propto \int_{z=0}^{z=P} \delta(\mathrm{Q}_z - \tfrac{2\pi n}{P})\, \mathrm{e}^{i\,\mathbf{Q}\cdot\mathbf{r}}\, dz$$

Using cylindrical coordinates the scalar product of the scattering vector $\mathbf{Q}$ and position $\mathbf{r}$ is

$$\begin{aligned}\mathbf{Q}\cdot\mathbf{r} &= \mathrm{Q}_\perp \cos(\Psi)\, R\cos(\tfrac{2\pi z}{P}) + \mathrm{Q}_\perp \sin(\Psi)\, R\sin(\tfrac{2\pi z}{P}) + \mathrm{Q}_z z \\ &= \mathrm{Q}_\perp R\cos(\tfrac{2\pi z}{P} - \Psi) + (\tfrac{2\pi z}{P})\, n\end{aligned}$$

It is convenient to rewrite this as

$$\mathbf{Q}\cdot\mathbf{r} = \xi\cos\varphi + n\varphi + n\Psi$$

with $\xi = \mathrm{Q}_\perp R$ and $\varphi = (2\pi z/P - \Psi)$. The scattering amplitude can then be written in the form

$$A(\mathbf{Q}) \propto \mathrm{e}^{i\,n\Psi} \int_0^{2\pi} \mathrm{e}^{i\,\xi\cos\varphi + i\,n\varphi}\, d\varphi$$

The n'th order Bessel function of the first kind is given in integral form by

$$J_n(\xi) = \frac{1}{2\pi i^n} \int_0^{2\pi} \mathrm{e}^{i\,\xi\cos\varphi + i\,n\varphi}\, d\varphi$$

It is then apparent that the scattering amplitude from a single helix assumes the form

$$A_1(\mathrm{Q}_\perp, \Psi, \mathrm{Q}_z) \propto \mathrm{e}^{i\,n\Psi}\, J_n(\mathrm{Q}_\perp R)$$

performed around the same time by Wilkins et al. [Wilkins et al., 1953] and Franklin and Gosling [Franklin and Gosling, 1953]. These experiments established the helical nature of the DNA molecule, and provided decisive structural parameters, such as its period and radius. The discovery of the double helix probably ranks as one of the most important scientific advances of the twentieth century. As the authors note: "It has not escaped our notice that the specific pairing we have postulated immediately suggests a possible copying mechanism for the genetic material". It is therefore more than worthwhile to describe the scattering from the double helix.

A photograph of the fibre diffraction pattern from DNA is shown in the top part of Fig. 4.31⋆. This is from the B phase, and is similar to one of the original patterns reported by Franklin and Gosling [Franklin and Gosling, 1953]. In contrast to the type of pattern recorded from a single crystal, this diffraction pattern arises from a large number of crystallites with random orientation about the chain axis. The reflections are spread into arcs because the alignment of the crystallites along the chain axis is not perfect. Although cylindrical averaging frequently occurs in fibres and results in overlap of systematically related reflections, the loss of information is not normally severe. This type of diffraction method has been used to determine the structures of 4 of the 5 principal DNA conformations. It is also used to study other biomolecules including filamentous viruses, cellulose, collagen, flagella etc. The use of neutron diffraction in combination with x-ray methods is especially powerful since it allows important insights into hydration of these molecules.

It is evident that the diffraction pattern from DNA possesses some of the features predicted by Cochran et al. for the scattering from a helix. In particular there is a characteristic cross of Bragg peaks. From the position of these peaks along the meridional (vertical) axis the period of the helix is found to be 34 Å, while from the angle of the cross it can be deduced that the radius of the helix is 10 Å. The double nature of the helix is only apparent from a detailed analysis of the pattern. Most tellingly the reflections from the 4'th order layer are missing on the film, although the 3'rd and 5'th order are clearly apparent. Indeed Rosalind Franklin herself was aware that this feature of the diffraction pattern could be explained naturally by assuming that DNA is formed from two intertwined helices as shown in the middle part of Fig. 4.31⋆. If the two helices are displaced along the common z axis by an amount Δ, then this corresponds to an azimuthal angle $\Psi = 2\pi(\Delta/P)$, and the scattering amplitude becomes

$$A_2(\mathrm{Q}_\perp, \Psi, \mathrm{Q}_z) \propto (1 + e^{in\left(\frac{2\pi}{P}\right)\Delta})\, J_n(\mathrm{Q}_\perp R) \tag{4.65}$$

with $\mathrm{Q}_z = n(2\pi/P)$. The waves scattered by the two helices interfere in such a way that the intensity of the 4'th layer reflections becomes vanishingly small when Δ/P=1/8, 3/8, 5/8, etc. In the bottom part of Fig. 4.31⋆ the intensity calculated from this equation is plotted. It can be seen that it accounts for most of the qualitative features of the central part of the diffraction pattern. To obtain better agreement it would obviously be necessary to specify the position of all the molecules in the structure and their scattering factors. One feature not accounted for by the simple model described here is the existence of strong, but diffuse reflections on the meridional axis close to the 10'th layer. These reflections arise from the fact that the double helix has 10 pairs of bases per period.

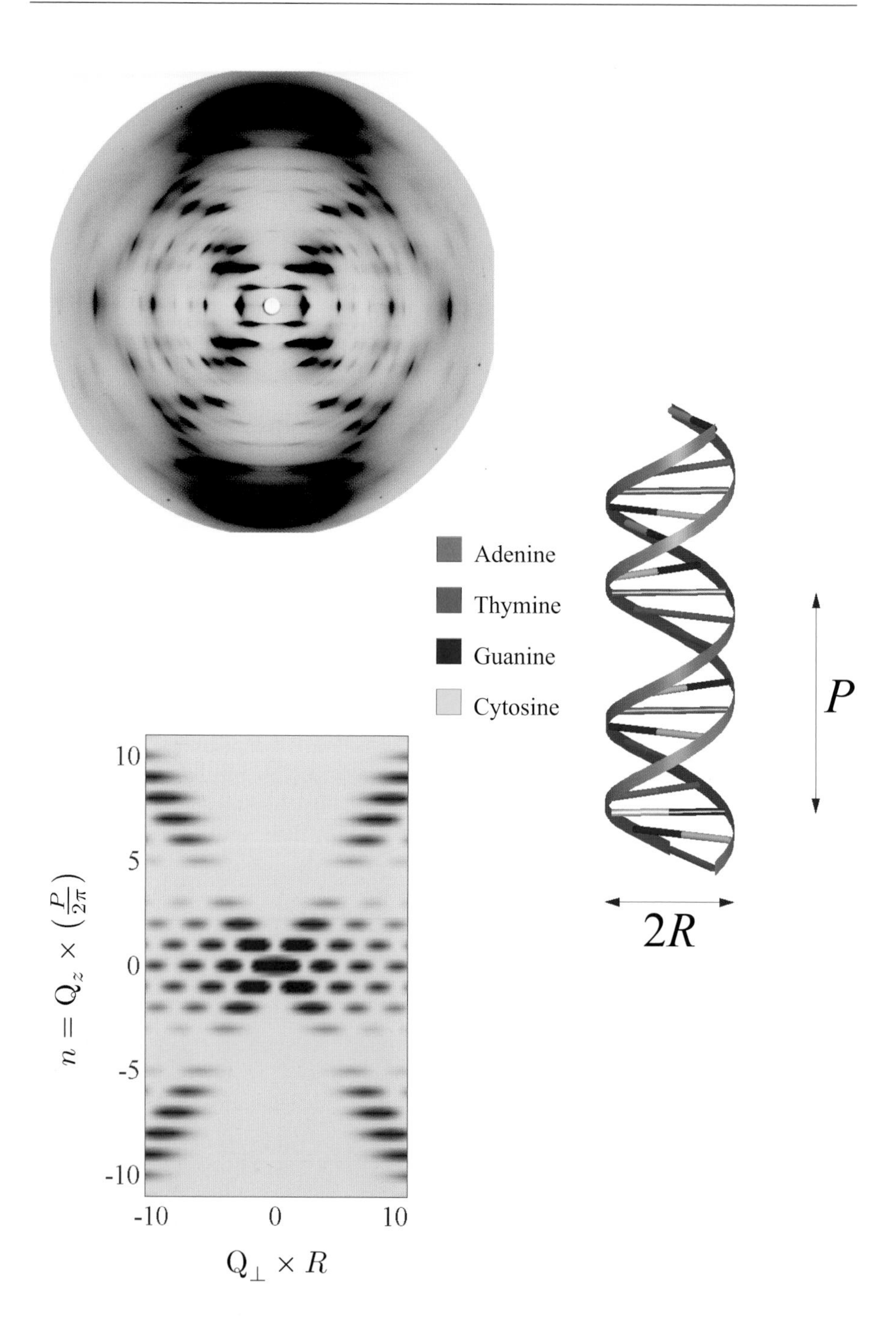

Figure 4.31: ⋆ The double helix of DNA. Top: Fibre diffraction data for the B conformation of DNA. (Image provided by Watson Fuller, University of Keele, UK.) Middle: The structure of DNA is formed from two intertwined helices displaced axially by 3/8 of a period. The backbone of the helices is formed from polypeptide chains, and the "steps" from a pairing of hydrogen bonded bases, adenine with thymine, and guanine with cytosine. Bottom: The intensity calculated from Eq. (4.65) for two helices displaced by 3/8'ths of a period.

4.6.3 Two-dimensional crystallography

One area that has benefited greatly from the high flux produced by synchrotron sources is the study of surfaces using X-ray scattering. Although X-rays are scattered only weakly by a monolayer of atoms or molecules, it transpires that the sensitivity is still high enough that it is possible to study the structure in great detail. Moreover, the fact that the scattering is weak simplifies the interpretation of the data considerably, as the kinematical approximation then applies. This contrasts with the case of strongly interacting probes, such as the electron, where the data analysis is complicated by the need to resort to a full multiple scattering theory.

X-ray scattering experiments from surfaces are usually performed with the angles of the incident and exit beams close to the critical angle, α_c, for total external reflection, as this limits the penetration depth of the beam which in turn reduces the background scattering from the bulk of the crystal. One consequence of this is that it is necessary to correct the above formula for refraction effects. In Section 3.4 it was shown that the transmittivity, $t(\alpha_i)$, of the incident beam is enhanced for angles close to α_c (Fig. 3.6). From Eq. (3.23) the amplitude transmittivity is

$$t(\alpha_i) = \frac{2\alpha_i}{\alpha_i + \alpha'}$$

where α' refers to the transmitted beam. It can be shown that similar arguments apply to the beam scattered from within the crystal, and this introduces a factor that depends on $t(\alpha_f)$, where α_f is the angle that the exit beam makes with the surface. Including these refraction effects, the integrated intensity given in Eq. (4.62) becomes

$$I_{\text{sc}}^{2D} \longrightarrow I_{\text{sc}}^{2D} \, |t(\alpha_i)|^2 \, |t(\alpha_f)|^2 \tag{4.66}$$

A free standing two-dimensional crystal is difficult to realize in nature, and quasi two-dimensional structures are instead studied on crystal surfaces. The structure may of course be the surface of the crystal itself, as the difference in the bonding of the surface atoms often leads to a reconstruction of the surface. Alternatively it may be the structure of an absorbed layer of atoms or molecules. In either case, experiments on surfaces are performed in two distinct steps. The first is to study the *in-plane* structure of the surface, in other words the positional coordinates of the topmost atoms within the plane of the surface. This is similar to a conventional crystallographic structure determination, and requires that the structure factor $|F_{hk}|$ of as many Bragg peaks as possible are determined, with $l \approx 0$. These can then be compared with calculations of the structure factor for different models of the surface. The in-plane structure is most readily determined when the Bragg peaks from the surface layer appear at different positions from those of the bulk. In other words when the surface has a different in-plane periodicity to that of the bulk. This is often found to be the case for either reconstructed surfaces, or absorbed layers. The second step is to study the *out-of-plane* structure. Here the intensity distribution of the scattering is studied as a function of l with the values of h and k set to coincide with a 2D Bragg peak. We have seen already (Section 4.4.6) that for an ideally terminated surface the scattering is extended along l to form the crystal

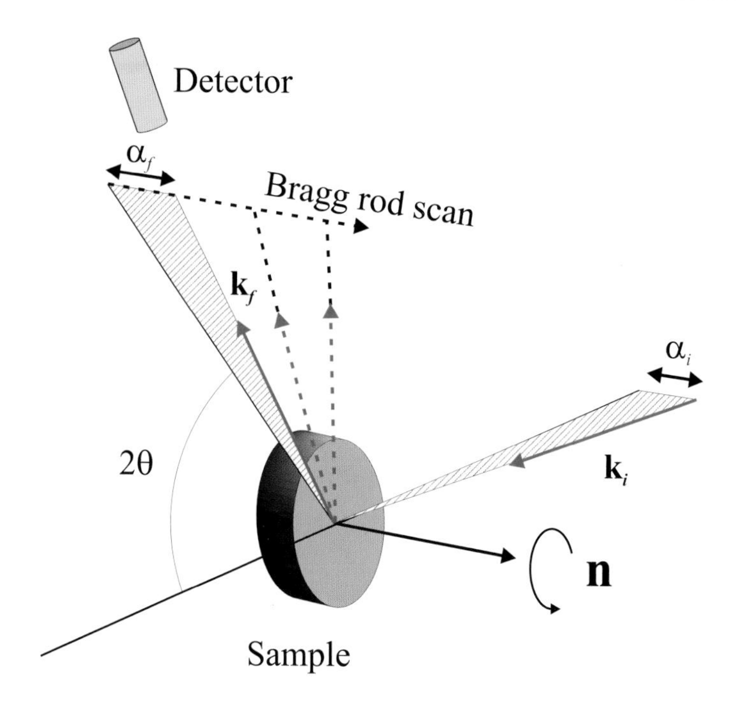

Figure 4.32: Geometry of a grazing incidence diffraction (GID) experiment from a solid single crystal surface at a synchrotron. The surface can be rotated around the surface normal $\mathbf{n}$, and the detector can be both rotated around $\mathbf{n}$ in the vertical plane, and moved perpendicular to it so as to scan along the crystal truncation rods.

truncation rods (CTR), and that the observed intensity distribution along the rods is sensitive to any modifications of the near-surface region.

A schematic of the experimental setup used to measure the scattering from a surface in grazing incidence geometry is shown in Fig. 4.32. The sample here is a disc with a surface normal $\mathbf{n}$ in the horizontal plane. The incident, monochromatic beam is horizontal and the incident wavevector $\mathbf{k}_i$ is at a glancing angle α_i to the surface. The incident glancing angle is close to the critical angle for total reflection, so that only a thin surface layer (thickness Λ) is exposed to the X-rays, as discussed in Section 3.4. When the sample is rotated around $\mathbf{n}$, the glancing incident angle should remain fixed which can be monitored by requiring a constant value of the specularly reflected beam. For a certain $(h, k,0)$ Bragg reflection with reciprocal lattice vector $\mathbf{G}_{hk}$ the Bragg angle θ is calculated from $2\mathrm{k}\sin\theta = \mathrm{G}_{hk}$. The detector arm is turned around $\mathbf{n}$ to the angle 2θ, and for $l \approxeq 0$ the detector is then translated a small angle α_f (of order α_i) along the surface normal. The sample lattice can be turned around $\mathbf{n}$ so that the reflecting (h, k) planes bi-sect the angle between $\mathbf{k}_i$ and $\mathbf{k}_f$. To obtain the intensity along the CTR's the detector is translated along the surface normal so that α_f becomes much larger than α_i.

Example: absorption of O on the Cu(110) surface

The example we shall use to illustrate many of the concepts introduced above is the structure of the (110) copper surface with oxygen atoms chemisorbed onto it [Feidenhans'l et al., 1990].

Copper crystallizes in a face centred cubic structure as shown in the top part of Fig. 4.33. The Cu atoms are located at the corners and at the centres of the faces of a cube. The copper crystal has been cut so that a (110) plane forms the surface. A front-view of the truncated bulk (110) surface is shown in the middle panel of Fig. 4.33. Cu atoms in the top layer are shown as large filled circles, whereas Cu atoms in the second layer are represented by smaller ones. Cu atoms in the third layer sit directly below those in the top layer, etc.

The cubic unit cell is not convenient for describing the positions of atoms in the surface layer. For the surface we rather choose the unit cell as shown shaded in the middle panel. Clearly the length of **a** is $\mathrm{a_c}/\sqrt{2}$ and the length of **b** is $\mathrm{a_c}$. The third axis of the unit cell, **c**, is perpendicular to the surface and has the length $a_c/\sqrt{2}$. With this choice there are two atoms per unit cell: one at (0,0,0) and one at (1/2,1/2,1/2). We define a full monolayer of Cu atoms as that of the top layer. Referring to this unit cell the reciprocal lattice ($\mathbf{a}^*$, $\mathbf{b}^*$) is shown to the right. The length of $\mathbf{a}^*$ and $\mathbf{b}^*$ are $2\pi/(a_c/\sqrt{2})$ and $2\pi/a_c$ respectively, so that a general reciprocal lattice vector in the plane of the surface is $\mathbf{G}_{hk} = h\mathbf{a}^* + k\mathbf{b}^*$.

If the bare Cu(110) surface terminated as the bulk structure, allowed (hk) reflections would require an even sum of h and k since $F_{hk} \propto 1+\mathrm{e}^{i\,\pi(h+k)}$. These reflections are indicated by diamonds in the reciprocal lattice shown in Fig. 4.33. However, when exposed to oxygen the surface undergoes a reconstruction from that of the truncated bulk lattice. By LEED (Low Energy Electron Diffraction) one can immediately see the symmetry of the surface unit cell, or rather that of the reciprocal lattice cell. For a certain dosage of oxygen, it turns out to be exactly like the truncated unit cell along the k-axis, but only half as large along the h-axis as indicated in the right panel by the h' index. This means that the unit cell in direct space must have doubled along the **a** direction as shown in the lower, left panel, and one refers to the reconstructed cell as a (2×1) cell. Furthermore, one can by other surface techniques (such as Scanning Tunnelling Microscopy, STM) determine that there is only half a monolayer of Cu atoms in the top layer and half a monolayer of oxygen. Finally, one knows that in the bulk structure of $\mathrm{Cu_2O}$, the oxygen is located midway between two Cu atoms with a Cu-O bond length of 1.852 Å, which is only slightly larger than $b/2 = \mathrm{a_c}/2$=1.8075 Å. It is therefore a good starting point to assume a model where every second row of Cu atoms along the **b** direction is missing (as this gives half a monolayer), and where half a monolayer of oxygen atoms is formed by the oxygen atoms occupying positions midway between neighbouring Cu atoms along the **b** direction. As Cu-O-Cu bonds are slightly larger than $\mathrm{a_c}/2$=1.8075 Å it is also likely that the O is displaced an amount z Å above or below the Cu-plane, with z given by $(1.8075^2 + z^2)=1.852^2$. As a refinement, one could further imagine that the Cu atoms in the next layer are pushed towards the missing row by an amount δ. This model unit cell is shown in the lower, left panel of Fig. 4.33.

The intensities observed in the experiment are given in Table 4.3$\star$, where corrections have been made for the Lorentz factor $1/\sin 2\theta$ as well as for the

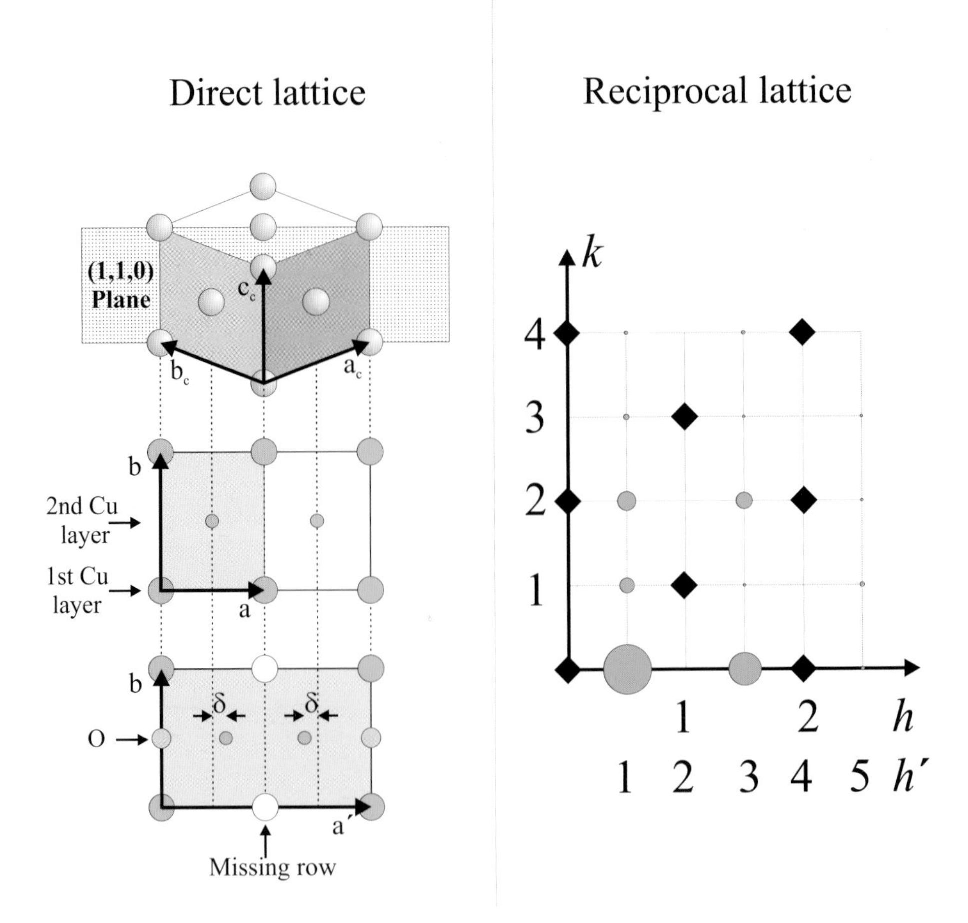

Figure 4.33: Top, left: The face centred cubic structure of Cu with a (110) plane indicated. The lattice vectors ($\mathbf{a}_c$,$\mathbf{b}_c$,$\mathbf{c}_c$) of the *conventional* unit cell are shown. Middle, left: The structure of the (110) surface layer, with the unit cell defined by **a** and **b**. Bottom, left: A model for the Cu surface exposed to O, with half of the Cu atoms in the surface layer missing along the **a** direction. Right: The reciprocal lattice of the Cu surface after exposure to O. The reflections arising from the doubling of the unit cell along the **a** direction (shown in the bottom, left panel) are represented by the filled circles, where their radius is proportional to the measured Bragg intensity. The diamonds indicate the allowed Bragg reflections from bulk Cu.

		k				
		0	1	2	3	4
	1	10.7 (10.7)	3.29 (3.85)	4.13 (4.23)	1.27 (1.01)	0.86 (0.75)
h'	3	7.09 (7.04)	0.39 (0.68)	3.58 (3.72)	0.32 (0.15)	0.84 (0.96)
	5	0.16 (0.52)	1.02 (1.23)	0.21 (0.23)	0.41 (0.48)	-

Table 4.3: ⋆ Observed intensities in arbitrary units for (h', k, l) reflections for O on Cu(110) with $l = 0$. The intensities from the model described in the text are given in the brackets.

crossed beam area, which is also proportional to $1/\sin 2\theta$. These data are also indicated in Fig. 4.33 as the shaded circles in reciprocal space. The area of each circle is proportional to the measured intensity when corrected for Lorentz factor and scattering area. In the reciprocal lattice the allowed bulk reflections are indicated by the diamonds.

Let us first discuss the in-plane data ($l \approx 0$) given in Table 4.3⋆. For h' odd and integer k, the structure factor for the first two layers (i.e. the top layer of unit cells) is the sum of three terms:

$$F_{h'k} = F^{\rm Cu1} + F^{\rm O} + F^{\rm Cu2}$$

where Cu1 and O refer to the first layer, and Cu2 to the Cu in the second layer. Using the unit cell shown in the bottom panel on the left of Fig. 4.33 the structure factors are

$$\begin{aligned}
F^{\rm Cu1} &= f^{\rm Cu}\,\mathrm{e}^{-M_{\rm Cu1}} \\
F^{\rm O} &= f^{\rm O}\,\mathrm{e}^{i\,2\pi(k/2)}\,\mathrm{e}^{-M_{\rm O}} = (-1)^k f^{\rm O}\,\mathrm{e}^{-M_{\rm O}} \\
F^{\rm Cu2} &= f^{\rm Cu}\,\mathrm{e}^{i\,2\pi(k/2)}\,\mathrm{e}^{i\,2\pi(h'/4)}\left[\mathrm{e}^{i\,2\pi h'\delta} - \mathrm{e}^{-i\,2\pi h'\delta}\right]\mathrm{e}^{-M_{\rm Cu2}} \\
&= f^{\rm Cu}\,\mathrm{e}^{i\,\pi k}\,\mathrm{e}^{i\,\pi h'/2}\,2i\,\sin(2\pi h'\delta)\,\mathrm{e}^{-M_{\rm Cu2}} \\
&= (-1)^{h'/2+k+1/2}\,f^{\rm Cu}\,2\,\sin(2\pi h'\delta)\,\mathrm{e}^{-M_{\rm Cu2}}
\end{aligned}$$

In evaluating the contribution from the second layer with Cu atoms we have utilized that for odd h', which we are dealing with in the table, $\mathrm{e}^{i\,2\pi(3h'/4)} = -\mathrm{e}^{-i\,2\pi(h'/4)}$. We have also taken into account that the thermal vibration (Debye-Waller) factor e^{-M} may be different for Cu atoms in the first layer and in the second layer, where the atoms are more tightly bound than in the first layer. The best fit to the data are obtained for δ=0.00606, corresponding to a displacement of 0.031 Å, and Debye-Waller factors of $B_T^{\rm Cu1} = 1.7 \pm 0.2$ Å^2, and $B_T^{\rm O} = 0 \pm 0.4$ Å^2. (For the second Cu layer $B_T^{\rm Cu2}$ was set equal to the value of 0.55 Å^2 for bulk copper.) The Debye-Waller factor is discussed in Section 4.4.7, and for the present example the relationship between the parameters M_X and B_T^X for element X is

$$M_X = B_T^X\left(\frac{\sin\theta}{\lambda}\right)^2 = B_T^X\left(\frac{\mathrm{G}_{h'k}}{4\pi}\right)^2 = B_T^X\left(\frac{1}{4\pi}\right)^2\left\{(h'a^*/2)^2 + (kb^*)^2\right\}$$

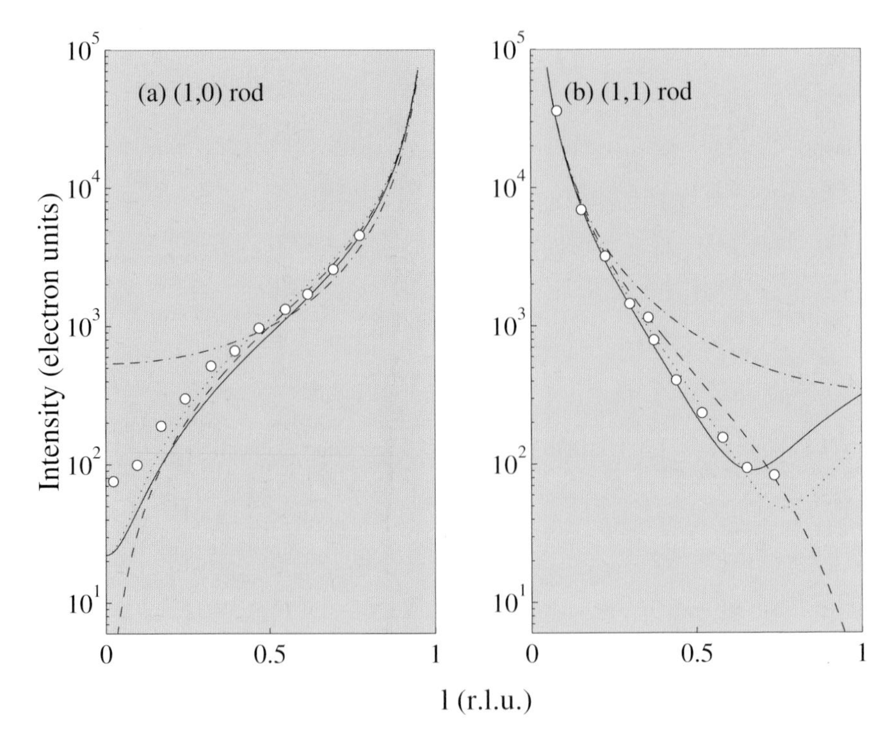

Figure 4.34: ⋆ Crystal truncation rods for (h =1, k =0) and (h =1, k =1) (see Fig. 4.33). The dashed-dotted curve corresponds to the expression for $F_{hkl}^{\rm CTR}(l)$ given in the text. The dashed curve corresponds to the missing row structure of Cu with z_0 = 0, i.e. no displacement. The solid curve is the best fit with the O 0.34 Å (z_1=0.0115) below the relaxed missing row, whereas the dotted curve has O located 0.34 Å (z_1=0.2775) above the missing row which is relaxed by z_0=0.1445 relative to the bulk.

With these parameters the observed and calculated intensities for the in-plane reflections (given in brackets in Table 4.3⋆) are seen to be in good agreement.

In addition to these measurements, CTR's along $(h = 1, k = 0, l)$ and $(h = 1, k = 1, l)$ were also measured with the results shown in Fig. 4.34⋆. The CTR data can be modelled by first considering the crystal to be ideally terminated, and then by adding to the complexity of the model until agreement with the data is reached. Successive layers along the surface normal direction have a phase factor of $\mathrm{e}^{i\,\Psi}$, where $\Psi = \pi(h + k + l)$. The structure factor of the CTR from the ideally terminated surface is then given by

$$F_{hk}^{\rm CTR}(l) = f^{\rm Cu}\mathrm{e}^{-M_{\rm Cu2}} \sum_{n=0}^{\infty} \mathrm{e}^{i\,n\Psi}$$

$$= f^{\rm Cu}\mathrm{e}^{-M_{\rm Cu2}} \frac{1}{1-\mathrm{e}^{i\,\Psi}} \qquad (4.67)$$

$$(4.68)$$

This is plotted as the dashed-dotted curve in Fig. 4.34⋆, and is in poor agreement with the data. What is missing from the model is a description of the surface

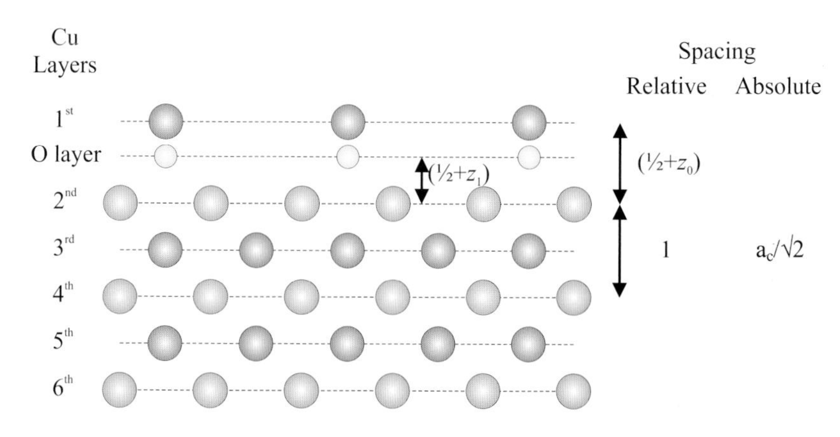

Figure 4.35: Schematic of the side view of O on Cu(110) surface showing the possible displacements perpendicular to the surface of the oxygen and copper atoms z_1 and z_0 respectively. The view is along the **b** axis of Fig. 4.33, and the displacements are not to scale.

structure, here denoted by $F^{\rm S}_{hk}(l)$, so that the total structure factor is

$$F^{\rm total}_{hk}(l) = F^{\rm CTR}_{hk}(l) + F^{\rm S}_{hk}(l)$$

The simplest modification to the model is to allow for the missing row structure by adding half a monolayer of Cu atoms by writing

$$F^{\rm S}_{hk}(l) = \frac{1}{2} f^{\rm Cu}\, {\rm e}^{-M_{\rm Cu1}}\, {\rm e}^{i\,\pi(h+k)}\, {\rm e}^{-i\,\pi l}$$

The last two phase factors arise because the origin in Eq. (4.67) was taken at $n = 0$, so that the next layer up is displaced in the negative z direction by half of a lattice unit, and is displaced in the plane by (1/2,1/2) (see middle left panel of Fig. 4.33). When this is added to $F^{\rm CTR}_{hk}$ it results in the dashed curve in Fig. 4.34⋆. Better agreement with the data is achieved by allowing the topmost Cu layer to relax outwards, and by including the oxygen (Fig. 4.35). The surface structure factor then becomes

$$F^{\rm S}_{hk}(l) = \frac{1}{2}\,{\rm e}^{i\,\pi(h+k)}\left(f^{\rm Cu}\,{\rm e}^{-M_{\rm Cu1}}\,{\rm e}^{-i\,2\pi(1/2+z_0)l} + f^{\rm O}\,{\rm e}^{i\pi k}\,{\rm e}^{-i\,2\pi(1/2+z_1)l}\right)$$

The best fit to the data is found with $z_0 = 0.1445$ and $z_1 = 0.0115$.

Further reading

Crystallography

An Introduction to X-ray Crystallography, M.M. Wolfson (Cambridge University Press, 1997)

X-ray diffraction, B.E. Warren (Dover Publications, 1990)

International Tables of Crystallography, (Kluwer Academic Publishers)

Solid State Physics

Solid State Physics, J.R. Hook and H.E. Hall (John Wiley and Sons, 1991)

Introduction to Solid State Physics, C. Kittel (John Wiley and Sons, 1996)

Surface Crystallography

Surface Structure Determination by X-ray Diffraction, R. Feidenhans'l, Surface Science Reports **10**, 105 (1989)

Surface X-ray Diffraction, I.K. Robinson and D.J. Tweet, Rep. Prog. Phys. **55**, 599 (1992)

5. Diffraction by perfect crystals

The X-ray beam from a synchrotron source is polychromatic. Typical values for the bandwidth in energy vary between a fraction of a keV for an undulator, up to a few hundred keV or so in the case of a bending magnet. Many experiments require a monochromatic beam, where both the energy and energy bandwidth can be set to convenient values. By far the most common type of monochromator is a crystal that Bragg reflects an energy band, or equivalently wavelength band, out of the incident beam. The band is centred around a wavelength λ given by Bragg's Law, $m\lambda = 2d\sin\theta$, where d is the lattice spacing, θ is the angle between the incident beam and the lattice planes, and m is a positive integer.

One requirement of a monochromator crystal is that it preserves the inherently good angular collimation of the synchrotron beam, which is of order 0.1 mrad. For this reason, perfect crystals, which are essentially free from any defects or dislocations, are often used. However, even a perfect crystal does not have an infinitely sharp response, but instead has an intrinsic width. This width may be defined in various ways, depending on the type of experiment imagined. Here we shall start out by considering the relative wavelength band, $\zeta = (\Delta\lambda/\lambda)$, a perfect crystal reflects out of a white, parallel incident beam.

Candidate monochromator materials must not only be perfect, but they should maintain this perfection when subjected to the large heat loads imposed by the incident white beam. In practice, few materials can meet these exacting requirements. Most monochromators are either fabricated from silicon, diamond or germanium, with each having its own advantages depending on the application, as discussed later.

To develop a theory of the diffraction of X-rays from perfect crystals it is necessary to go beyond the kinematical approximation used in Chapter 4. This approximation applies to imperfect crystals, formed from microscopic mosaic blocks (see Fig. 4.22 on page 149). The size of these blocks is taken to be small, in the sense that the magnitude of the X-ray wavefield does not change appreciably over the depth of the block[1]. The scattering amplitude is then evaluated by summing together the amplitude of the scattered waves, remembering to include the appropriate phase factors. Diffraction from macroscopic perfect crystals is fundamentally different from this scenario. As the incident wave propagates down into the crystal its amplitude diminishes, as a small fraction is reflected into the exit beam at each atomic plane. In addition there is a chance that the reflected beam will be re-scattered into the direction of the incident beam before it has left the crystal. The theory which has been developed to allow for these multiple scattering effects is known as dynamical diffraction theory.

At the outset it is important to specify exactly the scattering geometry, as

[1]The X-ray beam may of course be attenuated by absorption.

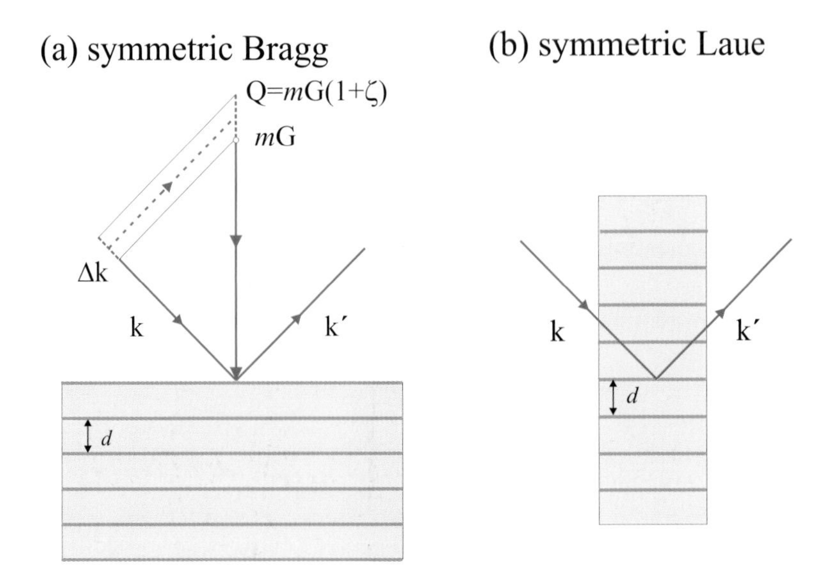

Figure 5.1: Diffraction by a crystal with a lattice spacing of d in (a) Bragg reflection and (b) Laue transmission geometries. Both cases are symmetric since the incident and exit beams form the same angle with respect to the physical surface. The incident beam is assumed to be parallel and white. The crystal reflects a width in wavevector given by $\Delta\mathrm{k}$. The small variable ζ is defined by $\zeta = \Delta\mathrm{G}/\mathrm{G} \equiv \Delta\mathrm{k}/\mathrm{k}$. The relative energy bandwidth or wavelength bandwidth is also equal to $\Delta\mathrm{k}/\mathrm{k}$.

it transpires that this has a profound influence on the diffraction profile from a perfect crystal. Diffraction may occur either in reflection or transmission, which are known as Bragg and Laue geometries respectively, as shown in Fig. 5.1. The angle that the physical surface makes with respect to the reflecting atomic planes is also an important factor. The reflection is said to be symmetric if the surface normal is parallel (perpendicular) to the reflecting planes in the case of Bragg (Laue) geometry. Otherwise it is asymmetric. Within the kinematical approximation the scattering is independent of the geometry. To take one striking example of how the diffraction profile from a perfect crystal is affected by the geometry, consider the symmetric Bragg and Laue cases (Fig. 5.1), and imagine that the incident beam is perfectly collimated and white, in the sense that it contains a continuous distribution of wavelengths. As we shall see, in the Bragg case the collimation of the beam is preserved, whereas the Laue geometry imparts an angular divergence to the reflected beam even though the incident beam is perfectly collimated. To start with we shall examine the symmetric Bragg case, and later explain how the results are modified with the crystal set in an asymmetric Bragg or Laue geometry.

The approach followed here is essentially the same as the one first developed by C. G. Darwin in 1914. In his method, Darwin treated the crystal as an infinite stack of atomic planes, each of which gives rise to a small reflected wave which may subsequently be re-scattered into the direction of the incident beam. An alternative approach was developed by Ewald (1916-1917), and later re-formulated by von Laue (1931). They treated the crystal as a medium with a periodic dielectric constant, and then solved Maxwell's equations to obtain results in agreement with those derived earlier by Darwin.

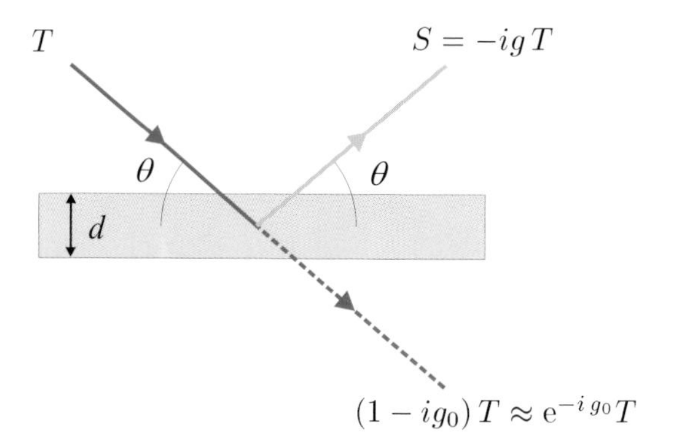

Figure 5.2: A wave T incident on a sheet of unit cells will be partly reflected and partly transmitted. The reflected wave is $-i\,gT$, and the transmitted wave is $T(1 - ig_0)$, where g and g_0 are small parameters given in the text.

Before deriving the dynamical theory, the reader is reminded of a few important results concerning the reflectivity from a thin slab. The kinematical diffraction from a stack of thin slabs is then calculated. This differs from the discussion in Chapter 4 as effects due to refraction are now included. The kinematical approximation is of importance, as any dynamical theory must give the same results in the limit of weak scattering.

5.1 One atomic layer: reflection and transmission

Consider an X-ray beam incident on a thin layer of electrons of density ρ and with a thickness d, such that $d \ll \lambda$, as shown in Fig. 5.2. The incident wave, T, is partly reflected specularly from the layer, and partly transmitted through it. From Eq. (3.33) in Chapter 3 we know that the reflected wave is phase shifted by $-\frac{\pi}{2}$ (i.e. by a factor of $-i$) with respect to the incident wave, and has an amplitude equal to

$$g = \frac{\lambda r_0 \rho\, d}{\sin\theta}$$

To generalize this expression so that it is applicable to the scattering from a layer of unit cells, the density ρ is replaced by $|F|/v_c$, where F is the structure factor of the unit cell, and v_c is its volume. This is a necessary step, as it can no longer be assumed that d is small compared to the wavelength, and hence destructive interference will reduce the scattering at higher scattering angles (see Section 4.2). Using Bragg's law, $m\lambda = 2d\sin\theta$ the above becomes

$$g = \frac{[2d\sin\theta/m]\, r_0\, (|F|/v_c)\, d}{\sin\theta} = \frac{1}{m}\left(\frac{2d^2 r_0}{v_c}\right)|F| \tag{5.1}$$

Since v_c is of order d^3, g is of order $r_0/d \simeq 10^{-5}$, and the reflectivity of even a thousand layers is order of 10^{-2} only. For simplicity the incident wave has been assumed to be polarized with its electric field perpendicular to the plane containing the incident and reflected waves, so that the polarization factor, P, is unity (see Eq. 1.7).

The transmitted wave in Fig. 5.2 may be written as $(1 - ig_0)T \approx \mathrm{e}^{-i\,g_0}T$, since g_0 is the small real number given by Eq. (3.9) on page 68. This equation may be recast in terms of g as

$$g_0 = \frac{|F_0|}{|F|}\, g \tag{5.2}$$

Here F_0 is the unit-cell structure factor in the forward direction, i.e. $\mathrm{Q} = \theta = 0$. We note that for forward scattering the polarization factor is always equal to unity, independent of the polarization of the incident beam.

5.2 Kinematical reflection from a few layers

A single layer of atoms reflects an X-ray beam only very weakly. It is straightforward to derive the reflectivity from a stack of N layers as long as the product of N and the reflectivity per layer, g, is small, *i.e.* $Ng \ll 1$. In this case we simply add the amplitude of rays reflected from layers at different depths in the stack, taking into account the phase factor $\mathrm{e}^{i\,\mathrm{Q}dj}$, where j labels the layer. This is the so-called kinematical approximation, and the *amplitude* reflectivity for N layers is

$$\begin{aligned} r_N(\mathrm{Q}) &= -ig \sum_{j=0}^{N-1} \mathrm{e}^{i\,\mathrm{Q}dj}\mathrm{e}^{-i\,g_0 j}\mathrm{e}^{-i\,g_0 j} \\ &= -ig \sum_{j=0}^{N-1} \mathrm{e}^{i\,(\mathrm{Q}\,d - 2g_0)j} \end{aligned} \tag{5.3}$$

The phase shift is $2g_0$ rather than just g_0 as each layer is traversed twice, once in the T direction, and once in the S direction.

The reciprocal lattice of a stack of layers with layer spacing d is a line of points in reciprocal space at multiples, m, of $\mathrm{G} = 2\pi/d$. In general we are interested in small, relative deviations of the scattering vector Q away from $m\mathrm{G}$, so that

$$\mathrm{Q} = m\mathrm{G}\,(1 + \zeta) \tag{5.4}$$

where ζ is then the small relative deviation (see Fig. 5.1). This is equivalent to the relative bandwidth in energy (or wavelength) since

$$\zeta = \frac{\Delta\mathrm{G}}{\mathrm{G}} = \frac{\Delta\mathrm{k}}{\mathrm{k}} = \frac{\Delta\mathcal{E}}{\mathcal{E}} = \frac{\Delta\lambda}{\lambda} \tag{5.5}$$

The phase factor appearing in the exponential of Eq. (5.3) can thus be re-written in terms of ζ as

$$\begin{aligned} \mathrm{Q}d - 2g_0 &= m\mathrm{G}(1+\zeta)\,\frac{2\pi}{\mathrm{G}} - 2g_0 \\ &= 2\pi\left(m + m\zeta - \frac{g_0}{\pi}\right) \end{aligned}$$

The sum then becomes

$$\begin{aligned} \sum_{j=0}^{N-1} \mathrm{e}^{i\,(\mathrm{Q}\,d-2g_0)j} &= \sum_{j=0}^{N-1} \mathrm{e}^{i\,2\pi m j}\mathrm{e}^{i\,2\pi(m\zeta - g_0/\pi)j} \\ &= \sum_{j=0}^{N-1} \mathrm{e}^{i\,2\pi(m\zeta - g_0/\pi)j} \end{aligned}$$

and this geometric series can be summed (see page 51) to yield

$$|r_N(\zeta)| = g\left[\frac{\sin(\pi N[m\zeta - \zeta_0])}{\sin(\pi[m\zeta - \zeta_0])}\right] \tag{5.6}$$

where ζ_0 is the displacement of the Bragg peak defined by

$$\boxed{\zeta_0 = \frac{g_0}{\pi} = \frac{2d^2|F_0|}{\pi m v_c}\, r_0} \tag{5.7}$$

From Eq. 5.6 the maximum amplitude reflectivity is Ng, and occurs when $\zeta = \zeta_0/m$. Therefore the reflectivity does not have its maximum at the reciprocal lattice points, but is displaced by an amount ζ_0/m. This displacement arises from the refraction of the incident wave as it enters the crystal, an effect that is usually neglected in the derivation of Bragg's law. The index of refraction is less than unity for X-rays, and inside the crystal the modulus of the X-ray wavevector has a smaller value than outside. For a fixed incident angle, the value of k outside of the crystal must be larger than $m\mathrm{G}/(2\sin\theta)$ in order to obtain maximal constructive interference, *i.e.* $\zeta_0 > 0$ as is also clear from Eq. 5.7.

For the kinematical approximation to be valid it is required that $Ng \ll 1$. Adding more and more layers with a reflectivity g per layer increases the peak reflectivity, but of course it can never exceed 100%. Close to $\zeta = \zeta_0/m$ the line shape starts to deviate from $|r_N(\zeta)|^2$ given by Eq. (5.6), and we enter into the dynamical diffraction as indicated by the shading in Fig. 5.3. However, when far enough away from the Bragg condition, *i.e.* when ζ is sufficiently different from ζ_0/m, the kinematical approximation is still valid, even for many layers. When N becomes large the side lobes of the function $|r_N(\zeta)|^2$ become closely spaced, and the rapidly varying numerator $\sin^2(N\pi[m\zeta - \zeta_0])$ can be approximated by its average value of 1/2 to obtain

$$|r_N(\zeta)|^2 \rightarrow \frac{g^2}{2\sin^2(\pi[m\zeta - \zeta_0])} \approx \frac{g^2}{2(\pi[m\zeta - \zeta_0])^2} \tag{5.8}$$

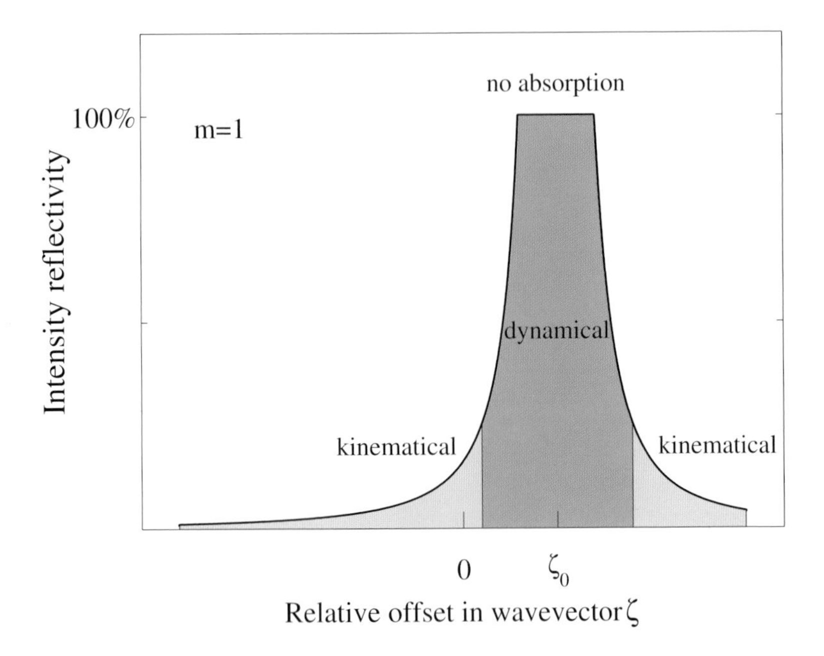

Figure 5.3: The intensity reflectivity from a stack of N atomic layers. The relative deviation from the Bragg condition is given by the parameter ζ. The reflectivity is peaked not at $\zeta = 0$, but at $\zeta_0 = g_0/(m\pi)$ due to refraction of the X-ray beam inside of the crystal. When $|\zeta - \zeta_0|$ is large the reflectivity is small, and we are in the kinematical regime as indicated by the lighter shading. As $|\zeta - \zeta_0| \to 0$ the kinematical approximation breaks down, and the reflectivity is then described by dynamical theory.

The significance of this result is that a correct dynamical diffraction theory must attain this limiting form for large values of $|\zeta - \zeta_0|$. In Fig. 5.3 the kinematical region is indicated by the lighter shading.

Dynamical reflection takes place near a reciprocal lattice point, and must join the kinematic reflection regime in a continuous way. It has been shown in Section 4.4.6 that the surface gives rise to rods of scattering in a direction perpendicular to the physical surface. Continuity then requires that the region of dynamical diffraction and these so-called crystal truncations rods must be connected in a continuous manner. This has the somewhat surprising consequence that the reflection from an asymmetric cut crystal is no longer specular – specular reflection only occurs from a symmetrically cut crystal in Bragg geometry. We shall return to this issue at the end of this chapter.

5.3 Darwin theory and dynamical diffraction

We now turn our attention to the problem of how to calculate the scattering from an infinite stack of atomic planes, where each one reflects and transmits the incident wave according to the equations given in Section 5.1. The planes are labelled by the index j, with the surface plane defined by $j = 0$ (Fig. 5.4(a)). The objective is to calculate the reflectivity, which is ratio of the total reflected wavefield S_0 to that of the incident field T_0.

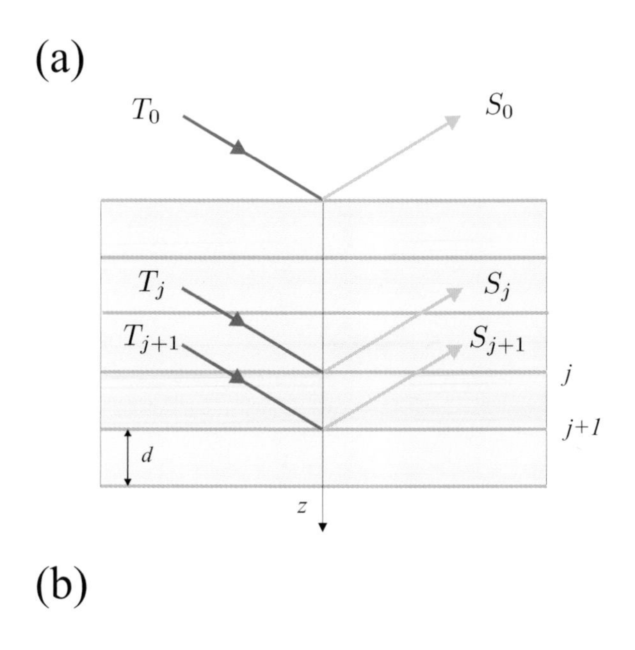

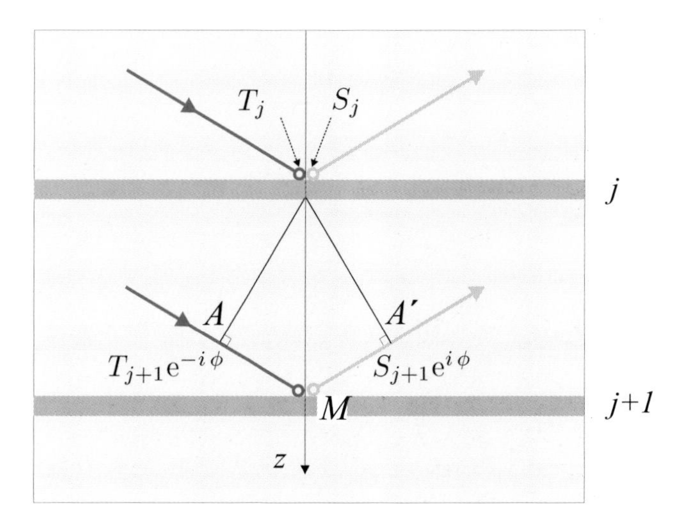

Figure 5.4: Definition of the T and S wavefields. (a) The amplitude reflectivity is given by S_0/T_0. (b) Schematic used to derive the difference equations. The S field at A' is related to the S field just above the atomic plane at M by the phase factor $e^{i\phi}$. The field $S_{j+1}e^{i\phi}$ is then the same as the S_j field just below layer j. Above layer j it gets an extra contribution $(-igT_j)$, from reflection of the T field. Similar arguments apply to the T field.

Within the crystal there are two wavefields: the T field propagating in the direction of the incident beam, and the S field in the direction of the reflected beam. These fields change abruptly when they pass through the atomic planes for two reasons. First, a small fraction, equal to $-ig$, of the wave is reflected. Second, the transmitted wave is phase shifted by an amount $(1 - ig_0)$. The derivation of Bragg's law relies on the fact that the reflected wave from layer $j+1$ is in phase with the one from layer j if the pathlength differs by an integer number of wavelengths. In Fig. 5.4(b) this corresponds to the requirement that the distance AMA' is equal to $m\lambda$, or equivalently that the phase shift in going from A to M is $m\pi$. As we are interested in deriving the (small) bandwidth of the reflecting region, the phase is restricted to small deviations about $m\pi$, and the phase is then given by

$$\phi = m\pi + \Delta$$

where Δ is a small parameter. The relative deviation in phase is therefore $\Delta/(m\pi)$, which must be equal to the corresponding relative deviation in scattering vector, ζ (Eq. (5.4)), so that

$$\Delta = m\pi\,\zeta \tag{5.9}$$

In our development of Darwin's theory Δ will be used as the independent variable in the first instance, and later when the algebra has been worked through, the results will be recast in terms of ζ, which is more useful for comparisons with experiment.

The fundamental difference equations

Let the T field just *above* layer j on the z axis be denoted T_j, and similarly for S_j. The S field just above layer $j+1$ is S_{j+1} on the z axis, that is the point M in Fig. 5.4. At point A' it is $S_{j+1}\mathrm{e}^{i\phi}$, and indeed it must have this value at any point on the wavefront through A', including the point on the z axis just below the j'th plane. On being transmitted through the j'th layer it changes its phase by the small amount $-ig_0$ so that the S field just above the j'th layer, which by definition is S_j, can be written as $(1-ig_0)S_{j+1}\mathrm{e}^{i\phi}$. To obtain the total field, we must also add the part due to the reflection of the wave T_j. In total then we have

$$S_j = -ig\,T_j + (1-ig_0)S_{j+1}\mathrm{e}^{i\phi} \tag{5.10}$$

Next consider the T field just below the j'th layer. This must be the same field that exits at M, except that its phase is shifted by an amount that corresponds to the distance from M to A, *i.e.* $T_{j+1}\mathrm{e}^{-i\phi}$. This field is composed of contributions from the field T_j after it has been transmitted through the j'th layer, and from the wave $S_{j+1}\mathrm{e}^{i\phi}$ after it has been reflected from the bottom of the j'th layer. This leads to the second difference equation

$$T_{j+1}\mathrm{e}^{-i\phi} = (1-ig_0)T_j - ig\,S_{j+1}\mathrm{e}^{i\phi} \tag{5.11}$$

The coupled T to S fields in Eq. (5.11) and (5.10) are separated in the following way.

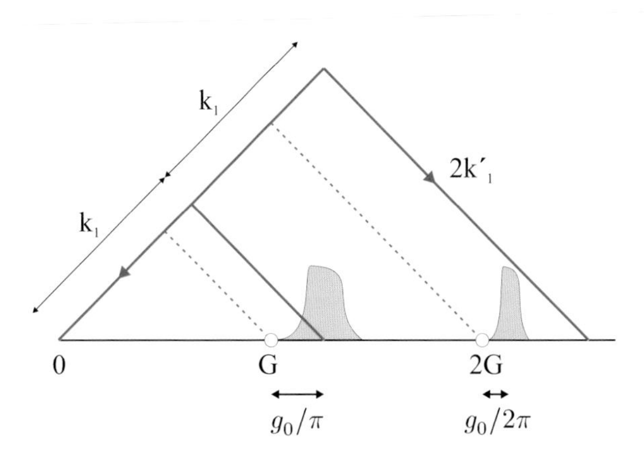

Figure 5.7: The reciprocal lattice of a stack of atomic planes separated by a distance d is a line of points at positions given by $\mathrm{G} = m2\pi/d$, where m is an integer. The centres of the Darwin reflectivity curves are offset from these points by an amount $\zeta_0/m = g_0/(m\pi)$, whereas the Darwin width ζ_{D} varies faster than $1/m^2$. (Note: the scale has been exaggerated.) The crystal is set to reflect a central wavelength of $\lambda_1 = 2\pi/\mathrm{k}_1$. If the incident spectrum also contains $\lambda_1/2 \equiv 2\mathrm{k}_1$, then the variation of ζ_{D} and ζ_0/m ensures that the reflectivity from this component is small.

implications when considering higher-order reflections from a monochromator crystal. According to Bragg's law a crystal illuminated by a collimated but white beam will not only reflect the desired wavelength $\lambda = 2d\sin\theta$, but also all multiples of λ/m. When performing experiments this can be a major source of irritation, and it is often necessary to take steps to reduce the higher-order contamination. For perfect crystals the situation is not as bad as it may first appear, as illustrated in Fig. 5.7. The variation of ζ_{D} and ζ_0 ensure that the contribution from the higher-order components is suppressed. One further way to reduce the higher-order contamination is to use a double crystal monochromator. Offsetting the angle of the second crystal by a small amount of order $g_0/(2\pi)$ will further reduce the reflectivity of the higher-order components without affecting significantly that of the fundamental wavelength.

From Eqs. 5.7 and 5.25 the explicit relationship between the refractive offset and the Darwin width is

$$\zeta^{\mathrm{offset}} = \frac{\zeta_0}{m} = \frac{\zeta_{\mathrm{D}}^{\mathrm{total}}}{2}\frac{|F_0|}{|F|} \tag{5.40}$$

It follows that the angular offset $\Delta\theta$ in degrees is

$$\Delta\theta = \frac{\zeta_{\mathrm{D}}^{\mathrm{total}}}{2}\frac{|F_0|}{|F|} \times \tan\theta \times \frac{360^\circ}{2\pi} \tag{5.41}$$

For example, the Si (111) reflection has a Darwin width of $w_{\mathrm{D}}^{\mathrm{total}} = 0.0020^\circ$, and a refractive offset of 0.0018°. An alternative expression for the offset $\Delta\theta$ in terms of δ, the difference of the refractive index from 1 (Eq. (3.1)), is

$$\Delta\theta = \frac{2\delta}{\sin 2\theta} \times \frac{360^\circ}{2\pi} \tag{5.42}$$

which can be shown to be equivalent to Eq. (5.41).

5.4.6 Effect of absorption

For a real crystal absorption has to be included in any calculation of the Darwin reflectivity curve. The way that this is achieved can be understood from a consideration of Fig. 5.2. If absorption is non negligible, then the transmitted wave not only undergoes a change in phase, proportional to g_0, but it is also attenuated. Thus absorption effects can be included by allowing g_0 to become complex, where the imaginary part of g_0 is proportional to the absorption cross-section. Similar considerations apply to the reflected wave. Equations (5.1) and (5.2) are therefore replaced by

$$g_0 = \left(\frac{2d^2 r_0}{m v_c}\right) F_0 \tag{5.43}$$

with

$$F_0 = \sum_j (Z_j + f_j' + i f_j'') \tag{5.44}$$

and

$$g = \left(\frac{2d^2 r_0}{m v_c}\right) F \tag{5.45}$$

with

$$F = \sum_j (f_j^0(\mathbf{Q}) + f_j' + i f_j'') \mathrm{e}^{i\,\mathbf{Q}\cdot\mathbf{r}_j} \tag{5.46}$$

where f_j' and f_j'' are the real and imaginary parts of the dispersion correction to the atomic scattering length $f^0(\mathbf{Q})$, and j labels the atoms in the unit cell.

With these alterations the formulae for the reflectivity are essentially the same, except that the variable x in Eq. 5.19 is now a complex number, x_c, which is given by

$$x_c = m\pi\frac{\zeta}{g} - \frac{g_0}{g} \tag{5.47}$$

with g_0 and g complex. To calculate the reflectivity curves for a given value of ζ one can plot the result versus the real part of x_c, so that the amplitude reflectivity may be written as

$$r(Re(x_c)) = \left(\frac{S_0}{T_0}\right) = \begin{cases} \dfrac{1}{x_c + \sqrt{x_c^2 - 1}} = x_c - \sqrt{x_c^2 - 1} & \text{for } Re(x_c) \geq 1 \\ \dfrac{1}{x_c + i\sqrt{1 - x_c^2}} = x_c - i\sqrt{1 - x_c^2} & \text{for } |Re(x_c)| \leq 1 \\ \dfrac{1}{x_c - \sqrt{x_c^2 - 1}} = x_c + \sqrt{x_c^2 - 1} & \text{for } Re(x_c) \leq -1 \end{cases}$$

and as usual the intensity reflectivity is obtained by taking the absolute square of $r(Re(x_c))$.

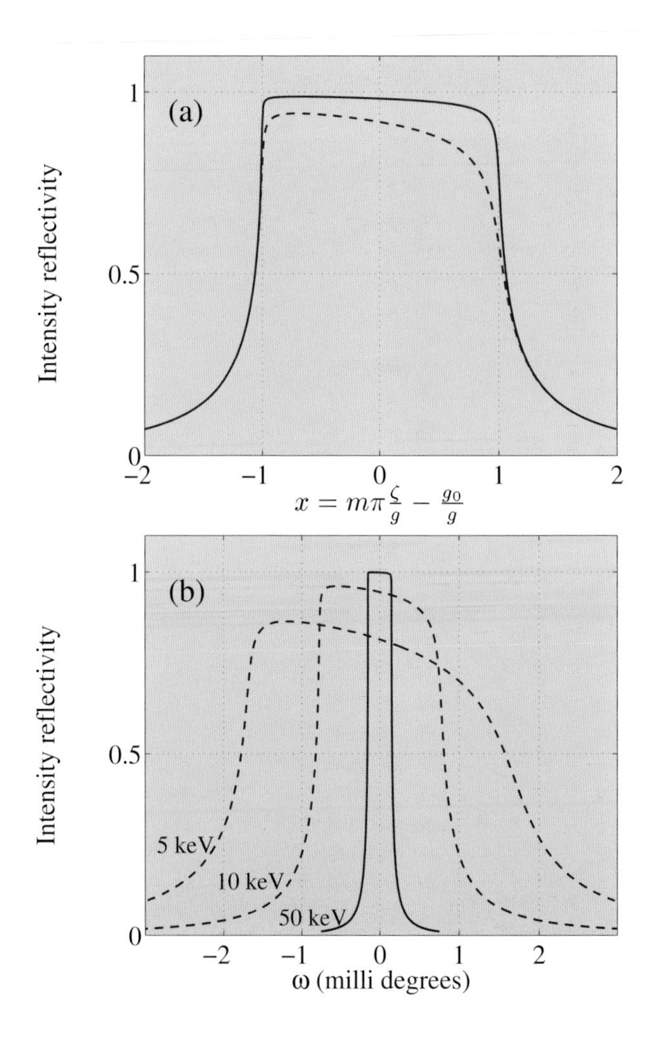

Figure 5.8: ⋆ Effect of absorption on the Darwin curve of the Si (111) reflection. (a) Plotted as a function of the variable x (see Eq. (5.19)). Solid line: λ=0.70926 Å, with $F_0 = 8 \times (14 + 0.082 - i\,0.071)$ and $F = 4|1-i| \times (10.54 + 0.082 - i\,0.071)$. Dashed line: λ=1.5405 Å, with $F_0 = 8 \times (14 + 0.25 - i\,0.33)$ and $F = 4|1-i| \times (10.54 + 0.25 - i\,0.33)$. (b) Plotted as a function of the rotation angle of the crystal in milli degrees (see Eq. (5.28)) at various energies.

The effects of absorption on the Darwin reflectivity curve are illustrated in Fig. 5.8(a)⋆, where the specific example of the Si (111) reflection has been taken. As expected the effect of absorption is more pronounced near $x \approx 1$ than at $x \approx -1$, because near $x \approx 1$ the X-ray wavefield is in phase with the position of the atomic planes. As the photon energy is increased the effect of absorption is diminished. In Fig. 5.8(b)⋆ the Darwin curves are plotted versus the rotation angle of the crystal in milli degrees at various energies. This part of the figure also serves to illustrate the point that, whereas the relative bandwidth ζ (which is proportional to x from Eq. (5.47)) is independent of energy, the angular Darwin width is not.

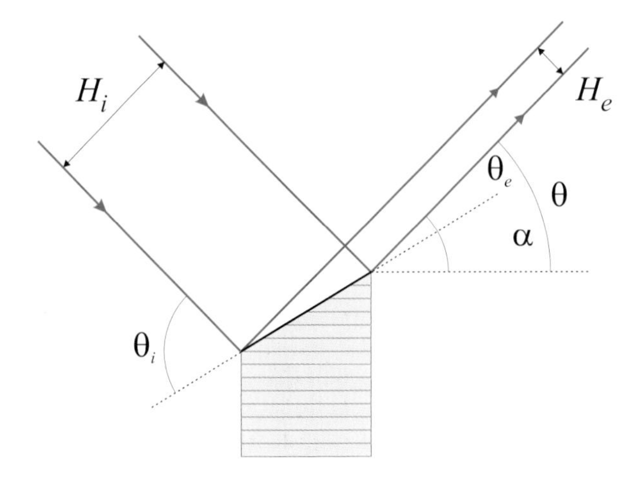

Figure 5.9: Asymmetric Bragg reflection. The surface of the crystal is at an angle α with respect to the reflecting atomic planes. The widths of the incident and scattered beams are then different. In this case the parameter b is greater than one.

5.4.7 Asymmetric Bragg geometry

In general the surface of a crystal will not be parallel to the atomic planes which reflect the incident beam, as shown in Fig. 5.9. Let α be the angle between the surface and the reflecting planes. The incident, θ_i, and exit, θ_e, glancing angles are then given by $\theta_i = \theta + \alpha$ and $\theta_e = \theta - \alpha$. For a reflection geometry it is required that both θ_e and θ_i are greater than zero, or in other words that α fulfills the condition $0 < |\alpha| < \theta$. In Fig. 5.9 α has been chosen to be greater than zero. This implies a compression of the width of the exit beam. The asymmetry parameter, b, is defined by

$$\boxed{\mathsf{b} \equiv \frac{\sin\theta_i}{\sin\theta_e} = \frac{\sin(\theta+\alpha)}{\sin(\theta-\alpha)}} \tag{5.48}$$

Symmetric Bragg diffraction corresponds to setting $\mathsf{b} = 1$. For the particular case shown in the Fig. 5.9, $\mathsf{b} > 1$. The widths of the incident, H_i, and exit, H_e, beams are related by the equation

$$H_i = \mathsf{b}\, H_e$$

It turns out that a compression in the width of the exit beam implies an increase in its angular divergence. This is a consequence of Liouville's theorem.[3] By the same reasoning the acceptance angle of the incident beam must decrease to compensate for the increase in the incident beam width. Let the angular acceptance of the incident beam be $\delta\theta_i$, and the reflected beam divergence be $\delta\theta_e$. We now assert that $\delta\theta_i$ and $\delta\theta_e$ are given in terms of the asymmetry parameter b and the Darwin width ζ_{D} by the equations

$$\delta\theta_e = \sqrt{\mathsf{b}}\,(\zeta_{\mathrm{D}} \tan\theta) \tag{5.49}$$

[3]Liouville's theorem states that for beams of particles, here photons, the product of beam width and divergence is a constant.

and

$$\delta\theta_i = \frac{1}{\sqrt{\mathsf{b}}} \left(\zeta_{\rm D} \tan\theta\right) \tag{5.50}$$

These formulae are certainly correct in the symmetric case with $\mathsf{b} = 1$ (see Eq. (5.28)). Moreover, since

$$\delta\theta_i \, H_i = \frac{1}{\sqrt{\mathsf{b}}} \left(\zeta_{\rm D} \tan\theta\right) \, \mathsf{b} H_e = \sqrt{\mathsf{b}} \left(\zeta_{\rm D} \tan\theta\right) \, H_e = \delta\theta_e \, H_e$$

the product of beam width and divergence is the same for the incident and exit beams, as required by Liouville's theorem.

An interesting application of asymmetric crystals is in the measurement of Darwin reflectivity curves. The angular Darwin width is small, typically of order of $\sim 0.002°$. Measurement of the reflectivity curve then requires a detector system that has a much better angular resolution than this value. This follows from the fact that the measured curve is the *convolution* of the Darwin reflectivity curve of the crystal and the angular resolution of the detector, or analyzer, system. So if the angular divergence of the analyzer is much smaller than that of the first crystal, then the measured curve is determined solely by the Darwin reflectivity of the first crystal. One way to achieve high angular resolution in the analyzer is to use an asymmetric crystal. From Eq. (5.49), its angular acceptance can be made arbitrarily small by decreasing the value of b. Double crystal spectrometers with two perfect crystals are discussed further in the next section on DuMond diagrams.

5.5 DuMond diagrams

An optical element inserted into an X-ray beam is supposed to modify some property of the beam such as its width, its divergence, or its wavelength band. It is useful to describe the modification of the beam by a transfer function. The transfer function relates the input parameters of the beam upstream from the optical element to the output parameters of the beam after the beam has passed the optical element.

When the optical element is a perfect crystal the relevant beam parameters are amongst other things the beam divergence and the wavelength band. The DuMond diagram is a graphical representation of the transfer function. In the diagram the horizontal axes are the beam divergence, with the input beam to the left and the output beam to the right. The vertical axis is common and is $\lambda/2d$, the wavelength normalized by twice the lattice spacing d. In the crudest approximation, where the finite width of the Darwin curve and refraction effects are neglected, only the points of the incident parameter space in the $(\theta_i, \lambda/2d)$ plane which satisfy $\lambda/2d = \sin\theta_i$ will be reflected. For a white incident beam that falls within an angular window $\theta_{i,\min} < \theta_i < \theta_{i,\max}$ the output side of the DuMond diagram consists of a line given by $\lambda/2d = \sin\theta_e$ with $\theta_{i,\min} < \theta_e < \theta_{i,\max}$.

One crystal

According to Bragg's law, constructive interference of waves scattered from an infinite crystal occurs if the angle of incidence, θ_B, and the wavelength, λ, are

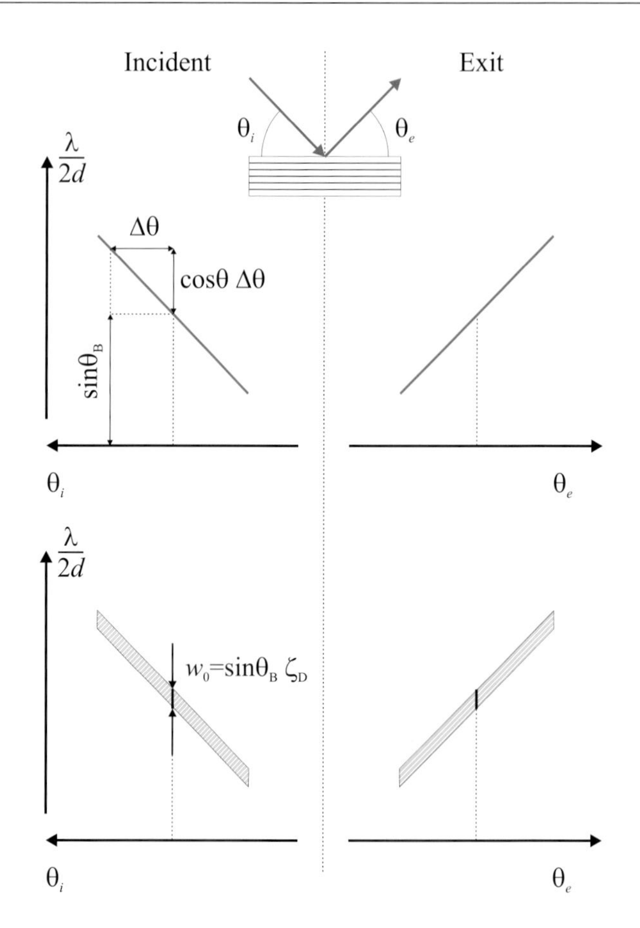

Figure 5.10: DuMond diagram for symmetric Bragg geometry. In this case the angles of the incident, θ_i, and exit, θ_e, beams relative to the crystal surface are the same. The DuMond diagram is a graphical representation of the Bragg reflection condition, where the axes are angle, relative to the Bragg angle θ_B, and $\lambda/2d$. In (a) the Darwin width has been neglected. The intensity is non-zero for points on the line only. (b) The finite Darwin width broadens the line into a band with a width along the ordinate of $w_0 = \sin\theta_B \zeta_{\rm D}$.

related exactly by

$$m\lambda = 2d\sin\theta_B$$

One way to represent this relationship is to plot a graph with $\lambda/2d$ on the ordinate and θ_B on the abscissa. Any point on the sinusoidal curve gives values of $\lambda/2d$ and θ_B that satisfy Bragg's law. Perfect crystals diffract over a small but finite range in angle and wavelength. When dealing with perfect crystals it is therefore necessary to consider deviations of the incident angle θ_i around θ_B, and deviations of wavelength around the value given by $2d\sin\theta_B$. For asymmetric crystals it is also necessary to consider the exit angle θ_e of the reflected beam.

The DuMond diagram is a graphical way to represent diffraction events, and is composed of two parts: one is a plot of $\lambda/2d$ against $\theta_i - \theta_B$, with θ_i increasing to the left; and the other is a plot of $\lambda/2d$ against $\theta_e - \theta_B$, with θ_e increasing to the right. For small deviations away from the Bragg condition the sinusoidal dependence of $\lambda/2d$ approximates to a straight line with slope $\cos\theta$. The top

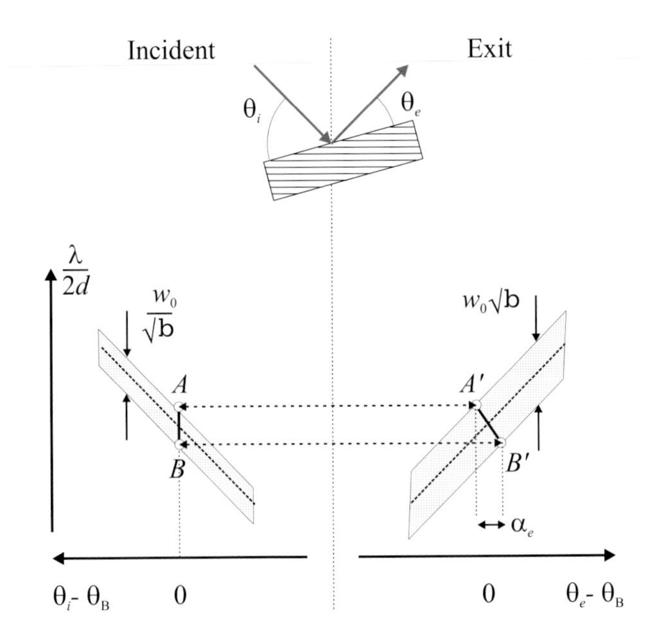

Figure 5.11: DuMond diagram for asymmetric Bragg geometry. The ratio of the widths of the incident and exit beams is given by the parameter b. This implies that the angular acceptance of the incident beam is reduced by a factor of $1/\sqrt{\mathsf{b}}$ while that of the exit beam is increased by the same amount. Thus in the DuMond diagram the incident bandwidth is reduced, and the exit one increased. Points A and B on the incident side are associated with points A' and B' on the exit side. This shows that an incident beam which is parallel and white acquires a finite angular divergence given by α_e when it has been diffracted by a crystal set in asymmetric Bragg geometry.

part of Fig. 5.10 shows the DuMond diagram for a crystal diffracting according to Bragg's law in a symmetric reflection geometry. When neglecting the finite Darwin width, the reflectivity is non-vanishing only on the line indicated, and the relative change in wavelength $\Delta\lambda/\lambda$ and the deviation $\Delta\theta$ from the Bragg angle are related by

$$\frac{\Delta\lambda}{\lambda} = \frac{\Delta\theta}{\tan\theta}$$

For the symmetric Bragg geometry assumed here the surface coincides with the reflecting planes: the reflection is specular, and the wavelength and exit angle are linked by the same condition as the one above.

The lower part of Fig. 5.10 shows the DuMond diagram for symmetric Bragg geometry, but now including the finite Darwin bandwidth: all wavelengths from a perfectly collimated white source within a relative bandwidth ζ_{D} have a reflectivity of 100%. Outside of this band, the reflectivity falls off quickly as we move from the dynamical to the kinematical regimes (see Fig. 5.3 and 5.5). In the latter the scattering is located along the crystal truncation rods which run parallel to the surface normal (Section 4.4.6). In terms of the DuMond ordinate $\lambda/2d$ the width of the central band is

$$w_0 = \frac{\Delta\lambda}{2d} = \left(\frac{\lambda}{2d}\right)\left(\frac{\Delta\lambda}{\lambda}\right) = \left(\frac{\lambda}{2d}\right)\zeta_{\mathrm{D}} = \sin\theta_B\,\zeta_{\mathrm{D}} \tag{5.51}$$

where $\zeta_{\rm D} = \Delta\lambda/\lambda$ is the Darwin width given by Eq. (5.25). As indicated, symmetry implies that a perfectly collimated incident beam is reflected to a perfectly collimated exit beam.

This is not the case for an asymmetric crystal, where the surface does not coincide with the reflecting planes, as is shown in Fig. 5.11. The exit beam width is now smaller than that of the incident beam. In Section 5.4.7 it has been shown that this implies that the bandwidth of the incident beam is reduced by a factor of $1/\sqrt{\mathsf{b}}$, while the bandwidth of the exit beam is increased by a factor of $\sqrt{\mathsf{b}}$. It is important to note that the crystal truncation rod is no longer parallel to the reciprocal lattice vector, since it runs perpendicular to the surface. The consequences of these considerations are illustrated in the lower part of Fig. 5.11. A perfectly collimated incident beam is reflected in the band AB. The scattering is elastic, so the point $A(B)$ is transferred to point $A'(B')$ on the exit part of the DuMond diagram. Since the points A' and B' have different abscissa, displaced by the amount α_e, a perfectly collimated incident beam acquires a finite divergence after Bragg reflection.

In the examples of the symmetric and asymmetric Bragg geometries there is an ambiguity left to resolve. This concerns the question of how to relate points on the DuMond diagram of the incident beam with those of the exit beam. For the asymmetric Bragg case, shown in the lower part Fig. 5.11, the point A on top of the incident band is shown connected to the point A' on top of the exit band. (The line runs at right angles to the $\lambda/2d$ axis since the scattering is elastic.) The reason for this is illustrated in Fig. 5.12. The transition from the dynamical to the kinematical regimes must be continuous. In the kinematical regime the scattering lies along the crystal truncation rods (CTR's). If the incident beam is white and parallel then the crystal reflects a band Δk out of the incident beam. A given wavevector in the incident beam, $\mathbf{k}_1$ say, is scattered to a final wavevector $\mathbf{k}_1'$, with $|\mathbf{k}_1| = |\mathbf{k}_1'|$. The direction of $\mathbf{k}_1'$ is found from where the Ewald sphere, indicated by the circular arc, crosses the CTR. For the asymmetric Bragg case the truncation rod does not lie along the direction of the wavevector transfer; it runs perpendicular to the physical surface. From Fig. 5.12(b) this implies that the scattering angle of the exit beam must increase as $|\mathbf{k}'|$ increases. This is consistent with the choice of associating B with B'. Continuity between the dynamical and kinematical regime also implies that the central band of the former does not lie along the wavevector transfer. In other words the reflection is not specular.

The same construction is shown for symmetric Laue geometry in Fig. 5.12(c). From this it is clear that a crystal diffracting in symmetric Laue geometry will impart a finite angular divergence to a parallel, white incident beam.

Two crystals in symmetric Bragg geometry

In Fig. 5.13 a white beam is incident on a crystal at a certain Bragg angle. To simplify the discussion it is assumed, as in the previous section, that the Darwin reflectivity curve may be approximated by a box function. The beam incident on the first crystal is thus the vertical, light shaded band with an angular width $2\Delta\theta_{in}$ in the DuMond diagrams in the lower part of the figure. A second crystal is set to reflect the central ray. This can be done in two ways.

If the Bragg planes in the second crystal are parallel to those in the first crystal, a ray deviating (dotted line) from the central ray by $\Delta\theta_{in}$ will be re-

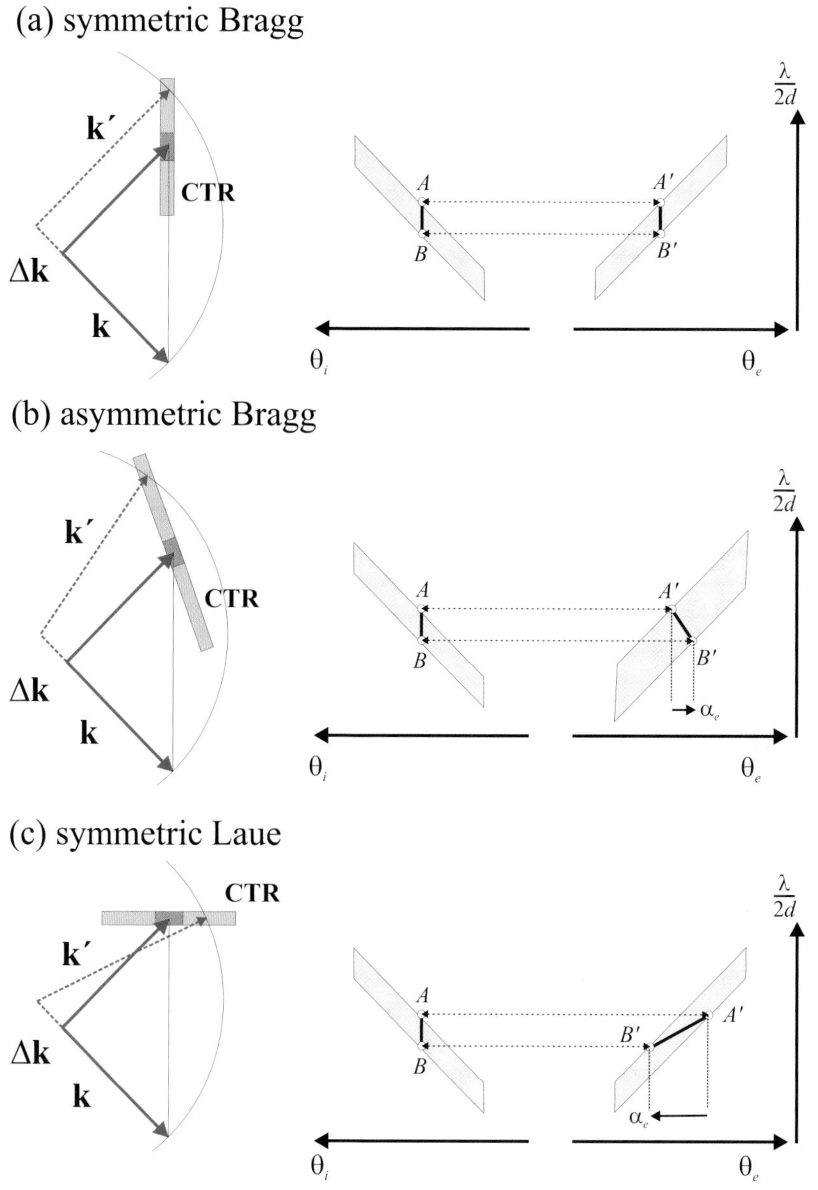

Figure 5.12: Scattering triangles (left) and DuMond diagrams (right) for (a) symmetric Bragg, (b) asymmetric Bragg and (c) symmetric Laue geometries. In the scattering triangles the crystal truncation rod (CTR) is represented by the rectangular box, with the darker shaded part being the central dynamical band. Continuity between the kinematical and dynamical regimes allows points A, B and A', B' in the DuMond diagrams of the incident and exit beams to be associated with each other in an unambiguous way.

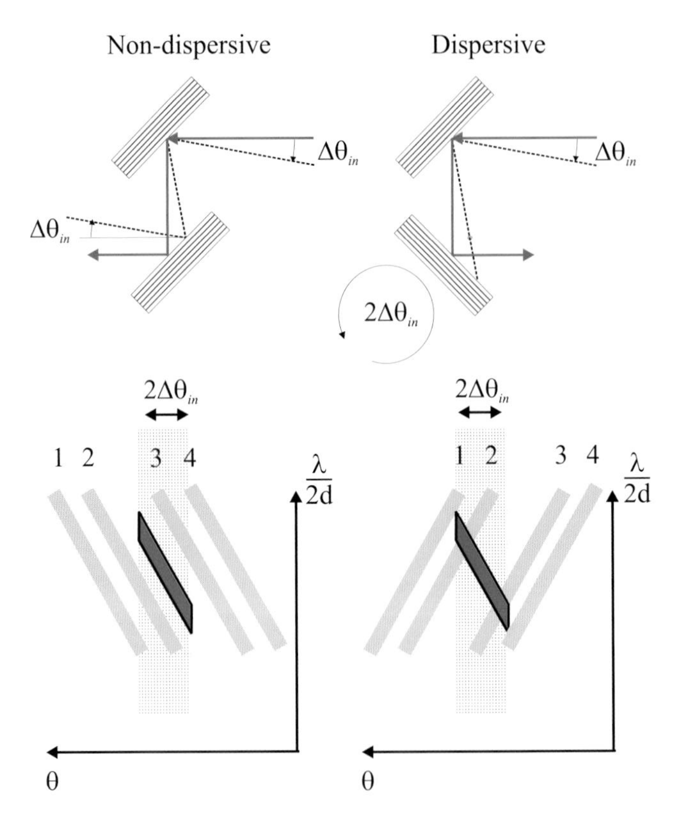

Figure 5.13: Non-dispersive geometry (left): X-rays from a white source are incident on two crystals aligned in the same orientation. The central ray (full line) will be Bragg reflected by both crystals and will emerge parallel to the original ray. A ray incident at a higher angle than that of the central ray will only be Bragg reflected if it has a longer wavelength. The angle of incidence this ray makes with the second crystal is the same as that it made with the first, and will be Bragg reflected. The DuMond diagram in the lower part shows that a scan of the second crystal has a width equal to the convolution of the Darwin widths of the two crystals, independent of the incident angular divergence. Dispersive geometry (right): A ray incident at a higher angle than the central ray at the first crystal will be incident at a lower angle at the second crystal. The second crystal must be rotated by the amount $2\Delta\theta_{in}$ for Bragg's law to be fulfilled. The geometry is therefore wavelength dispersive.

flected at the same setting as the central ray. In a DuMond diagram this means that the response band of the second crystal is parallel to that of the first crystal. In scanning the angle of the second crystal there is no overlap with the intensity provided by the first crystal for any of the four settings shown. Only when the angular setting of the second crystal is in between those labelled 2 and 3 will there be scattered intensity after the second crystal. Since the bands are assumed to be box-like, the intensity versus angle will be triangular with a FWHM equal to the angular Darwin width w_1 of one crystal, independent of the incident angular bandwidth. Furthermore, the reflected wavelength band from the second crystal equals that after the first crystal and is determined by the angular spread $2\Delta\theta_{in}$. This orientation is therefore termed non-dispersive.

On the other hand, in the alternative orientation, the response band of the second crystal has the opposite slope to that of the first crystal. The angular width is now dependent of the incident width – in the limit of a very small Darwin width w_1 it is actually equal to that of the incident width. In the DuMond diagram there will be scattering after the second crystal in all of the positions 2 through 4. The wavelength bandwidth after the second crystal in position 3 is now much smaller than the wavelength bandwidth after the first crystal, and the orientation is termed dispersive. It is clear that this qualitative discussion of the dispersive setting can be sharpened to a quantitative estimate of both the angular and wavelength bandwidths after scattering from the second crystal.

Further reading

X-ray diffraction, B.E. Warren (Dover Publications, 1990) Chapter 14

The Optical Principles of the Diffraction of X-rays, R.W. James (Ox Bow Press) Chapter 2

X-ray monochromators, T. Matsushita and H. Hashizume, in Handbook of Synchrotron Radiation, Vol. 1b, ed. E.E. Koch (North holland, 1983) p.261

6. Photoelectric absorption

Almost everyone has benefited in one way or another from the characteristics of the X-ray absorption cross-section. For example, most people have had the experience at the dentist of holding a piece of photographic film inside of their mouth during the few seconds or so it takes to record a shadow picture of a suspicious tooth. The ability to take shadow pictures, or radiographs, relies on two basic aspects of the absorption process. The first is that X-ray absorption has a pronounced dependence on the atomic number Z, varying approximately as the fourth power of Z. This feature provides the necessary contrast between materials of different densities, such as skin, bone, etc. The second relates to the penetrating power of the X-ray beam, which for a given element varies approximately as the reciprocal of the photon energy $\mathcal{E}$ to the third power. By adjusting the energy of the beam it is thus possible to obtain a suitable penetration depth into the material of interest.

The absorption cross-section per atom, σ_a, is an easy quantity to measure. In a transmission experiment the ratio of beam intensities is recorded with (I) and without (I_0) the sample. For a sample of thickness z the transmission, T, is given by

$$T = \frac{I}{I_0} = \mathrm{e}^{-\mu z} \tag{6.1}$$

The absorption coefficient μ is related to σ_a through

$$\mu = \left(\frac{\rho_m N_{\mathrm{A}}}{A} \right) \sigma_a$$

where N_{A}, ρ_m and A are Avogadro's number, the mass density, and the atomic mass number, respectively (Eq. (1.14)).

An instructive way to illustrate the stated dependencies of σ_a on Z and $\mathcal{E}$ is shown in Fig. 6.1. Here the experimentally determined values of σ_a have been scaled by dividing by Z^4 and multiplying by $\mathcal{E}^3$. Five elements have been selected, with values of Z ranging between 13 and 82, providing a range of $(82/13)^4 \approx 1500$ for the dependence on Z. The energy range covered is one decade, so that altogether the total span for the Z and $\mathcal{E}$ dependencies is more than six decades. For $Z < 47$ (Ag), and for $\mathcal{E} \geq 25$ keV, all of the scaled cross-sections collapse onto a single curve with a value of approximately 0.02 barn (1 barn $\equiv 10^{-24}$ cm^2). Below certain characteristic energies (approximately 25 keV for Ag, 14 keV for Kr, 7 keV for Fe) the scaled cross-section drops to another value, approximately one decade lower, where it joins the level that the heaviest element Pb has for $\mathcal{E} > 16$ keV. The element-specific energies of the discontinuous jumps in the absorption cross-section are called *absorption edges*, and the physical reason for their appearance is quite simple to understand. Electrons are bound in atoms with discrete energies. For example, the K electrons in Kr

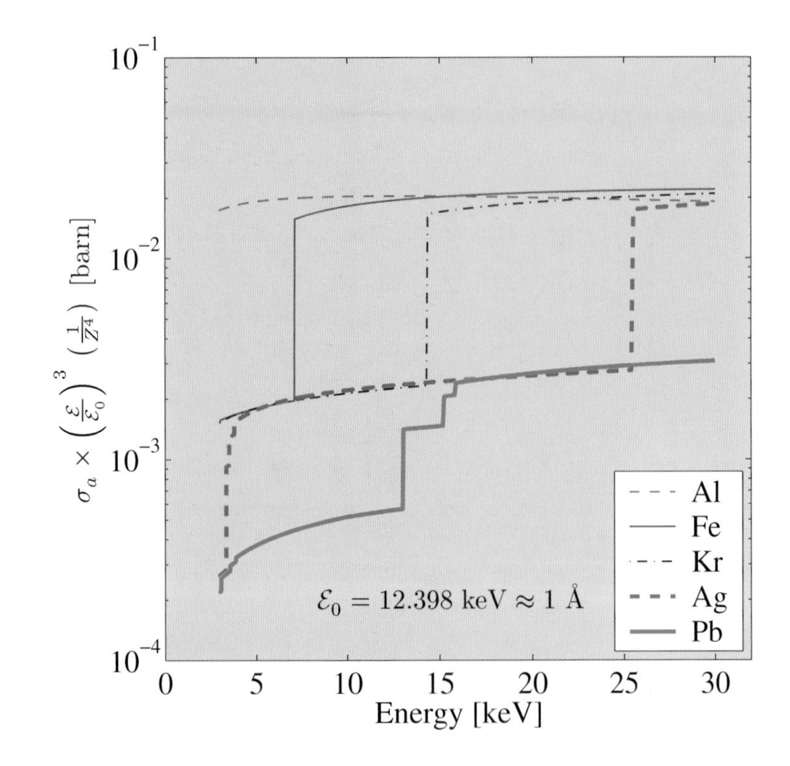

Figure 6.1: The scaled absorption cross-section as a function of photon energy for a selection of elements. The absorption cross-section per atom, σ_a, has been scaled by dividing it by the atomic number Z to the fourth power, and multiplying it by the photon energy $\mathcal{E}$ to the third power.

have a binding energy of 14.32 keV. At photon energies greater than 14.32 keV there is the possibility that the photon can interact with the atom, removing one of the K electrons in the process, with the photon being annihilated at the same time. This is known as *photoelectric absorption.* When the photon energy drops below the threshold value of 14.32 keV this particular process is no longer energetically possible, one of the channels for photoelectric absorption closes, and therefore the absorption cross-section falls by a certain amount.

The K edge for Pb is 88 keV, beyond the range of energies plotted in Fig. 6.1. However, three other discontinuities are apparent for Pb in the range 13-16 keV. These are the L edges, which correspond to the removal of electrons from the L shell. The structure evident in the L edges arises from the fact that the degeneracy of the electron energy in the L shell is lifted by two mechanisms. First, due to screening of the nuclear charge by the inner K electrons, the self-consistent one-electron potential drops faster than the pure Coulomb potential, with the consequence that the energy of the 2s electrons is lower than that of the 2p electrons. By convention the 2s energy is labelled L_I. Furthermore the 2p level is split by spin-orbit coupling into levels denoted L_{II} and L_{III}. The nomenclature used to label the absorption edges is summarized in Fig. 6.2.

Photoelectric absorption is sometimes referred to as true absorption. This is to distinguish it from other processes, such as Thomson or Compton scattering,

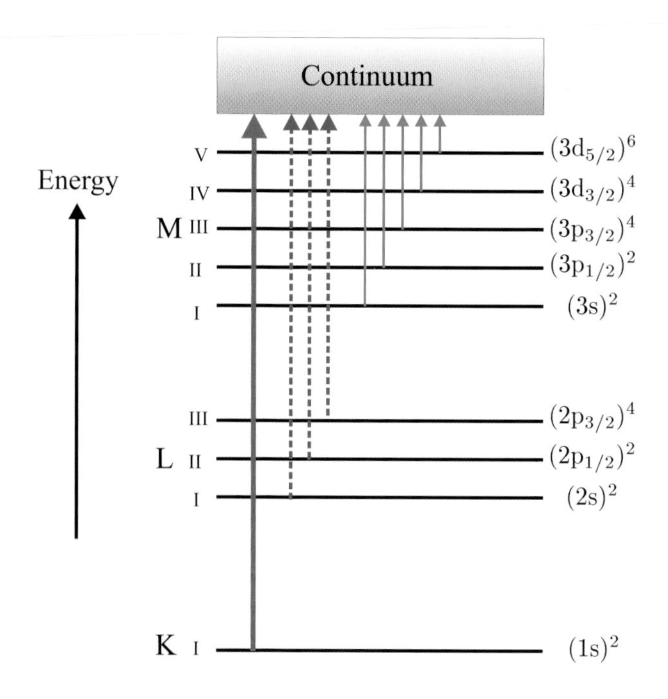

Figure 6.2: A summary of the nomenclature used to label the absorption edges of the elements. The K edge corresponds to the energy required to remove an electron from the 1s shell to the continuum of free states, etc. The electronic shells are labelled as $(n\ell_j)^{2j+1}$, where n, ℓ and j are the principal, orbital angular momentum, and total angular momentum quantum numbers, respectively, of the single-electron states. The multiplicity is $2j+1$.

that also act to reduce the intensity of a beam of photons. The photoelectric process is the dominant contribution to the absorption cross-section whenever the photon energy is much less than the rest mass of the electron, mc^2=511 keV, and is the subject of this chapter. Our main objectives here are twofold:

1. Investigate whether from a first-principles calculation of the absorption cross-section, σ_a, it is possible to account not only for the observed dependences on Z and $\mathcal{E}$, but also for the absolute magnitude of the jump in σ_a at the K edge.

2. Outline the theoretical description of the oscillations in σ_a which are observed just above an absorption edge. These are known as Extended X-Ray Absorption Fine Structure (EXAFS), and in Chapter 1 the example was given of crystalline Kr in Fig. 1.11 on page 20.

In contrast to Thomson scattering, photoelectric absorption cannot be explained by classical physics, and instead it is necessary to invoke a quantum mechanical description of both the X-ray field and the photoelectron. Readers who are unfamiliar with this approach are referred to Appendix C.

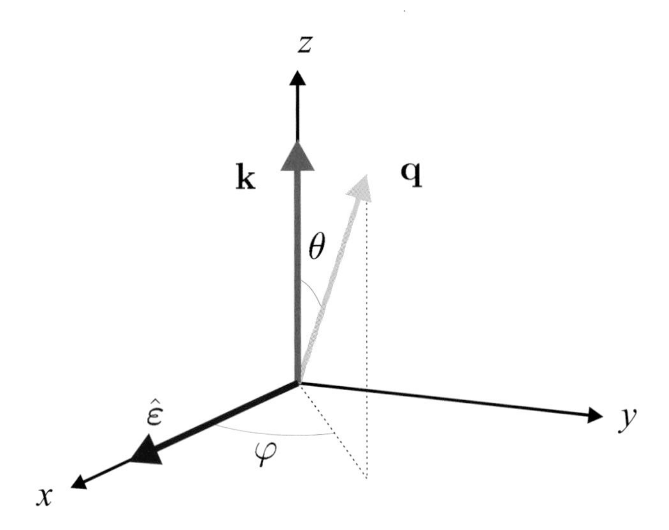

Figure 6.3: The coordinate system relating the angles (θ, φ) to the wavevector of the photoelectron $\mathbf{q}$ and the wavevector and polarization of the incident photon $(\mathbf{k}, \hat{\varepsilon})$. In this case the photon propagates along the z direction and has its electric field polarized along x.

6.1 X-ray absorption from an isolated atom

For the sake of definiteness an absorption process is chosen where a K electron is expelled from an absorbing atom, although the calculation would be essentially the same if an electron in another shell were to be considered.

Our starting point is the formula (Eq. (A.8)) for the absorption cross-section derived in Appendix A:

$$\sigma_a = \frac{2\pi}{\hbar c}\frac{V^2}{4\pi^3}\int \left|M_{if}\right|^2 \delta(\mathcal{E}_f - \mathcal{E}_i)\,\mathrm{q}^2 \sin\theta\, d\mathrm{q} d\theta d\varphi \tag{6.2}$$

This equation comes directly from first-order perturbation theory. In the absorption process an X-ray photon specified by $\mathbf{k}, \hat{\varepsilon}$ (where $\mathbf{k}$ and $\hat{\varepsilon}$ are the wavevector and polarization) is annihilated from the initial state $|i\rangle$, and a photoelectron is expelled into the continuum, where it ends in the final state $|f\rangle$ with a momentum $\mathbf{p} = \hbar\mathbf{q}$ and energy $\mathcal{E}_f = \hbar^2\mathrm{q}^2/2m$. As the photoelectron may be expelled into any direction, it is necessary to integrate over the entire solid angle, 4π, with the value of q^2 restricted to obey energy conservation by the introduction of the delta function in the integrand. The angles (θ, φ) relate the direction of $\mathbf{q}$ to $(\mathbf{k}, \hat{\varepsilon})$ as shown in Fig. 6.3. It is also recalled that in order to normalize the wavefunctions the system is confined to a box of volume V.

The crucial quantity in the formula for σ_a is the matrix element $M_{if}=\langle f|\mathcal{H}_I|i\rangle$, where $\mathcal{H}_I$ is the interaction Hamiltonian that produces transitions between the initial $|i\rangle$ and final $|f\rangle$ states. Here we refer to Appendix C where it is shown that $\mathcal{H}_I$ is conveniently expressed in terms of the vector potential $\mathbf{A}$ of the incident photon field. Both the electric and magnetic fields may be derived directly from **A**, and hence quantizing the electromagnetic field amounts to quantizing the vector potential **A**. For a plane wave the time-independent operator representing the vector potential is

$$\mathbf{A} = \hat{\varepsilon}\sqrt{\frac{\hbar}{2\epsilon_0 V\omega}}\left[a_{\mathbf{k}}\,\mathrm{e}^{i\mathbf{k}\cdot\mathbf{r}} + a_{\mathbf{k}}^{\dagger}\,\mathrm{e}^{-i\mathbf{k}\cdot\mathbf{r}}\right] \tag{6.3}$$

where $a_{\mathbf{k}}$ and $a_{\mathbf{k}}^{\dagger}$ are the annihilation and creation operators. They act on the eigenstates of the photon field, and either destroy or create a photon specified by $(\mathbf{k}, \hat{\varepsilon})$.

Neglecting any magnetic interactions, the interaction Hamiltonian, $\mathcal{H}_I$, contains two terms, one linear in **A** and one that varies as A^2 (Eq. (C.7)). As shall now be shown, in first-order perturbation theory the linear term gives rise to absorption, whereas the squared term produces, amongst other things, Thomson scattering of the photon (see Appendix C). The explicit form of the matrix element of the linear term is

$$M_{if} = \langle f|\frac{e}{m}\mathbf{p}\cdot\mathbf{A}|i\rangle \tag{6.4}$$

This matrix element is evaluated by first neglecting the Coulomb interaction between the photoelectron and the positively charged ion that is left behind. In other words the photoelectron is assumed to be free. Its wavefunction must be normalized, and is thus proportional to $V^{-1/2}$. This together with the $V^{-1/2}$ dependence of **A** makes M_{if} proportional to V^{-1}, and hence, according to Eq. (6.2), σ_a is independent of V as required.

6.1.1 Free-electron approximation

In the initial state $|i\rangle$ there is one photon, specified by its wavevector and polarization, $(\mathbf{k}, \hat{\varepsilon})$, and one K electron in its ground state. The initial state $|i\rangle$ is a product of the photon and electron states, and is hence written as $|i\rangle = |1\rangle_{\mathrm{x}}|0\rangle_e$. Similarly the final state is given by $|f\rangle = |0\rangle_{\mathrm{x}}|1\rangle_e$, where the photon has been annihilated and the photoelectron expelled from the atom. From Eqs. (6.4) and (6.3) the matrix element for the absorption process is

$$M_{if} = \frac{e}{m}\sqrt{\frac{\hbar}{2\epsilon_0 V\omega}}\left[{}_e\langle 1|_{\mathrm{x}}\langle 0|\,(\mathbf{p}\cdot\hat{\varepsilon})\,a\,\mathrm{e}^{i\mathbf{k}\cdot\mathbf{r}} + (\mathbf{p}\cdot\hat{\varepsilon})\,a^{\dagger}\,\mathrm{e}^{-i\mathbf{k}\cdot\mathbf{r}}|1\rangle_{\mathrm{x}}|0\rangle_{\mathrm{e}}\right]$$

To facilitate the evaluation of this matrix element the operators are allowed to act to the left on the final state. The advantage is that the final state of the electron is free, and is therefore an eigenfunction of the momentum operator of the electron, **p**, with an eigenvalue $\hbar\mathbf{q}$. For the photon part it is recalled that when operating to the left an annihilation operator transforms to become a creation operator, etc., with the result that ${}_{\mathrm{x}}\langle n|a = (\sqrt{n+1})\,{}_{\mathrm{x}}\langle n+1|$ and ${}_{\mathrm{x}}\langle n|a^{\dagger} = (\sqrt{n})\,{}_{\mathrm{x}}\langle n-1|$, where n is the number of photons. The terms of interest are

$${}_e\langle 1|_{\mathrm{x}}\langle 0|(\mathbf{p}\cdot\hat{\varepsilon})\,a = \hbar(\mathbf{q}\cdot\hat{\varepsilon})\,{}_e\langle 1|_{\mathrm{x}}\langle 1|$$

and

$${}_e\langle 1|_{\mathrm{x}}\langle 0|(\mathbf{p}\cdot\hat{\varepsilon})\,a^{\dagger} = 0$$

since in the second case there are no photons in the final state to annihilate. It follows that the absorption matrix element simplifies to become

$$\begin{aligned} M_{if} &= \frac{e}{m}\sqrt{\frac{\hbar}{2\epsilon_0 V\omega}}\left[(\hbar\mathbf{q}\cdot\hat{\varepsilon})\,{}_e\langle 1|_x\langle 1|\,\mathrm{e}^{i\mathbf{k}\cdot\mathbf{r}}|1\rangle_x|0\rangle_e + 0\right] \\ &= \frac{e\hbar}{m}\sqrt{\frac{\hbar}{2\epsilon_0 V\omega}}(\mathbf{q}\cdot\hat{\varepsilon})\,{}_e\langle 1|\mathrm{e}^{i\mathbf{k}\cdot\mathbf{r}}|0\rangle_e \\ &= \frac{e\hbar}{m}\sqrt{\frac{\hbar}{2\epsilon_0 V\omega}}(\mathbf{q}\cdot\hat{\varepsilon})\int \psi^*_{e,f}\,\mathrm{e}^{i\mathbf{k}\cdot\mathbf{r}}\psi_{e,i}\,d\mathbf{r} \end{aligned}$$

The integral is over the position $\mathbf{r}$ of the photoelectron, and involves the plane wave $\mathrm{e}^{i\,\mathbf{k}\cdot\mathbf{r}}$ of the incident photon field (Eq. (6.3)). Here the initial wavefunction of the electron, $\psi_{e,i}$, is taken to be that of the 1s bound state, while the final wavefunction, $\psi_{e,f}$, is assumed to be that of a free electron, which are written respectively as

$$\psi_{e,i} = \psi_{1s}(\mathbf{r})$$

and

$$\psi_{e,f} = \frac{1}{\sqrt{V}}\,\mathrm{e}^{i\,\mathbf{q}\cdot\mathbf{r}}$$

The matrix element for the photoelectric absorption process is thus

$$M_{if} = \frac{e\hbar}{m}\sqrt{\frac{\hbar}{2\epsilon_0 V\omega}}(\mathbf{q}\cdot\hat{\varepsilon})\int \frac{\mathrm{e}^{-i\,\mathbf{q}\cdot\mathbf{r}}}{\sqrt{V}}\,\mathrm{e}^{i\,\mathbf{k}\cdot\mathbf{r}}\psi_{1s}(\mathbf{r})\,d\mathbf{r} \tag{6.5}$$

With the wavevector transfer defined by $\mathbf{Q} = \mathbf{k} - \mathbf{q}$, the integral is written as

$$\phi(\mathbf{Q}) = \int \psi_{1s}(\mathbf{r})\,\mathrm{e}^{i\,(\mathbf{k}-\mathbf{q})\cdot\mathbf{r}}\,d\mathbf{r} = \int \psi_{1s}(\mathbf{r})\,\mathrm{e}^{i\,\mathbf{Q}\cdot\mathbf{r}}\,d\mathbf{r}$$

This is nothing other than the Fourier transform of the wavefunction of the electron in its initial state. The modulus squared of the matrix element for the particular process where the photoelectron is expelled into the direction specified by the polar angles (θ, φ) is

$$\left|M_{if}\right|^2 = \left(\frac{e\hbar}{m}\right)^2 \frac{\hbar}{2\epsilon_0 V^2\omega}\,(\mathrm{q}^2\sin^2\theta\cos^2\varphi)\phi^2(\mathbf{Q})$$

since $(\mathbf{q}\cdot\hat{\varepsilon}) = \mathrm{q}\sin\theta\cos\varphi$, as can be seen from Fig. 6.3.

The absorption cross-section per K electron is found by substituting the above matrix element into Eq. (6.2) to obtain

$$\begin{aligned} \sigma_a &= \frac{2\pi}{\hbar c}\frac{V^2}{4\pi^3}\left(\frac{e\hbar}{m}\right)^2 \frac{\hbar}{2\epsilon_0 V^2\omega} I_3 \\ &= \left(\frac{e\hbar}{m}\right)^2 \frac{1}{4\pi^2\epsilon_0 c\omega} I_3 \end{aligned} \tag{6.6}$$

where the three-dimensional integral I_3 is defined by

$$I_3 = \int \phi^2(\mathbf{Q})\, \mathrm{q}^2 \sin^2\theta \cos^2\varphi \; \delta(\mathcal{E}_f - \mathcal{E}_i)\, \mathrm{q}^2 \sin\theta \, d\mathrm{q} d\theta d\varphi \qquad (6.7)$$

To proceed it is required to specify an explicit form for $\phi(\mathbf{Q})$ and hence also for $\psi_{1s}(\mathbf{r})$. Here it is taken to be that of the 1s state of the hydrogen atom, but with a nuclear charge of Z. In this case the wavefunction is

$$\psi_{1s}(\mathbf{r}) = \frac{2}{\sqrt{4\pi}} \kappa^{\frac{3}{2}} \, \mathrm{e}^{-\kappa \mathrm{r}} \qquad (6.8)$$

where $\kappa = Z/a_0$, and a_0 is the Bohr radius. The Fourier transform of ψ_{1s} may be evaluated using the method described on page 113 for the Fourier transform of $|\psi_{1s}|^2$. The result is

$$\phi(\mathbf{Q}) = \int \psi_{1s}(\mathbf{r})\, \mathrm{e}^{i\,\mathbf{Q}\cdot\mathbf{r}}\, d\mathbf{r} = \frac{4\sqrt{4\pi}\,\kappa^{\frac{5}{2}}}{[\mathrm{Q}^2 + \kappa^2]^2}$$

We are now in a position to evaluate the integral I_3 defined in Eq. (6.7). The integration over φ is straightforward: the integral over one period of $\cos^2\varphi$ is equal to π. Next, consider the integral over the delta function. The energy of the initial state is $\mathcal{E}_i = \hbar\omega - \hbar\omega_{\mathrm{K}}$, i.e. equal to the difference between the incident photon energy and $\hbar\omega_{\mathrm{K}}$, the binding energy of the K electron. The energy of the final state is equal to the kinetic energy of the photoelectron, $\mathcal{E}_f = \hbar^2\mathrm{q}^2/2m$. It is convenient to introduce $\tau = \mathrm{q}^2$ as the integration variable, rather than to use q itself. The differential element $d\mathrm{q}$ then becomes $d\mathrm{q} = d\tau/(2\mathrm{q}) = d\tau/(2\sqrt{\tau})$. As far as the θ integration is concerned, the substitution $\mu = \cos\theta$ is made to obtain

$$I_3 = \pi \int \phi^2(\mathbf{Q})\, \tau^2\, (1-\mu^2)\; \delta\left(\left(\frac{\hbar^2}{2m}\right)\tau - (\hbar\omega - \hbar\omega_{\mathrm{K}})\right) \frac{1}{2\sqrt{\tau}}\, d\tau d\mu$$

Integrating over τ is achieved using the properties of the delta function (see the box on page 147). This results in a factor of $(2m/\hbar^2)$, with the integrand evaluated at

$$\tau = \tau_0 = \left(\frac{2m}{\hbar^2}\right)[\hbar\omega - \hbar\omega_{\mathrm{K}}] \qquad (6.9)$$

In terms of the integration variables the square of $\phi(\mathbf{Q})$ is

$$\phi^2(\mathbf{Q}) = \frac{64\pi\,\kappa^5}{[\mathrm{k}^2 + \tau - 2\mathrm{k}\sqrt{\tau}\mu + \kappa^2]^4}$$

since $\mathrm{Q}^2 = (\mathbf{k}-\mathbf{q})\cdot(\mathbf{k}-\mathbf{q}) = \mathrm{k}^2 + \mathrm{q}^2 - 2\mathbf{k}\cdot\mathbf{q} = \mathrm{k}^2 + \tau - 2\mathrm{k}\sqrt{\tau}\mu$. The three-dimensional integral I_3 therefore reduces to a one-dimensional integral I_1, with

$$I_3 = 32\pi^2 \left(\frac{2m}{\hbar^2}\right) I_1(\tau_0, \kappa)$$

and

$$I_1(\tau_0, \kappa) = \int_{-1}^{1} \frac{\kappa^5 (1-\mu^2)\, \tau_0^{\frac{3}{2}}}{\left[\mathrm{k}^2 + \tau_0 - 2\mathrm{k}\sqrt{\tau_0}\mu + \kappa^2\right]^4}\, d\mu \qquad (6.10)$$

Evaluation of integral I_1 and its limit when $\hbar\omega \gg \hbar\omega_\mathrm{K}$

The integral I_1 given in Eq. (6.10) may be expressed in the form

$$I_1 = g\int_{-1}^{1} \frac{(1-\mu^2)}{(a\mu - b)^4}\, d\mu = \left(\frac{4}{3}\right)\frac{g}{(a^2-b^2)^2}$$

with the parameters g, a and b defined as

$$\begin{aligned}
g &= \tau_0^{\frac{3}{2}}\kappa^5 &&= c^{-3}\left[\omega_c(\omega-\omega_\mathrm{K})\right]^{\frac{3}{2}} c^{-5}\omega_A^5 &&\rightarrow c^{-8}\omega_A^5[\omega_c\omega]^{\frac{3}{2}}\\
a &= 2\mathrm{k}\sqrt{\tau_0} &&= 2c^{-2}\omega\left[\omega_c(\omega-\omega_\mathrm{K})\right]^{\frac{1}{2}} &&\rightarrow 2c^{-2}\omega[\omega_c\omega]^{\frac{1}{2}}\\
b &= \mathrm{k}^2+\tau_0+\kappa^2 &&= c^{-2}\left[\omega^2+\omega_c(\omega-\omega_\mathrm{K})+\omega_A^2)\right] &&\rightarrow c^{-2}[\omega_c\omega]
\end{aligned}$$

In the second equation we have introduced the following definitions of energies (or cyclic frequencies): $\tau_0 = \frac{2m}{\hbar}(\omega-\omega_\mathrm{K}) = c^{-2}\omega_c(\omega-\omega_\mathrm{K})$, $\hbar\omega_c = 2mc^2$ and $\hbar\omega_A = \hbar c\kappa = Z\hbar c/a_0$ (see Fig. 6.4). Furthermore, the arrows indicate the limit when $\hbar\omega \gg \hbar\omega_\mathrm{K}$, but still with $\hbar\omega \ll \hbar\omega_c$. In this limit $b \gg a$, and we find that

$$I_1 \rightarrow \left(\frac{4}{3}\right)\frac{g}{b^4} = \left(\frac{4}{3}\right)\left[\frac{\omega_A^2}{\omega\omega_c}\right]^{\frac{5}{2}}$$

Collecting the above results together the expression for the absorption cross-section given in Eq. (6.6) becomes

$$\begin{aligned}
\sigma_a &= \left(\frac{e\hbar}{m}\right)^2 \frac{1}{4\pi^2\epsilon_0 c\omega}\, 32\pi^2\left(\frac{2m}{\hbar^2}\right) I_1(\tau_0,\kappa)\\
&= \frac{e^2}{4\pi\epsilon_0 mc\omega}\, 32\,(2\pi)I_1(\tau_0,\kappa)
\end{aligned}$$

This can be simplified by noting that the Thomson scattering length is $r_0 = e^2/(4\pi\epsilon_0 mc^2)$, and that $\omega = 2\pi c/\lambda$. Hence the atomic absorption cross-section per K electron is

$$\boxed{\sigma_a = 32\lambda r_0\, I_1(\tau_0,\kappa)} \tag{6.11}$$

with τ_0 given by Eq. (6.9) and $\kappa = Z/a_0$. At this point it is interesting to note the following:

1. If the dimensionless integral I_1 turns out to be of order unity, as indeed it does at the edge, then the absorption cross-section per K electron is much larger than the scattering cross-section, which is of order r_0^2, since $\lambda \gg r_0$.

2. The dimension of the absorption cross-section is length squared, as expected.

3. The volume, V, of the box introduced for normalization purposes has disappeared from the final formula, also as expected.

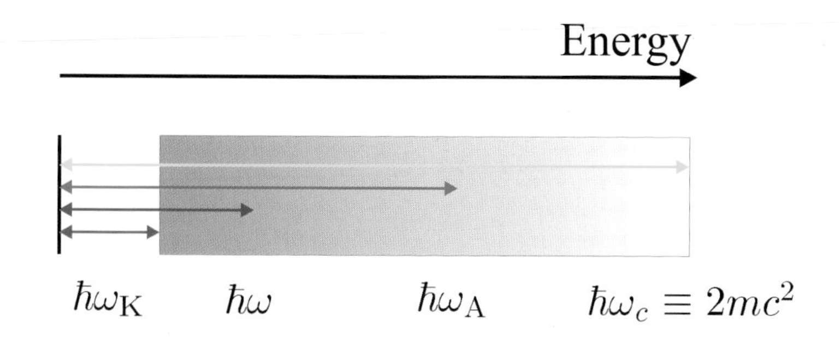

Figure 6.4: A schematic of the different energy scales involved in calculating the absorption cross-section. The energy of the absorption edge is $\hbar\omega_K$, and is proportional to Z^2 in a simple hydrogen-like model of the atom. The energy $\hbar\omega_A$ is related to κ, the inverse length scale of the wavefunction ψ_{1s} (Eq. (6.8)). The relationship is $\hbar\omega_A \equiv \hbar c\kappa$, and thus $\hbar\omega_A$ is proportional to Z. The highest characteristic energy is $\hbar\omega_c$ and is defined to be twice the rest mass energy of the electron. i.e. 2×511 keV.

The evaluation of the integral $I_1(\tau_0, \kappa)$ is given in the box on the preceding page, along with its asymptotic behaviour in the limit that the photon energy is much greater than the binding energy of the K electron, but still much smaller than the rest mass energy of the electron. With energies as defined in Fig. 6.4 one finds

$$\sigma_a = 32\lambda r_0 \left(\frac{4}{3}\right)\left[\frac{\omega_A^2}{\omega\omega_c}\right]^{\frac{5}{2}} \quad \text{for } \hbar\omega_K \ll \hbar\omega \ll \hbar\omega_c \tag{6.12}$$

It is apparent that σ_a varies as Z^5 *via* ω_A, and as $\omega^{-7/2}$ *via* the dependence of I_1 on $\omega^{-5/2}$ and the factor of $\lambda = 2\pi c/\omega$. This behaviour is somewhat different from the experimental findings, summarized in Fig. 6.1, where σ_a is approximately proportional to Z^4 and ω^{-3}. The reason for this discrepancy is the approximation made at the beginning of this section, where we neglected the Coulomb interaction between the photoelectron and the positively charged ion. The benefit is that we have been able to obtain, with moderate effort, an analytical expression for σ_a. However, the price to be paid for this is apparently high, as the result is not sufficiently accurate. It is therefore necessary to consider a treatment of the problem beyond the free-electron approximation.

6.1.2 Beyond the free-electron approximation

Here the full calculation of the correct wavefunction of the photoelectron in the Coulomb field of the ion is not given. Instead the result which was derived in the 1930s by Stoppe is stated without proof. Stoppe introduced the dimensionless photon energy variable

$$\xi = \sqrt{\frac{\omega_K}{\omega - \omega_K}}$$

and conveniently enough, his result can be written as a correction factor $f(\xi)$ to the asymptotic expression for the integral I_1 given above in Eq. (6.12). The absorption cross-section per K electron allowing for the Coulomb interaction between the photoelectron and the ion is then

$$\sigma_a = 32\lambda r_0 \left(\frac{4}{3}\right) \left[\frac{\omega_A^2}{\omega\omega_c}\right]^{\frac{5}{2}} f(\xi) \tag{6.13}$$

Stoppe's correction function, $f(\xi)$, depends on both Z and $\hbar\omega$. When it is included in the formula for σ_a it transpires that there is good agreement between the experimental and theoretical dependences on Z and $\hbar\omega$, and reasonable agreement with the absolute value of the cross-section. The explicit form of the correction function is

$$f(\xi) = 2\pi\sqrt{\frac{\omega_{\mathrm{K}}}{\omega}} \left(\frac{\mathrm{e}^{-4\xi \operatorname{arccot}\xi}}{1 - \mathrm{e}^{-2\pi\xi}}\right)$$

Two limits are particularly illuminating to consider, namely when the photon energy is much greater than the binding energy, $\hbar\omega \gg \hbar\omega_{\mathrm{K}}$ or equivalently $\xi \to 0$, and when the photon energy approaches the threshold energy from above, $\hbar\omega \to \hbar\omega_{\mathrm{K}}^{+}$ or $\xi \to \infty$. At high photon energies we have $\operatorname{arccot}\xi \to \pi/2$, so that $\mathrm{e}^{-4\xi \operatorname{arccot}\xi} \to \mathrm{e}^{-2\pi\xi}$. Thus the high-energy limit of the correction factor is

$$f(\xi) \to 2\pi\xi \left(\frac{\mathrm{e}^{-2\pi\xi}}{1 - \mathrm{e}^{-2\pi\xi}}\right) \to 1 \qquad \text{for } \hbar\omega \gg \hbar\omega_{\mathrm{K}}$$

This result makes physical sense. When the photon energy is high, so is the energy of the photoelectron, and it makes little difference whether the photoelectron is free or moves in the relatively weak attractive field of the positive ion. For photon energies approaching the threshold ($\hbar\omega \to \hbar\omega_{\mathrm{K}}^{+}$, or $\xi \to \infty$) we have that $\operatorname{arccot}\xi \to 0$ as $1/\xi$, so that the product $\xi \operatorname{arccot}\xi \to 1$, and

$$f(\xi) \to \left(\frac{2\pi}{\mathrm{e}^4}\right) \qquad \text{for } \hbar\omega \to \hbar\omega_{\mathrm{K}}^{+}$$

At threshold we therefore find a discontinuous jump in the absorption cross-section *per* K electron of

$$\sigma_a = 32\lambda r_0 \left(\frac{4}{3}\right) \left[\frac{\omega_A^2}{\omega_{\mathrm{K}}\omega_c}\right]^{\frac{5}{2}} \left(\frac{2\pi}{\mathrm{e}^4}\right) \tag{6.14}$$

In order to evaluate either the energy dependence of σ_a (Eq. (6.13)) or the height of the step in σ_a at the K edge (Eq. (6.14)) it is necessary to know how to calculate ω_A and ω_{K}. The simplest approach is to take the model of a hydrogen atom as the starting point. The K shell ionization energy, $\hbar\omega_{\mathrm{K}}$, of an atom with Z electrons is then approximately the binding energy of the hydrogen atom times Z^2. We thus can write $\hbar\omega_{\mathrm{K}} = Z^2 e^2/(4\pi\epsilon_0 2a_0)$. The energy $\hbar\omega_A$ that we have introduced is given by $Z\hbar c/a_0$, and hence scales with Z. Within the model of the hydrogen-like atom the ratio $\omega_A^2/(\omega_{\mathrm{K}}\omega_c)$ is independent of Z, and as $a_0 = 4\pi\hbar^2\epsilon_0/(me^2)$, the ratio is equal to unity. The edge jump per K electron is therefore

$$\begin{aligned}\sigma_a(\lambda_\mathrm{K}) &\approxeq 32\lambda_\mathrm{K}r_0\left(\frac{4}{3}\right)\left(\frac{2\pi}{\mathrm{e}^4}\right)\\ &= \left(\frac{256\pi}{3\mathrm{e}^4}\right)\lambda_\mathrm{K}r_0\end{aligned} \tag{6.15}$$

From Eq. (6.13), this approximation also allows the energy dependence of σ_a *per* K electron to be written in a particularly convenient form as

$$\boxed{\sigma_a \approxeq 32\lambda r_0\left(\frac{4}{3}\right)\left[\frac{\omega_\mathrm{K}}{\omega}\right]^{\frac{5}{2}} f(\xi)} \tag{6.16}$$

Comparison with experiment

As an example we have chosen the absorption cross-sections of the noble gas elements Ar ($Z = 18$), Kr ($Z = 36$) and Xe ($Z = 54$). The energy dependences of σ_a are shown in Fig. 6.5 in the vicinity of the K edges at 3.20, 14.32, and 34.56 keV for the three elements respectively. The dot-dashed lines are obtained from state-of-the-art calculations performed within the self-consistent Dirac-Hartree-Fock framework [Chantler, 1995] (See also Henke et al. [1993].).

Here a comparison is made with our simpler model given in Eq. (6.16). This was derived using the hydrogen-like model of the atom, and has only one free parameter, the energy of the K edge, $\hbar\omega_\mathrm{K}$. For a hydrogen-like atom $\hbar\omega_\mathrm{K}$ is given by Z^2 times the binding energy of the hydrogen atom, 13.60 eV. For the three elements chosen here the K edges are calculated to be 4.41, 17.63, and 39.66 keV. These are significantly greater than the experimental values. Slight adjustments to our approach, such as replacing Z by $(Z-1)$, which allows for the shielding of the nuclear charge by one of the K electrons, brings little improvement to the estimate of $\hbar\omega_\mathrm{K}$, although Moseley showed in 1915 that is does give good agreement with the K_α fluorescence energies, $\hbar(\omega_\mathrm{K} - \omega_L)$. The hydrogen-like model probably works better in the case of fluorescence than absorption as the former involves differences in energies of the inner shell electrons, while the latter depends on being able to calculate the absolute value of the binding energy correctly. In either case the model is expected to become less appropriate as Z increases due to multi-electron and relativistic effects.

An alternative approach to calculating $\hbar\omega_\mathrm{K}$ is to assume that Z scaling is approximately valid, and to treat $\hbar\omega_\mathrm{K}$ as an experimentally determined parameter to be put into the theory. This approach has been adopted for Ar, Kr and Xe, and the photoelectron cross-section σ_a given by Eq. (6.16) with $\hbar\omega_\mathrm{K}$ equal to the experimental values are represented by the solid lines in Fig. 6.5. In all three cases the theory developed here is seen to reproduce the energy dependence of σ_a very well. Agreement with the absolute value of σ_a is reasonable in the case of Ar, but becomes progressively worse for Kr and Xe. This is in line with the expectation that the hydrogen-like model becomes a poorer approximation the higher the value of Z. However, our intention here was not to derive exact values of σ_a, but rather to show how a relatively simple model of the photoelectron absorption process is capable of accounting for the main

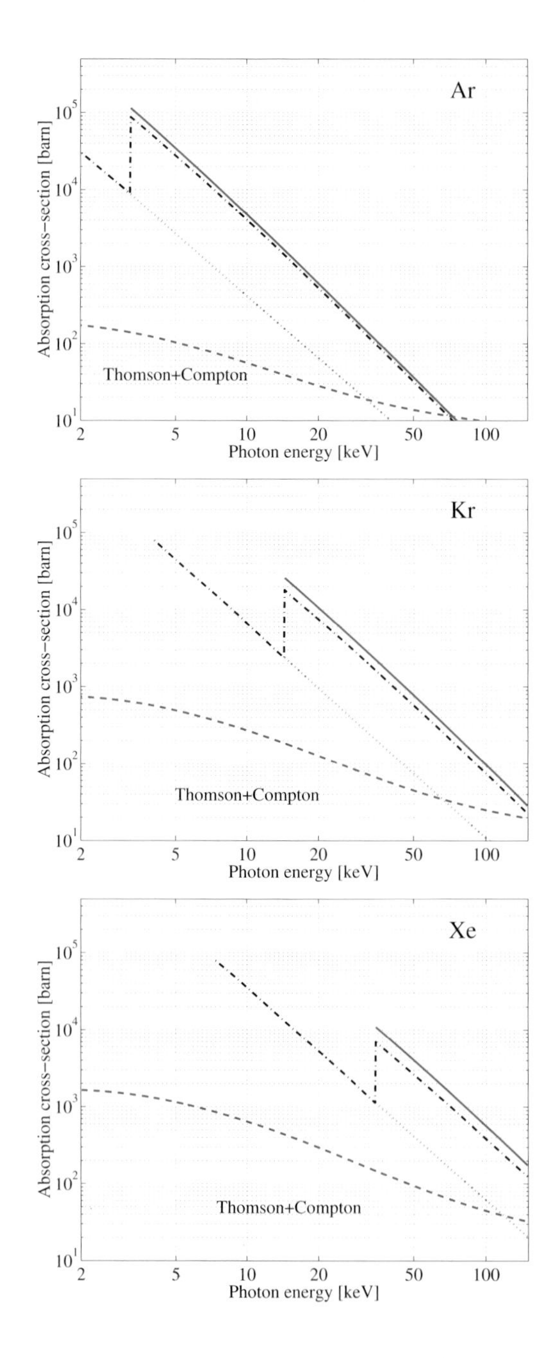

Figure 6.5: The photoelectric absorption cross-sections of Ar, Kr and Xe plotted on a double logarithmic scale for energies in the vicinity of their K edges. The dot-dashed lines represent the results of calculations within the self-consistent Dirac-Hartree-Fock framework [Chantler, 1995]. The solid lines are calculated from Eq. (6.16) with $\hbar\omega_K$ equal to the experimentally observed values of 3.20, 14.32 and 34.56 keV respectively. In each case the result was multiplied by a factor of 2 to allow for the two K electrons, and then the extrapolated contribution from the L electrons (dotted lines) was added to it to produce the final result. For completeness the cross-sections for Thomson and Compton scattering are plotted as the dashed lines. The L edges of Kr ($\sim$ 2 keV) and Xe ($\sim$ 5 keV) have been omitted for clarity.

experimental features. Accurate methods for obtaining values of σ_a required for analyzing experimental data are described and tabulated in a number of places, including the International Tables of Crystallography.

For completeness the cross-section from processes other than the photoelectric effect have also been included in Fig. 6.5. At the energies shown, these include Thomson and Compton scattering: at higher energies pair production becomes important. It is interesting to compare the limiting behaviour of the contribution made by these scattering processes for the different elements. In the case of Ar, the electrons can be considered to be effectively free for the highest photon energies shown, since the photon energy is much greater than that binding the electrons to the atom. The cross-section should then approach the value expected for a gas of Z electrons, i.e. $18\times0.667= 12$ barn per atom, similar to the value shown in Fig. 6.5. (The total scattering cross-section per electron is 0.667 barn, see Eq. (B.5) on page 271.) The other extreme is Xe at low energies, where the electrons are tightly bound. The wavevector transfers, $\mathbf{Q}$, accessible at these energies are small, and hence to a good approximation the atomic form factor squared is equal to Z^2 (Eq. (4.8)). The limiting cross-section in this case should therefore be $54\times54\times0.667= 1945$ barn per atom, again close to the value shown.

6.2 EXAFS and near-edge structure

In Fig. 1.11 on page 20 the absorption cross-section is plotted for atomic Kr in various environments. Comparing the cross-section for Kr in its gaseous phase with its behaviour when bound to the surface of graphite it is clear that the absorption depends on the environment of the absorbing atom. The oscillations evident in Fig. 1.11 extend for several hundred eV above the edge and are referred to as the extended X-ray absorption fine structure, or EXAFS for short. Very near the edge the actual absorption cross-section may appear to overshoot the step like behaviour shown in Fig. 6.5. For historical reasons this is known as the "white line" as this is the way that it appeared on photographic films used in early X-ray experiments. Physically it corresponds to transitions of core electrons to unfilled bound states just below the continuum of free electron states. As the density of such bound states close to the edge may be higher than the density of unbound states the absorption has a peak. For higher photon energies a photoelectron is liberated, which propagates from the source atom as a spherical wave. This outgoing wave is back scattered by neighbouring atoms. EXAFS oscillations are produced by the interference between the outgoing and back scattered waves as shown schematically in Fig. 6.6.

In the analysis of EXAFS spectra it is customary to introduce the dimensionless quantity $\chi(\mathrm{q})$, defined by

$$\boxed{\chi(\mathrm{q}(\mathcal{E})) = \frac{\mu_\chi(\mathcal{E}) - \mu_0(\mathcal{E})}{\mu_0(\mathcal{E})}} \tag{6.17}$$

Here $\mu_0(\mathcal{E})$ is the absorption coefficient of the isolated atom (which obviously does not display EXAFS), and $\mu_\chi(\mathcal{E})$ is the absorption coefficient of the atom

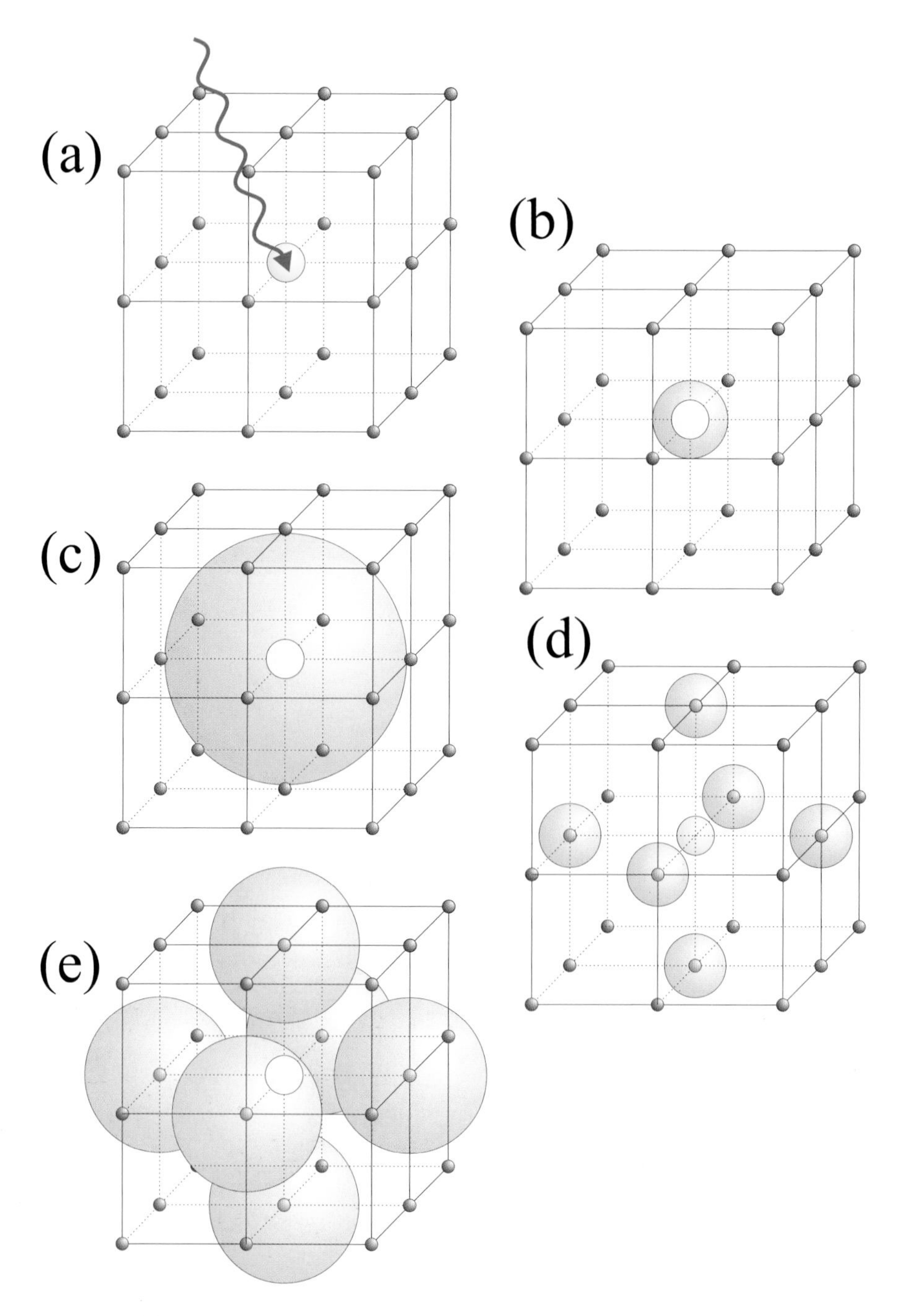

Figure 6.6: Schematic of the EXAFS process. (a): An X-ray photon is incident on an atom located on a lattice. The energy of the photon is high enough that it liberates an electron from a core state in the atom, and the photon is absorbed in the process. (b)-(c): The outgoing wavefunction of the photoelectron propagates from the absorbing atom as a spherical wave until it reaches one of the neighbouring atoms. (d)-(e): The photoelectron wavefunction is scattered by the neighbouring atoms, which then gives rise to a back scattered wave. The interference between the outgoing and back scattered wavefunctions gives rise to EXAFS oscillations in the absorption cross-section [After Stern, 1976].

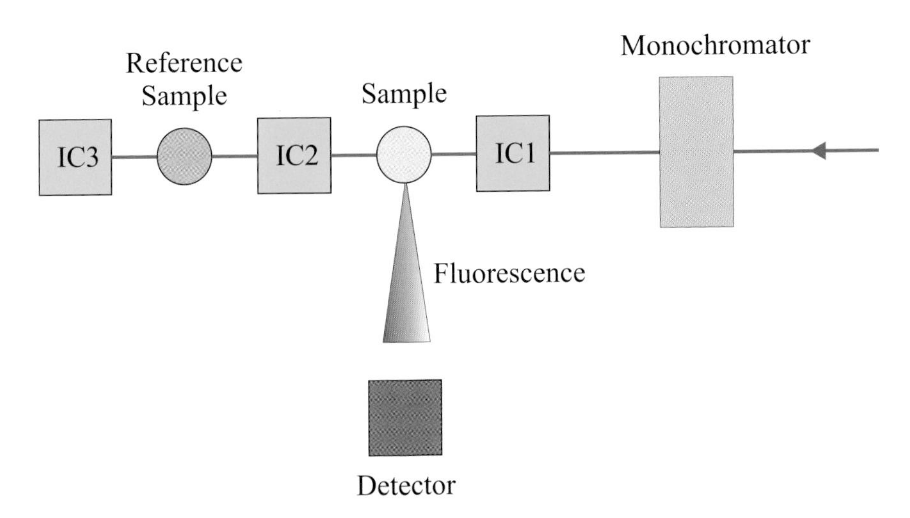

Figure 6.7: Schematic layout of an EXAFS experiment in plan view. The energy of the incident beam is defined by a double-crystal monochromator. Incident and transmitted beam intensities are recorded by ionization chambers IC1 and IC2. It is also possible to measure the EXAFS signal by measuring the fluorescence yield with an energy sensitive detector.

in the material of interest. Rather than the photon energy $\mathcal{E}$, the photoelectron wavenumber q is used as the independent variable:

$$\frac{\hbar^2 \mathrm{q}^2}{2m} = \mathcal{E} - \hbar\omega_{\mathrm{K}} \tag{6.18}$$

The typical apparatus required for an EXAFS experiment is sketched in Fig. 6.7. A double crystal monochromator (as described in Chapter 5) is used to produce a monochromatic beam from the "white" synchrotron beam. For relatively low X-ray energies it is found that the energy resolution provided by the Si(111) reflection is mostly adequate, whereas at higher energies the Si(311) or (511) may be needed. For all of these reflections, the second-order is forbidden, but it is important to ensure that higher orders in the beam are removed, either by offsetting slightly the angle of the second crystal as described on page 190, or by the use of mirrors.

The absorption spectrum can be measured in a transmission geometry, as indicated in Fig. 6.7. The transmission, defined as the ratio of intensities before, I_0, and after, I_1, the sample, is related to the absorption coefficient $\mu(\mathcal{E})$ at photon energy $\mathcal{E}$ by

$$T = \frac{I_1}{I_0} = \mathrm{e}^{-\mu(\mathcal{E})d}$$

where d is the sample thickness (see Eq. (6.1)). The absorption coefficient $\mu(\mathcal{E})$ is then obtained from the measured transmission as a function of $\mathcal{E}$ from the above equation.

The measured absorption coefficient can be partitioned into contributions from the atoms of interest, $\mu_x(\mathcal{E})$, and that due to all the other atoms in the sample, $\mu_A(\mathcal{E})$, so that we may write

$$\mu(\mathcal{E}) = \mu_A(\mathcal{E}) + \mu_\chi(\mathcal{E}) = \mu_A(\mathcal{E}) + \mu_0(\mathcal{E})\left[1 + \chi(\mathrm{q})\right] \tag{6.19}$$

Both $\mu_A(\mathcal{E})$ and $\mu_0(\mathcal{E})$ vary smoothly as a function of $\mathcal{E}$ in the EXAFS region of interest, and can, by a combination of theoretical knowledge and numerical spline methods, be subtracted from the data in order to obtain $\chi(\mathrm{q})$. It is often of advantage to measure simultaneously a reference sample, as indicated in Fig. 6.7.

A second possible way to determine $\chi(\mathrm{q})$ is to measure the fluorescent radiation, which is emitted after the photoelectric absorption (Fig. 6.7). In this case an energy sensitive detector is preferred, as this allows the fluorescent radiation (which is monochromatic) to be isolated. The unwanted contribution from scattering processes can be further minimized by placing the detector at 90° to the incident beam in the horizontal plane. The polarization factor P for Thomson scattering is then identically zero (see Eq. (1.6)). With the detector subtending a solid angle $\Delta\Omega$, and the fluorescent yield relative to the Auger electron processes given by the parameter ϵ, the fluorescent intensity is

$$I_{\mathrm{f}} = I_0 \epsilon \left(\frac{\Delta\Omega}{4\pi}\right) \frac{\mu_\chi(\mathcal{E})}{\mu(\mathcal{E}) + \mu(\mathcal{E}_{\mathrm{f}})} \left[1 - \mathrm{e}^{-(\mu(\mathcal{E}) + \mu(\mathcal{E}_{\mathrm{f}}))d}\right] \tag{6.20}$$

This formula applies to the specific case where the detector is set at 90° to the incident beam, and the normal to the sample surface bisects this angle in two. The last two factors may be understood through the following arguments. First consider an infinitely thick sample, where the the incident beam is normal to the surface, and where the fluorescent radiation is detected in backscattering at 180° relative to the incident beam. The last factor in Eq. (6.20) is then unity. The incident beam at depth x is attenuated by a factor of $\mathrm{e}^{-\mu x}$. Fluorescent radiation originating at that depth is created in a sheet of thickness dx with a probability $\mu_\chi(\mathcal{E})\,dx$, and has to get out of the sample in order to be detected. It follows that the intensity of the observed fluorescent radiation must be proportional to

$$\int_0^\infty \mu_\chi(\mathcal{E})\mathrm{e}^{-\mu(\mathcal{E})x}\mathrm{e}^{-\mu(\mathcal{E}_{\mathrm{f}})x}\,dx = \frac{\mu_\chi(\mathcal{E})}{\mu(\mathcal{E}) + \mu(\mathcal{E}_{\mathrm{f}})}$$

By symmetry, the same argument holds when the incident and fluorescent beams have the same path length in the sample. Hence the result is valid when the sample normal is at 45° to the incident beam. If the sample has a finite thickness d, the upper limit in the integration should not be infinity but d, and this leads to the last factor in Eq. (6.20). Using Eq. (6.20) one can obtain $\mu_\chi(\mathcal{E})$ from the measured fluorescent yield, and the analysis proceeds as in the transmission case described above.

6.2.1 Theoretical outline

In this chapter it has been explained how a relatively simple model of the photoelectron process is capable of accounting for the main features of the absorption cross-section of an *isolated* atom. The key ingredient in this model was the matrix element $\langle f|\mathcal{H}_I|i\rangle$ of the interaction Hamiltonian, Eq. (6.4). The interaction Hamiltonian $\mathcal{H}_I$ is fundamental, and does not depend on the details of the neighbouring atoms. The initial state $|i\rangle$ describes the innermost electrons in the absorbing atom, and also cannot depend greatly on the environment of the atom. It follows that the EXAFS oscillations must arise from modification of the final state. This should come as no surprise. We have already seen that the assumption of a truly free photoelectron was only asymptotically correct in the high-energy limit: good agreement with experiment was found for a final state where the electron is *unbound*, moving in the attractive field of the ionized atom.

Let the relatively small modification to the final state $|f_0\rangle$ of the free atom due to neighbouring atoms be $|\Delta f\rangle$, so that the final state becomes $|f_0 + \Delta f\rangle$. The modulus squared of the matrix element is then

$$
\begin{aligned}
|\langle f_0 + \Delta f\,|\mathcal{H}_I|\,i\rangle|^2 &= [\langle f_0\,|\mathcal{H}_I|\,i\rangle + \langle \Delta f\,|\mathcal{H}_I|\,i\rangle]\,[\langle f_0\,|\mathcal{H}_I|\,i\rangle + \langle \Delta f\,|\mathcal{H}_I|\,i\rangle]^* \\
&\approxeq |\langle f_0\,|\mathcal{H}_I|\,i\rangle|^2 + \{\langle f_0\,|\mathcal{H}_I|\,i\rangle^* \langle \Delta f\,|\mathcal{H}_I|\,i\rangle + \text{c.c.}\} \\
&= |\langle f_0\,|\mathcal{H}_I|\,i\rangle|^2 \left(1 + \left\{ \frac{\langle f_0\,|\mathcal{H}_I|\,i\rangle^* \langle \Delta f\,|\mathcal{H}_I|\,i\rangle}{|\langle f_0\,|\mathcal{H}_I|\,i\rangle|^2} + \text{c.c.} \right\}\right)
\end{aligned}
$$

where c.c. refers to the complex conjugate. By comparison with Eq. (6.19) it can be seen that the first term describes the absorption coefficient of the free atom, $\mu_0(\mathcal{E})$. It can also be inferred that the second term must represent the EXAFS oscillations, with

$$
\chi(\mathrm{q}) \propto \langle \Delta f\,|\mathcal{H}_I|\,i\rangle \tag{6.21}
$$

The initial state wavefunction is strongly localized within the absorbing atom, with an extension given approximately by the Bohr radius, a_0=0.53 Å divided by Z. So as far as the modification is concerned, the initial wavefunction of the electron is highly localized and can be approximated by a delta function. We denote the change in the photoelectron wavefunction due to the neighbouring atoms by $\psi_{\text{back.sc.}}(\mathbf{r})$. Physically, the EXAFS modification is due to *back scattering* of the photoelectron by the neighbouring atoms, as sketched in Fig. 6.6. Referring back to Eq. (6.5) it can be seen that the appropriate form of the matrix element is

$$
\langle \Delta f\,|\mathcal{H}_I|\,i\rangle \propto \int \psi_{\text{back.sc.}}(\mathbf{r}) e^{i\,\mathbf{k}\cdot\mathbf{r}} \delta(\mathbf{r})\, d\mathbf{r} = \psi_{\text{back.sc.}}(0)
$$

In comparison to Eq. (6.5) the plane wave of the photon field $e^{i\,\mathbf{k}\cdot\mathbf{r}}$ has been retained, but the wavefunction of the electron in its initial state, $\psi_{1\text{s}}(\mathbf{r})$, has been simplified to be a delta function $\delta(\mathbf{r})$, and $\psi_{\text{back.sc.}}$ has been inserted for the perturbation of the final state $\langle \Delta f|$. We thus assert that

$$
\chi(\mathrm{q}) \propto \psi_{\text{back.sc.}}(0)
$$

and an expression for $\psi_{\text{back.sc.}}$ is now developed in a step-by-step procedure.

The wavefunction of the photoelectron emitted from the absorbing atom is an outgoing spherical wave, i.e. of the form $(e^{i\,\mathrm{qr}}/\mathrm{r})$, where r is measured from the centre of the absorbing atom. Let us first assume that there is only one neighbouring atom at a distance R from the absorbing atom. This neighbouring atom will scatter the incoming wave into a new spherical wave, with an amplitude proportional to the amplitude of the incident wave, and to a scattering length $t(q)$. Altogether then the back scattered wave at r=0 will be proportional to $t(\mathrm{q})$ $(\mathrm{e}^{i\,\mathrm{q}R}/R) \times(\mathrm{e}^{i\,\mathrm{q}R}/R)$, or $t(\mathrm{q})(\mathrm{e}^{i\,2\mathrm{q}R}/R^2)$. The free electron wavefunction, $\mathrm{e}^{i\,\mathrm{qr}}/\mathrm{r}$, used in this argument neglects the electrostatic potential between the negatively charged electron and the ions of the lattice. Formally, such a potential can be taken into account by a phase shift $\delta(\mathrm{q})$, with the result that the wavefunction is modified to be of the form $\mathrm{e}^{i[\mathrm{qr}+\delta(\mathrm{q})]}/\mathrm{r}$. The calculation of such a phase shift is a central problem in the branch of solid state physics which is concerned with electrons moving in the periodic potential of ions on a lattice, and we shall see an example of the result of such a calculation in the following section, Fig. 6.10. In the present context of EXAFS one must distinguish between the phase shift produced by the absorbing atom, $\delta_{\mathrm{a}}(\mathrm{q})$, and that coming from the back scattering atoms, $\delta_{\mathrm{back.sc.}}(\mathrm{q})$. The total phase shift, $\delta(\mathrm{q})$, is of course the sum of the two. Thus as a first step in obtaining an expression for the back-scattered wavefunction we write

$$\psi^{(1)}_{\mathrm{back.sc.}}(0) = t(\mathrm{q})\frac{\mathrm{e}^{i\,(2\mathrm{q}R+\delta)} + \mathrm{c.c.}}{\mathrm{q}R^2}$$
$$\propto \frac{t(\mathrm{q})\sin(2\mathrm{q}R+\delta)}{\mathrm{q}R^2}$$

Following convention a factor of q has been included in the denominator, if for no other good reason than to obtain a dimensionless expression for $\psi^{(1)}_{\mathrm{back.sc.}}(0)$.

The neighbouring atom is of course not stationary, but vibrates about its equilibrium position. If the r.m.s. value of the displacement parallel to $\mathbf{q}$ is σ, then the amplitude of the back scattered wave is reduced by the Debye-Waller factor of $\mathrm{e}^{-\mathrm{Q}^2\sigma^2/2}$ (see Section 4.4.7). For a scattering vector of $\mathrm{Q} = 2\mathrm{q}\sin 90^\circ = 2\mathrm{q}$ we have

$$\psi^{(2)}_{\mathrm{back.sc.}}(0) \propto \frac{t(\mathrm{q})\sin(2\mathrm{q}R+\delta(\mathrm{q}))}{\mathrm{q}R^2}\,\mathrm{e}^{-2(\mathrm{q}\sigma)^2}$$

It is the vibrations of the back scattering atom *relative* to that of the absorbing atom that is taken into account in this way. Since the two atoms are close neighbours, acoustic, long wavelength phonons will not contribute to σ, so it is smaller than determined from crystallography.

The state of a photoelectron and an atom left behind with a hole in its K shell is not a steady state – it has a finite lifetime. The discussion we have given so far tacitly assumed that the back scattered electron will find the atom in the initial state, but due to the lifetime there is a certain probability that the hole in the K shell has been filled in the meantime. In addition, the photoelectron may be scattered by other electrons in its round trip, so we introduce a phenomenological mean-free pathlength Λ to obtain

$$\psi^{(3)}_{\mathrm{back.sc.}}(0) \propto \frac{t(\mathrm{q})\sin(2\mathrm{q}R+\delta(\mathrm{q}))}{\mathrm{q}R^2}\,\mathrm{e}^{-2(\mathrm{q}\sigma)^2}\mathrm{e}^{-2R/\Lambda}$$

Finally, we assume that the absorbing atom is surrounded by shells of neighbours, with N_j atoms in the j'th shell at a distance R_j. The shells may have different types of atoms, so the back scattering amplitude $t(\mathrm{q})$ also needs a suffix j, and we therefore write

$$\mathrm{q}\chi(\mathrm{q}) \propto \sum_j N_j \frac{t_j(\mathrm{q}) \sin(2\mathrm{q}R_j + \delta_j(\mathrm{q}))}{R_j^2} \mathrm{e}^{-2(\mathrm{q}\sigma_j)^2} \mathrm{e}^{-2R_j/\Lambda} \tag{6.22}$$

This is the standard expression used for analyzing EXAFS data. The goal is to extract the radii of the neighbouring shells, R_j, and their occupation number, N_j. The q dependence of the back scattering amplitude $t_j(\mathrm{q})$, and of the phase shift $\delta_j(\mathrm{q})$ is a subtlety that complicates the analysis. This is usually overcome by a combination of theory, and use of reference samples where R_j, N_j and σ_j are known.

6.2.2 Example: CdTe nano-crystals

EXAFS is now established as a powerful method for determining the structures of materials. It should be emphasized that EXAFS is a *local* probe, in the sense that information on only the first few neighbouring shells is obtained. This in itself should not be seen as a severe limitation of the technique, as it means that EXAFS can not only be used to study well-ordered single crystals, but also disordered materials such as glasses. Diffraction techniques discussed in Chapter 4 can also be applied to both ordered and disordered materials, and in this way EXAFS and diffraction are complementary to one another. In the example considered here this complementarity has been exploited in a study of very small crystals, so-called nano-crystals, of the semiconducting material CdTe. This example has been chosen for a number of reasons. First, the data turns out to be easier to interpret than most EXAFS data. The reason is that in this case only the EXAFS signal from the nearest-neighbour shell is significant, and the complication of the summation over neighbouring shells in Eq. (6.22) is avoided. Second, the solid state physics of CdTe nano-crystals is interesting, and may turn out to be of technological importance.

CdTe is a II-VI semiconductor compound. (The corresponding III-V compound is InSb.) The homologous compound CdS may be known to the reader from its use as photoelectric cells in cameras, and diodes in electronic circuitry. Perhaps the most important feature of CdTe nano-crystals is that the electronic band-gap, which determines the photo-sensitivity in the visible part of the electromagnetic spectrum, depends strongly on the size of the crystal when the size of the crystal is reduced to the nano-meter scale. This is simply due to quantum mechanical confinement of the electrons within the nano-crystal. The nano-crystal may be regarded as a large molecule with discrete electronic energy levels, rather than the continuous band of allowed electronic energies found in bulk crystals. Nano-crystals of CdTe with a well-defined size have been produced by chemical methods [Rogach et al., 1996]. The core of the nano-crystal is formed from a tetrahedron of Cd and Te atoms packed in the cubic zinc-blende structure, with an organic part, SCH_2CH_2OH, attached to the Cd atoms on the surface of the tetrahedra. The absorption spectrum of these CdTe nano-crystals in the UV part of the electromagnetic spectrum (see the original article for de-

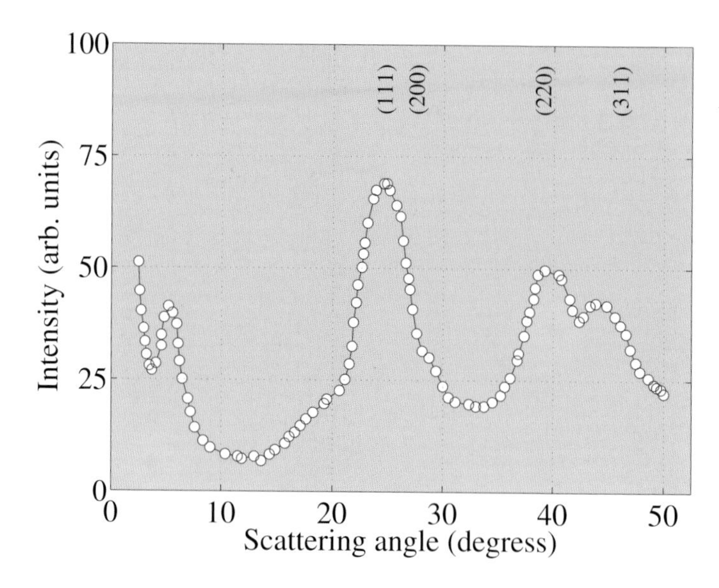

Figure 6.8: Powder diffraction data for CdTe nano-crystals with an average diameter of around 15 Å. The X-ray wavelength was 1.54 Å. The graph was prepared by digitising the data of Rogach et al. [1996]

2θ (Degrees)	Relative intensity	Wavevector Q (Å^{-1})	Peak Width FWHM (Degrees)	No. of unit cells N	Miller indices (h,k,l)	$\frac{Q}{\sqrt{h^2+k^2+l^2}}$
5.2	–	0.37016	3.17	–	–	–
24.6	1	1.7383	6.3	3.5	(1,1,1)	1.0
39.0	0.7	2.7239	8.5	4	(2,2,0)	0.96
46.0	0.5	3.1884	9.5	4	(3,1,1)	0.96

Table 6.1: Analysis of the diffraction data shown in Fig. 6.8 for CdTe nano-crystals. The data were taken from Rogach et al. [1996].

tails) exhibits two distinct peaks at 2.9 eV (425 nm) and 2.7 eV (460 nm). In contrast the band-gap of the bulk CdTe is 1.5 eV (827 nm).

The powder diffraction spectrum from CdTe nano-crystals is shown in Fig. 6.8. Here a slight digression from the main subject of this chapter is made to discuss this diffraction pattern. It serves to illustrate nicely several of the subjects that have already been treated in Chapter 4, and is relevant for interpreting the EXAFS spectra that will be considered a little later. The diffraction data were taken with an X-ray wavelength of 1.54 Å, and are tabulated in Table 6.1.

The first peak in the diffraction pattern occurs at a scattering angle of $2\theta = 5.2°$. This small-angle scattering feature is the interference peak from particles a distance R apart (see Section 4.1). The average distance R is given by $QR = 2\pi$. (Here recall that for elastic scattering the modulus of the wavevector transfer is related to the scattering angle by $Q=(4\pi/\lambda)\sin\theta$.) In this particular case with the peak at 5.2°, and for an X-ray wavelength of 1.54 Å, this means that the nano-crystals are approximately 17 Å apart. If the nano-particles are roughly spherical in morphology, then this is also the diameter of each sphere.

Now consider the three peaks at higher scattering angles. These correspond to the powder diffraction peaks from randomly oriented CdTe nano-crystals. To establish this we first need to index the powder pattern, i.e. assign Miller indices to each of the high-angle diffraction peaks. Bulk CdTe has the zinc blende structure, which in order of increasing scattering angle has strong Bragg reflections with Miller indices of (1,1,1), (2,2,0) and (3,1,1). The modulus of the wavevector transfer is related to the d spacing for a given set of (h, k, l) planes by $d = a/\sqrt{h^2 + k^2 + l^2}$, where a is the lattice parameter. As Q=$2\pi/d$, this means that the ratio of Q to $\sqrt{h^2 + k^2 + l^2}$ should be a constant, equal to $2\pi/a$, if the assignment of Miller indices given in Table 6.1 is correct. It is apparent that this ratio is indeed almost constant with a value of $\approx$1.0 within error. Thus the lattice parameter of the CdTe forming the nano-crystals is $a \approx 2\pi/1.0 \approx 6.3$ Å, somewhat smaller than the bulk value of 6.48 Å. It is also important to compare the intensities of the high-angle peaks with what would be expected for a powder of bulk CdTe. Using Eq. (4.63) the relative intensities of the peaks from bulk CdTe are evaluated to be in the ratios 1:0.78:0.42, similar to the ratios given in Table 6.1.

It is apparent from the data shown in Fig. 6.8 that the diffraction peaks appear to be broad. They are in fact much broader than the instrumental resolution. The width results from the fact that each nano-crystal is built up from so few Cd and Te atoms. The width of a diffraction peak is inversely proportional to the number N of unit cells that scatter coherently. From the box on page 51 we know that the relative width (FWHM) is equal to $0.88/N$, and thus from Table 6.1 the number of unit cells in the nano-crystal is $\approx$4. From this it can be expected that the size of the nano-crystal is about 4 times the unit cell length, i.e. 4$\times$6.3=25.2 Å, a value reasonably close to the size of the nano-crystal estimated from the interference peak at small scattering angles.

Now to the EXAFS data. The absorption spectra, in the vicinity of the K edge of Te, for bulk and nano-crystalline CdTe are shown in the top row of Fig. 6.9. The first step in the data analysis is to obtain the EXAFS part of the signal, χ(q), from its definition in Eq. (6.17). The smooth part of the absorption coefficient, μ_0, is derived from the inverse transmission curve in the top row, and is represented by the dotted lines. The next step is to locate the energy of the K edge (here 31.813 keV) which allows the photon energy scale in keV to be converted into electron wavenumber in Å^{-1} using Eq. (6.18). It is then possible to generate χ(q) from Eq. (6.17). The result is shown in the second row of Fig. 6.9, where it has been weighted somewhat arbitrarily by q^3. Without performing any further analysis it is clear that there is a difference between the nano-crystal and the bulk forms of CdTe. The nano-crystal has one dominant frequency (or wavelength) which means that the EXAFS is dominated by the distance to the nearest-neighbour shell. In contrast, the bulk data for q$^3\chi$(q) have a superposition of at least two frequencies. Clearly to understand the bulk data it is necessary to take into account more than the nearest-neighbour shell. These observations can be placed on a more quantitative footing by taking the Fourier transform of q$^3\chi$(q). This is shown in the bottom panel of Fig. 6.9. These radial distribution functions have peaks corresponding to the position of the shells neighbouring a Te atom. The nano-crystal has one shell at a distance of 2.79 Å. This should be compared to the Te-Cd distance in bulk CdTe of 6.48$\times\sqrt{3}/4$=2.806 Å. The small contraction of the nano-crystal is presumably due to the epitaxial strain from the interaction between the Cd ions on the

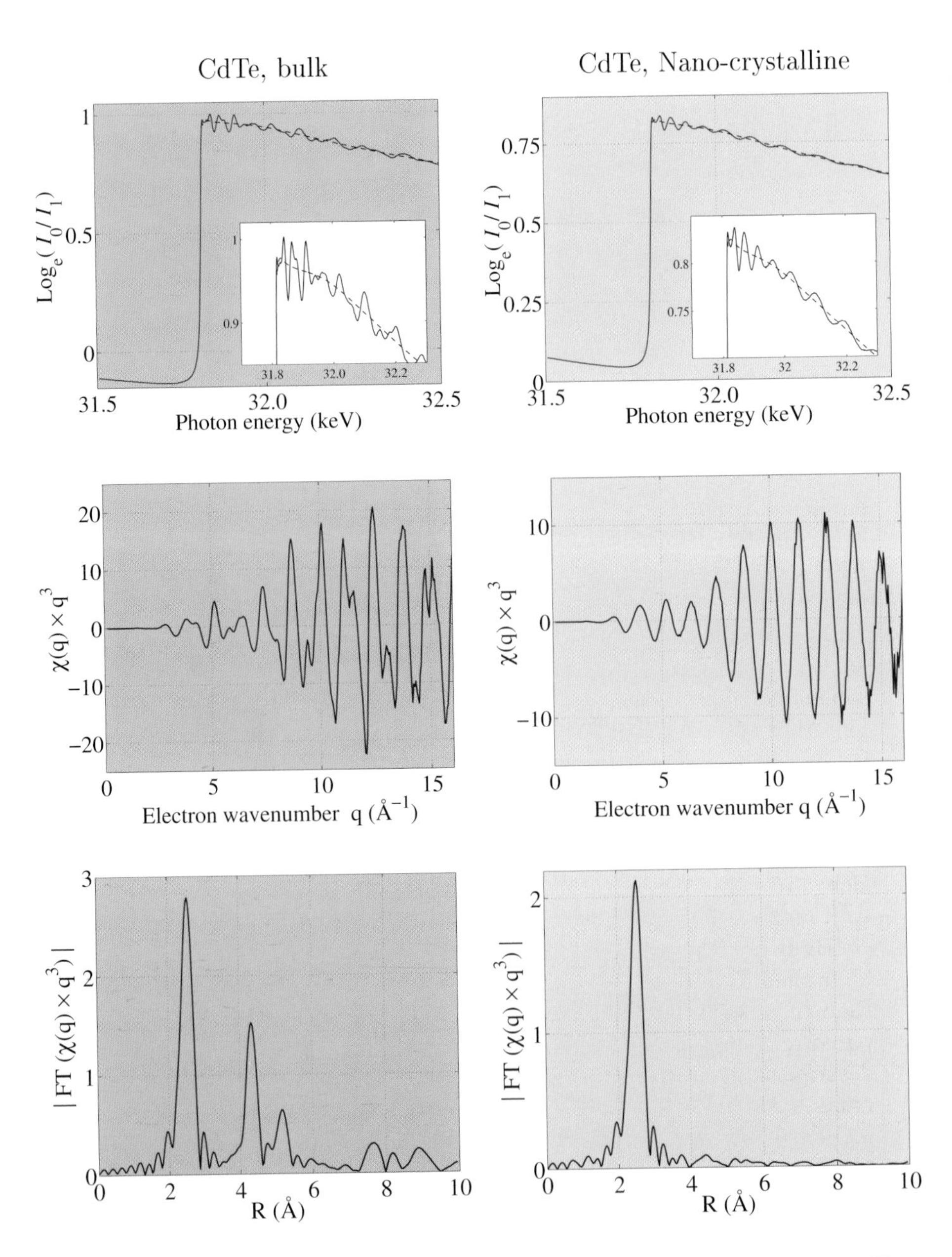

Figure 6.9: A comparison of the EXAFS spectra from bulk and nano-crystalline CdTe. The data were taken near the K edge of Te (31.813 keV) at a temperature of 8K. Top row: The absorption spectra. The dotted line indicates the smooth signal that would be obtained from an isolated Te atom. Middle row: $\chi(q)$ multiplied by q^3 as a function of the electron wavenumber q. Bottom row: The Fourier transform of the data shown in the row above. The resulting radial distribution function has peaks corresponding to the position of successive shells of atoms centred on a Te atom. The nano-crystal has one such shell at 2.79 Å. (Data supplied by J. Rockengerger, L. Tröger, A.L Rogach, M. Tischer, M. Grundmann, A. Eychmüller, and H. Weller [Rockengerger et al., 1998]).

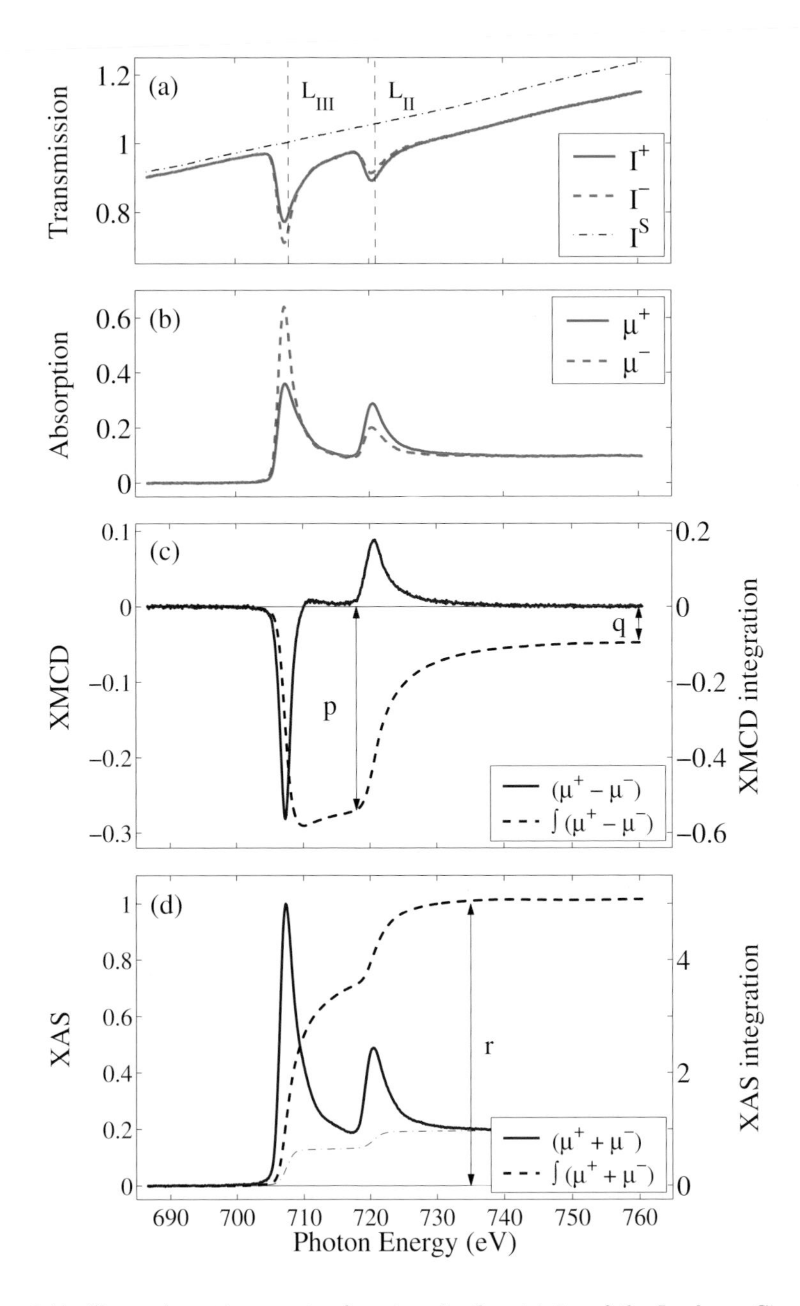

Figure 6.14: X-ray absorption spectra from iron in the vicinity of the L edges. Circularly polarized light was used, and the spectra were recorded with the spin of the incident photon parallel (I^+, solid curve) and antiparallel (I^-, dashed curve) to the spin of the Fe 3d electrons. (a) Transmission spectra of Fe/parylene thin films, and of the parylene substrate alone, taken at opposite saturation magnetizations; (b) the X-ray absorption spectra calculated from the transmission data shown in (a); (c) the XMCD spectra; (d) the summed X-ray absorption spectra. In (c) and (d) the values of the integrals p, q and r which appear in the sum rules are given by the dashed lines. The dot-dashed line in (d) indicates two steps in the absorption cross-section at the L_{III} and L_{II} edges. These were removed from the data before integration of the spectra. (Taken from Chen et al. [1995].)

where n_{3d} is the number of electrons in the 3d state, and p, q and r are given by

$$\begin{aligned} p &= \int_{\mathrm{L}_{III}} (\mu^+ - \mu^-)\, d\mathcal{E} \\ q &= \int_{\mathrm{L}_{III}+\mathrm{L}_{II}} (\mu^+ - \mu^-)\, d\mathcal{E} \\ r &= \int_{\mathrm{L}_{III}+\mathrm{L}_{II}} (\mu^+ + \mu^-)\, d\mathcal{E} \end{aligned} \tag{6.26}$$

For the spin sum rule the expression given in Eq. (6.25) is only approximate as we have neglected the so-called $\langle T_z \rangle$ term, which in the case of the 3d metals introduces an error of a few %.

The validity of the sum rules has been established through a number of experiments. In Fig. 6.14 we show the data for iron obtained by Chen et al. [Chen et al., 1995], who have performed one of the most exacting tests of the sum rules to date. The L edges for the 3d elements fall in the soft part of the X-ray spectrum. This usually means that the dichroic signal has to be inferred by measuring either the fluorescent radiation or the photoelectron yield. Both of these approaches introduce systematic errors which are difficult to correct for properly. By studying iron thin films grown on a parylene substrate, Chen et al. were able to perform their experiments in a transmission geometry, hence avoiding these complications. The absorption spectra for the parallel, μ^+, and anti-parallel, μ^-, configurations are shown in part (b). Strong white lines are evident at the positions of the L_{III} and L_{II} edges, corresponding to the transitions ($2\mathrm{p}_{3/2} \rightarrow$ 3d) and ($2\mathrm{p}_{1/2} \rightarrow$ 3d), respectively. The dichroism signal is shown in part (c), which also indicates the values of the integrals p, q and r appearing in the sum rules, Eqs. (6.24) and (6.25). In part (d) the X-ray absorption spectra for the sum of the μ^+ and μ^- is plotted. With the values of p, q and r indicated, and with n_{3d}=6.61 taken from theory, good agreement (within 7%) was found with other experimental techniques and theory for the values of the spin and orbital moments.

In addition to XMCD, materials may also exhibit X-ray magnetic linear dichroism (XMLD) [van der Laan et al., 1986]. This is the analogue of the Faraday rotation effect in the X-ray region. Although in principle it can be used to extract similar information to XMCD it is somewhat more demanding from an experimental point of view, and has not yet gained widespread use.

It is worthwhile to consider briefly what implications follow from the optical theorem and the observation of XMCD. The optical theorem states that absorption is proportional to the imaginary part of the scattering length in the forward direction (Eq. 3.15). This can of course be turned on its head, so that a particular type of absorption must imply an imaginary component to the scattering length: scattering and absorption are two sides of the same coin. As explained in the next chapter, the Kramers-Kronig relations state that there then must also be a contribution to the real part of the scattering length. Thus XMCD implies that there exists enhanced or resonant magnetic scattering at certain absorption edges. This takes us beyond the scope of this book, although a few remarks on resonant magnetic scattering are made at the end of the next chapter.

Further reading

E.A. Stern, *The Analysis of Materials by X-ray Absorption*, Scientific American, **234** No. 4, p. 96 (1976).

R. Loudon, *The Quantum Theory of Light*, (Oxford University Press, 1983)

7. Resonant scattering

In earlier chapters the scattering of X-rays has been discussed in terms of the classical Thomson scattering from an extended distribution of free electrons. Within this approximation the scattering length of an atom is written as $-r_0 f^0(\mathbf{Q})$, where $f^0(\mathbf{Q})$ is the atomic form factor, and r_0 is the Thomson scattering length of a single electron. The atomic form factor is nothing other than the Fourier transform of the charge distribution, and is hence a real number. We have also seen in Chapter 3 that with absorption processes included, the atomic scattering length must be generalized to be complex, the imaginary part being proportional to the absorption cross-section, σ_a (see Section 3.3 on page 68). It seems clear therefore that in order to pursue a classical model for the scattering, and at the same time require a complex scattering amplitude, a more elaborate model than that of a cloud of free electrons must be invoked. An obvious extension is to allow for the fact that electrons may be bound in atoms. In a classical picture they will respond to the driving field of the X-ray as damped harmonic oscillators, with an associated resonant frequency ω_s and a damping constant Γ.

As we shall see, the forced oscillator model does indeed give an imaginary component to the atomic scattering length, and in addition produces a correction to the real part. Altogether the scattering amplitude of the atom, in units of $-r_0$, can be written in the form

$$f(\mathbf{Q},\omega) = f^0(\mathbf{Q}) + f'(\omega) + i\,f''(\omega) \tag{7.1}$$

where f' and f'' are the real and imaginary parts of the *dispersion corrections*. It should be clear that the dispersion corrections are energy (or equivalently frequency) dependent. As they take on their extremal values at the absorption edges they are also known as the resonant scattering terms. At one time it was also common to refer to them as the anomalous scattering corrections, but since they are now mostly understood, it is generally agreed that there is nothing really anomalous about them. The dispersion corrections are dominated by electrons in the K shell, except perhaps for the heavier elements, where the L and M shells become important. The electrons in these shells are so spatially confined that the $\mathbf{Q}$ dependence can be neglected, and this explains why it has been omitted in Eq. (7.1). The Thomson term, $f^0(\mathbf{Q})$, on the other hand, does not depend on the photon energy, but only on the scattering vector $\mathbf{Q}$. The $\mathbf{Q}$ dependence is due to the fact that the non-resonant scattering is produced by *all* atomic electrons, which have a spatial extent of the same order of magnitude as the X-ray wavelength (see the discussion of the atomic form factor on page 114).

It is important to emphasize that the resonant scattering is elastic. That is, the scattered X-ray has the same energy as that of the incident one. In a quantum mechanical picture of the resonant scattering the incident photon

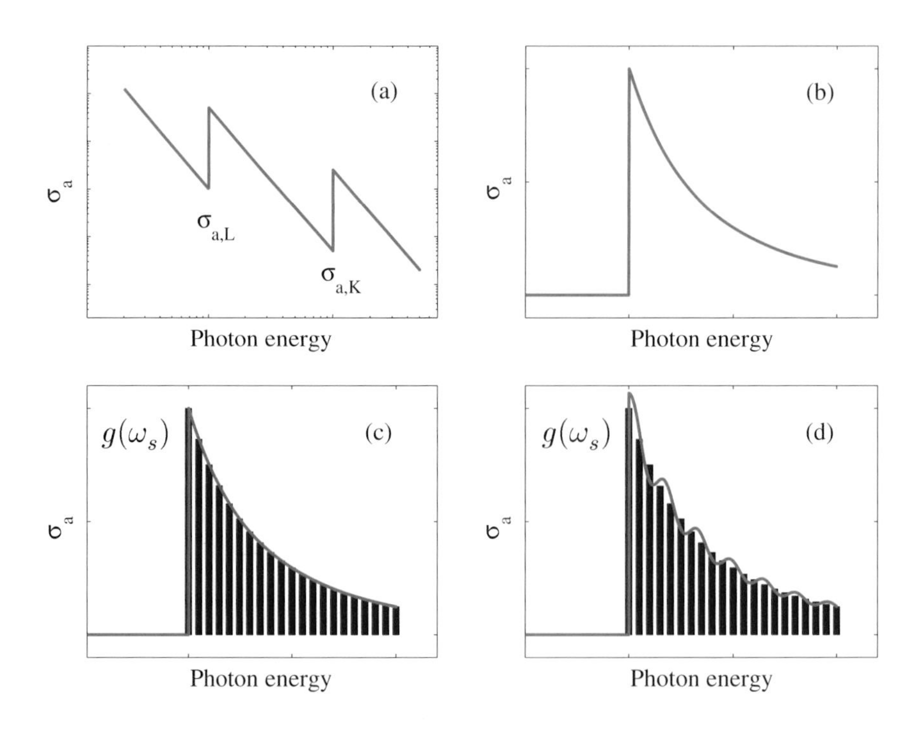

Figure 7.1: (a) Double logarithmic plot of the absorption cross-section as a function of photon energy. The absorption cross-section σ_a has characteristic edges. In between these edges σ_a varies approximately as the inverse cube of the energy. (b) The absorption cross-section for a K electron on a linear scale. (c) The absorption cross-section for an isolated atom may be modelled by a series of harmonic oscillators described by a smooth weighting function $g(\omega_s)$. (d) This often is not an adequate approach, as it does not take into account near-edge structure, such as the white line, or EXAFS oscillations produced by neighbouring atoms.

excites an electron to a higher lying level. The electron then decays back to the initial state by emitting a photon of the same energy as the incident one. This type of process, emission via some intermediate state, requires second-order perturbation theory to describe it, and the resonant behaviour then arises from the energy denominator present in the theory.

At this stage it is worth pausing to anticipate one of the limitations of the single oscillator model. The imaginary part of the dispersion correction f'' represents the dissipation in the system, or in other words the absorption. Indeed the explicit relationship between f'' and σ_a has already been given in Eq. (3.15) on page 70. It is known from elementary considerations that the imaginary part of the response of a forced harmonic oscillator displays a resonance when the driving frequency is close to the natural frequency, and that the width of this resonance is small for light damping. It follows that the single oscillator model can be expected to yield at best a peak in f'', and hence one also in σ_a. This clearly does not resemble the absorption cross-section of an atom sketched in Fig. 7.1(a). This has a discontinuous jump at an absorption edge, followed by a ω^{-3} fall-off, as discussed in Chapter 6 on absorption. In order to model this behaviour, one must instead assume a superposition of oscillators with relative weights, so-called oscillator strengths, $g(\omega_s)$, proportional to $\sigma_a(\omega = \omega_s)$.

The resonant scattering terms are of particular importance in the crystallography of complex systems, such as in the determination of the structure of macromolecules. The reason is that it is often difficult, if not impossible, to solve uniquely the structure of a unit cell that may contain thousands of atoms by collecting data at a single wavelength. It turns out that a solution may be found by recording data sets at several photon energies around the absorption edge of one of the atoms (usually a heavy atom) in the structure. The technique that exploits this approach is known as MAD, for Multi-wavelength Anomalous Diffraction. In fact it is possible to solve the phase problem in crystallography using MAD. The success of this technique depends on an accurate knowledge of the dispersion corrections. In this chapter we shall explain the basic principles behind how f' and f'' are determined. In the following section the expressions for the dispersion corrections are derived by treating the atomic electrons as harmonic oscillators. This is obviously a crude approximation, but one that nonetheless allows us to explore more general aspects of the relationship of f' and f'' to each other, and to the absorption cross-section.

7.1 The forced charged oscillator

Consider a classical model of an electron bound in an atom. Let the electron be subject to the electric field of an incident X-ray beam, $\mathbf{E}_{\text{in}} = \hat{\mathbf{x}}\,\mathrm{E}_0 \mathrm{e}^{-i\,\omega t}$, linearly polarized along the x axis, with amplitude E_0 and frequency ω. The equation of motion of the electron is

$$\ddot{x} + \Gamma\,\dot{x} + \omega_s^2\,x = -\left(\frac{e\,\mathrm{E}_0}{m}\right)\,\mathrm{e}^{-i\,\omega t} \tag{7.2}$$

The velocity-dependent damping term, $\Gamma\dot{x}$, represents dissipation of energy from the applied field, primarily due to re-radiation. The damping constant, Γ, which has the dimension of frequency, is usually much smaller than the resonant frequency, ω_s. By substituting the trial solution $x(t) = x_0 \mathrm{e}^{-i\,\omega t}$ into the above we obtain the following expression for x_0, the amplitude of the forced oscillation:

$$x_0 = -\left(\frac{e\,\mathrm{E}_0}{m}\right)\frac{1}{(\omega_s^2 - \omega^2 - i\,\omega\Gamma)} \tag{7.3}$$

The radiated field strength at distance R and time t is as usual (see Eq. (1.1) on page 7) given by the acceleration $\ddot{x}(t - R/c)$ at the earlier time $t' = t - R/c$:

$$\mathrm{E}_{\text{rad}}(R,t) = \left(\frac{e}{4\pi\epsilon_0 Rc^2}\right)\,\ddot{x}(t - R/c)$$

Inserting $\ddot{x}(t - R/c) = -\omega^2 x_0 \mathrm{e}^{-i\,\omega t}\mathrm{e}^{i\,(\omega/c)R}$, with x_0 given by Eq. (7.3), leads to

$$\mathrm{E}_{\text{rad}}(R,t) = \frac{\omega^2}{(\omega_s^2 - \omega^2 - i\,\omega\Gamma)}\left(\frac{e^2}{4\pi\epsilon_0 mc^2}\right)\mathrm{E}_0\mathrm{e}^{-i\,\omega t}\left(\frac{\mathrm{e}^{i\,\mathrm{k}R}}{R}\right)$$

or equivalently

$$\frac{\mathrm{E}_{\text{rad}}(R,t)}{\mathrm{E}_{\text{in}}} = -r_0\frac{\omega^2}{(\omega^2 - \omega_s^2 + i\,\omega\Gamma)}\left(\frac{\mathrm{e}^{i\,\mathrm{k}R}}{R}\right)$$

The atomic scattering length, f_s, is defined to be the amplitude of the outgoing spherical wave, $(\mathrm{e}^{i\,\mathrm{k}R}/R)$. In units of $-r_0$ this is

$$f_s = \frac{\omega^2}{(\omega^2 - \omega_s^2 + i\,\omega\Gamma)} \tag{7.4}$$

where the subscript "s" is there to remind us that the result is for a single oscillator. For frequencies large compared to the resonant frequency, $\omega \gg \omega_s$, the electron can be considered to be free, and the Thomson scattering expression is recovered, i.e. $f_s = 1$. Here we recall that the *total* cross-section for the scattering of an electromagnetic wave by a single free electron is

$$\sigma_{\mathrm{T}} = \left(\frac{8\pi}{3}\right) r_0^2$$

as derived on page 271 in Appendix B. It follows from Eq. (7.4) that the free electron result can be generalized to the case of a bound electron by writing

$$\sigma_{\mathrm{T}} = \left(\frac{8\pi}{3}\right) \frac{\omega^4}{(\omega^2 - \omega_s^2)^2 + (\omega\Gamma)^2}\, r_0^2$$

In the limit that $\omega \ll \omega_s$ and $\Gamma \rightarrow 0$ the cross-section becomes

$$\sigma_{\mathrm{T}} = \left(\frac{8\pi}{3}\right) \left(\frac{\omega}{\omega_s}\right)^4 r_0^2$$

This is the limiting form appropriate for the scattering of electromagnetic radiation in the visible part of the spectrum, and is known as Rayleigh's law[1]. For this reason the scattering of X-rays from atoms is also sometimes referred to as Rayleigh scattering.

The expression for f_s given in Eq. (7.4) can be rearranged in the following way:

$$\begin{aligned} f_s &= \frac{\omega^2 - \omega_s^2 + i\,\omega\Gamma + \omega_s^2 - i\,\omega\Gamma}{(\omega^2 - \omega_s^2 + i\,\omega\Gamma)} \\ &= 1 + \frac{\omega_s^2 - i\,\omega\Gamma}{(\omega^2 - \omega_s^2 + i\,\omega\Gamma)} \\ &\approx 1 + \frac{\omega_s^2}{(\omega^2 - \omega_s^2 + i\,\omega\Gamma)} \end{aligned} \tag{7.5}$$

where the last line follows from the fact that Γ is usually much less than ω_s. By writing it in this form it is clear that the second term is the dispersion correction to the scattering factor. The dispersion correction is then written as $\chi(\omega) = f_s' + i f_s''$, so that

$$\chi(\omega) = f_s' + i\,f_s'' = \frac{\omega_s^2}{(\omega^2 - \omega_s^2 + i\,\omega\Gamma)} \tag{7.6}$$

[1]The ω^4 dependence of the cross-section explains, amongst other things, why the sky is blue in the middle of the day, and turns red at sunrise and sunset. Blue light has a shorter wavelength than red, and is more strongly scattered. Light from the sun is scattered from particles in the atmosphere. During the day the light reaching an observer comes partly from diffuse scattering, and is hence dominated by the blue part of the spectrum. When the sun is viewed close to the horizon the blue light is scattered out of the direct sun light, producing a red hue to the sky.

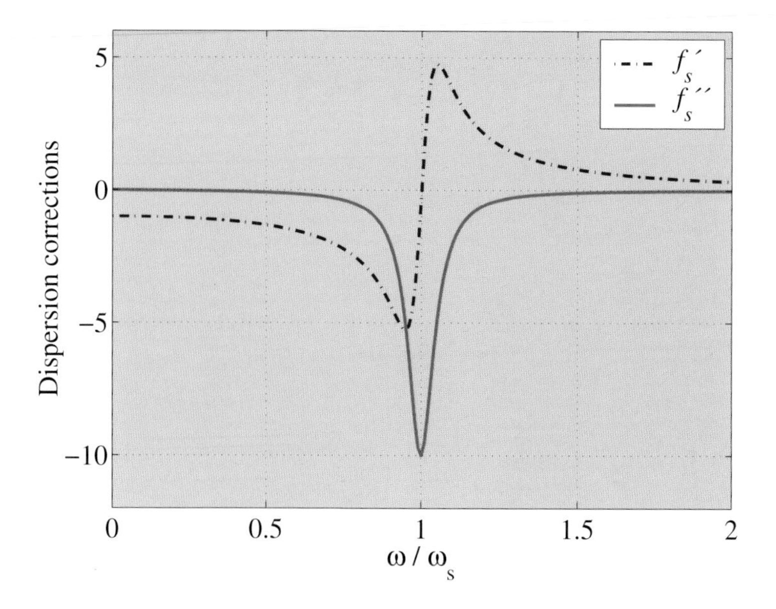

Figure 7.2: Single oscillator model. The real, f'_s, and imaginary, f''_s, parts of the dispersion corrections as a function of the driving frequency ω relative to the resonant frequency ω_s. In the example shown here the damping Γ has been chosen to be equal to $0.1\ \omega_s$.

with the real part given by

$$f'_s = \frac{\omega_s^2(\omega^2 - \omega_s^2)}{(\omega^2 - \omega_s^2)^2 + (\omega\Gamma)^2} \tag{7.7}$$

and the imaginary part by

$$f''_s = -\frac{\omega_s^2\omega\Gamma}{(\omega^2 - \omega_s^2)^2 + (\omega\Gamma)^2} \tag{7.8}$$

The frequency dependence of the dispersion corrections for the single oscillator model are shown in Fig. 7.2.

Comparing the above expression for f''_s to Eq. (3.15), and noting that $\omega/\mathrm{k} = c$, the absorption cross-section for a single oscillator model becomes

$$\sigma_{\mathrm{a,s}}(\omega) = 4\pi r_0 c \frac{\omega_s^2\,\Gamma}{(\omega^2 - \omega_s^2)^2 + (\omega\Gamma)^2} \tag{7.9}$$

As the damping constant Γ is small compared to the resonance frequency ω_s, the absorption cross-section has a sharp peak at $\omega = \omega_s$, the peak width being $\Delta\omega_{\mathrm{FWHM}} \approx \Gamma$. The effective absorption cross-section may thus be represented by a delta function centred at $\omega = \omega_s$:

$$\sigma_{\text{a,s}}(\omega) = 4\pi r_0 c \, \frac{\pi}{2} \, \delta(\omega - \omega_s) \tag{7.10}$$

where the factor of $\frac{\pi}{2}$ ensures that when Eq. (7.10) is integrated over ω it gives the same result as integrating Eq. (7.9).

7.2 The atom as an assembly of oscillators

In Fig. 7.1(a) a schematic plot is shown of the atomic X-ray absorption cross-section as a function of photon energy, where it is seen to exhibit characteristic absorption edges. For example, an X-ray photon with energy greater than the K edge energy can expel an electron from the K shell of the atom. This opens up a new channel for absorption, and produces an abrupt increase in the cross-section. In Chapter 6 on absorption we have shown how the magnitude of the absorption edge may be calculated from first principles, and also how the absorption between the edges varies approximately as ω^{-3}. The K absorption edge shown on a linear plot in Fig. 7.1(b) is clearly not the simple line spectrum of a single oscillator predicted by Eq. (7.10), and instead a more elaborate model is required.

If there were only one discrete quantum state that the electron could be excited into, then the classical line spectrum of a single oscillator would be an adequate description of the re-radiation. However, there is a continuum of free states above the absorption edge that the electron can be excited into. A different characteristic frequency ω_s can be associated with each of these states. Explicitly, the absorption cross-section given in Eq. (7.10) is generalized to

$$\sigma_{\text{a}}(\omega) = 2\pi^2 \, r_0 \, c \sum_s g(\omega_s) \, \delta(\omega - \omega_s)$$

where $g(\omega_s)$ is the relative weight of each transition, and where the narrow absorption lines have been approximated by delta functions (Fig. 7.1(c)). The expression for the real part of the dispersion correction, f', then also becomes the weighted superposition of single oscillators:

$$f'(\omega) = \sum_s g(\omega_s) \, f'_s(\omega_s, \omega)$$

7.3 The Kramers-Kronig relations

When trying to interpret experimental data it is sometimes better not to rely on theoretical values of the dispersion corrections. The reason is simply that they may not be accurate enough. A more serious difficulty is that it is not straight forward to allow for effects, such as the existence of a white line, or EXAFS oscillations in σ_a which depend on the particular environment of the resonantly scattering atom (see Fig. 7.1(d)). Instead a method has been developed for obtaining $f'(\omega)$ indirectly from the absorption cross-section $\sigma_a(\omega)$. The starting point is to determine $\sigma_a(\omega)$ experimentally, from which it is possible to obtain

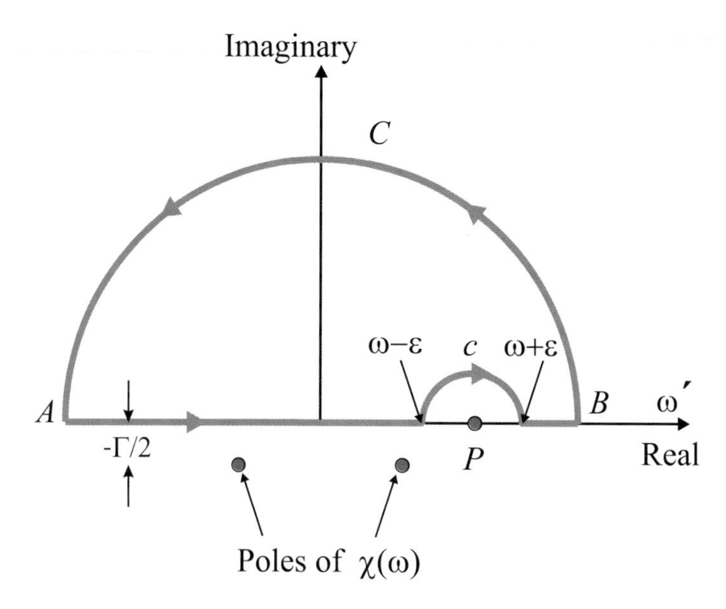

Figure 7.3: The Kramers-Kronig relation can be derived from Cauchy's theorem using the contour integral in the complex plane shown here.

$f''(\omega)$ through[2]

$$f''(\omega) = -\left(\frac{\omega}{4\pi\, r_0\, c}\right)\, \sigma_a(\omega) \tag{7.11}$$

(see Eq. (3.15)). The next step is to exploit the general relationships that exist between f' and f''. These are written as

$$f'(\omega) = \frac{1}{\pi}\mathcal{P}\int_{-\infty}^{+\infty} \frac{f''(\omega')}{(\omega' - \omega)} d\omega' = \frac{2}{\pi}\mathcal{P}\int_{0}^{+\infty} \frac{\omega'\, f''(\omega')}{(\omega'^2 - \omega^2)} d\omega' \tag{7.12}$$

$$f''(\omega) = -\frac{1}{\pi}\mathcal{P}\int_{-\infty}^{+\infty} \frac{f'(\omega')}{(\omega' - \omega)} d\omega' = -\frac{2\omega}{\pi}\mathcal{P}\int_{0}^{+\infty} \frac{f'(\omega')}{(\omega'^2 - \omega^2)} d\omega' \tag{7.13}$$

and are known as the Kramers-Kronig relations. The meaning of these relations is that if the energy dependence of the absorption cross-section is known, then $f''(\omega)$ can be found from Eq. (7.11), and with this substituted into Eq. (7.12) it is possible to derive the associated real part of the dispersion correction to the scattering amplitude. This method of obtaining f' from σ_a is illustrated in the next section by considering the simple model introduced in Chapter 6 for the variation of σ_a in the vicinity of a K edge.

The Kramers-Kronig equations relating f' to f'' require further comment. First, the $\mathcal{P}$ in front of the integral stands for "principal value". This means that the integration over ω' is actually performed by integrating from $-\infty$ to $(\omega - \epsilon)$ and from $(\omega + \epsilon)$ to $+\infty$, and then the limit $\epsilon \to 0$ is taken. Second,

[2] f'' is negative since σ_a is a positive real number. In other texts the sign convention is sometimes such that f'' is positive.

Kramers-Kronig relations

The derivation of the Kramers-Kronig relations is based on Cauchy's theorem concerning the contour integral taken in the counter clockwise direction of an analytic function $F(z)$ in the complex plane, z. If $F(z)$ has a *simple* pole at z_0 which is encompassed by the contour, then the value of the contour integral is equal to $2\pi i$ times the *residue*, which for a simple pole is equal to $(z - z_0)F(z_0)$.

Let us apply this theorem to the function $\chi(z)/(z - \omega)$ with $\chi(z) = \omega_s^2/(z^2 - \omega_s^2 + iz\Gamma)$ given by Eq. (7.6). It is straightforward to show that the poles of $\chi(z)$ are in the lower half of the complex plane at $Im(z) = -\Gamma/2$ as shown in Fig. 7.3. We shall then consider the contour comprising the path with $z = \omega'$ on the real axis from $A(\rightarrow -\infty)$ to $(\omega - \epsilon)$, then clockwise along the semi-circle c, then again along the real axis from $(\omega + \epsilon)$ to $B(\rightarrow +\infty)$, and then finally back to A along the large semi-circle C.

On the real axis, our function $\chi(z)/(z - \omega)$ has a pole at $\omega' = \omega$ with a residue equal to $\chi(\omega)$. As the path did not encompass this pole, the entire contour integral must be equal to zero. The contribution from the large semi-circle C is also zero, because for large z our function decays as $|z|^{-3}$, whereas the path length is only proportional to $|z|$. Altogether then, the sum of the principal integral, $\mathcal{P}\int \chi(z)/(z - \omega)d\omega'$, and the integral along c, $\int_c \chi(z)/(z - \omega)d\omega'$, is zero.

The clockwise integral along the small *half* circle is $-\pi i\,\chi(\omega)$, since a contour integral counter clockwise along the *full* circle around the pole P must, according to Cauchy's theorem, be equal to $+2\pi i\,\chi(\omega)$. Splitting $\chi(z)$ into its real and imaginary components (Eqs. (7.7) and (7.8)) finally yields

$$i\pi\left(f_s'(\omega) + if_s''(\omega)\right) = \mathcal{P}\int_{-\infty}^{\infty} \frac{f_s'(\omega') + if_s''(\omega')}{\omega' - \omega}d\omega' \tag{7.14}$$

Identifying the real and imaginary parts on the left and right hand sides leads to the Kramers-Kronig relations for a single oscillator. Since $f'(\omega)$ and $f''(\omega)$ are linear superpositions of single oscillators, the Kramers-Kronig relations also apply to them.

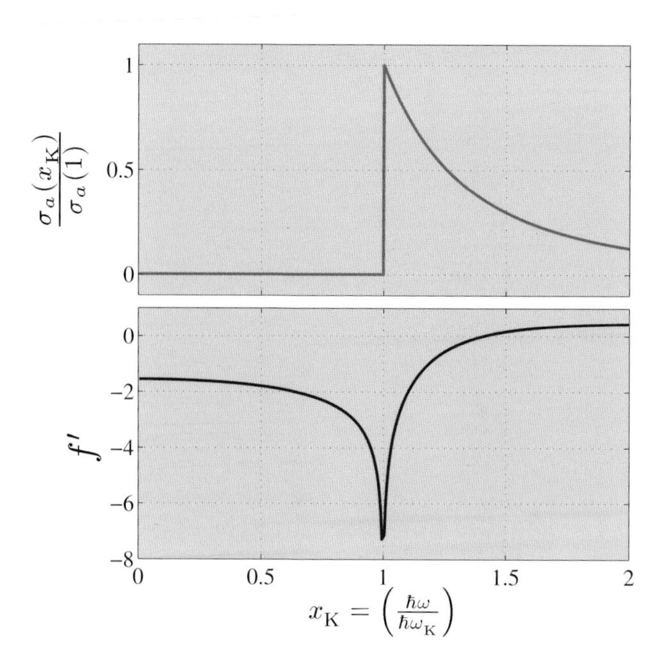

Figure 7.4: Estimate of the dispersion corrections around the K edge. Top panel: The imaginary part of the dispersion correction f'' is proportional to the absorption cross-section σ_a, here assumed to vary as $1/\omega^3$ above the edge. Bottom panel: Numerical estimate of the real part of the dispersion correction f' for two K shell electrons in an atom. The curve for f' is given by Eq. (7.17) which has been derived from the Kramers-Kronig transform of f''.

the alternative form of the expressions for $f'(\omega)$ and $f''(\omega)$ have been obtained by multiplying the numerator and denominator by $(\omega' + \omega)$, and by utilizing the fact that $f'(\omega')$ is an even function, and $f''(\omega')$ an odd function according to Eq. (7.7) and Eq. (7.8). The validity of the Kramers-Kronig relations for the single oscillator expression given in Eqs. (7.7) and (7.8) may be established either by direct substitution, or by the more general derivation given in the box on the preceding page.

7.4 Numerical estimate of f'

In this section an estimate is made of f', the real part of the dispersion correction, for photon energies near the K absorption edge. Equation (7.12) can be used together with (7.11) to relate f' to the energy dependence of the absorption cross-section:

$$
\begin{aligned}
f'(\omega) &= \frac{2}{\pi}\mathcal{P}\int_0^{+\infty} \frac{\omega' f''(\omega')}{(\omega'^2 - \omega^2)} d\omega' \\
&= -\frac{2}{\pi}\frac{1}{(4\pi\, r_0\, c)}\mathcal{P}\int_0^{+\infty} \frac{\omega'^2\,\sigma_a(\omega')}{(\omega'^2 - \omega^2)} d\omega' \qquad (7.15)
\end{aligned}
$$

To evaluate this integral the frequency ω is normalized to that of the K edge by introducing $x_{\mathrm{K}} = \omega/\omega_{\mathrm{K}}$ as the independent variable, so that the integration variable becomes $x{=}\omega'/\omega_{\mathrm{K}}$. The energy dependence of the absorption cross-section was discussed in Chapter 6, and this allows us to write $\sigma_a(\omega')$ in the form

$$\sigma_a\left(\frac{\omega'}{\omega_{\mathrm{K}}}\right) \approxeq \begin{cases} \sigma_a\left(\frac{\omega'}{\omega_{\mathrm{K}}}=1\right)\left(\frac{\omega'}{\omega_{\mathrm{K}}}\right)^{-3} & \text{for } \left(\frac{\omega'}{\omega_{\mathrm{K}}}\right) \geq 1, \\ 0 & \text{for } \left(\frac{\omega'}{\omega_{\mathrm{K}}}\right) < 1 \end{cases}$$

This is plotted in the top panel of Fig. 7.4. With this functional form for $\sigma_a(\omega')$ the principal value integral given in Eq. (7.15) is

$$\begin{aligned} \mathcal{P}\int_0^{\infty} \frac{\omega'^2\sigma_a(\omega')}{(\omega'^2-\omega^2)}\,d\omega' &\approxeq \sigma_a(1)\,\mathcal{P}\int_1^{\infty}\frac{x^2x^{-3}}{(x^2-x_{\mathrm{K}}^2)}\,\omega_{\mathrm{K}}\,dx \\ &= \sigma_a(1)\,\omega_{\mathrm{K}}\,\mathcal{P}\int_1^{\infty}\frac{1}{x(x^2-x_{\mathrm{K}}^2)}\,dx \\ &= \sigma_a(1)\,\omega_{\mathrm{K}}\,I(x_{\mathrm{K}}) \end{aligned}$$

where the integral $I(x_{\mathrm{K}})$ is defined by

$$I(x_{\mathrm{K}}) = \mathcal{P}\int_1^{\infty}\frac{1}{x(x+x_{\mathrm{K}})(x-x_{\mathrm{K}})}\,dx \tag{7.16}$$

Collecting these results together, the expression for the real part of the dispersion correction becomes

$$\begin{aligned} f'(\omega) &= -\frac{2}{\pi}\frac{1}{(4\pi\,r_0\,c)}\,\sigma_a(1)\,\omega_{\mathrm{K}}\,I(x_{\mathrm{K}}) \\ &= -\frac{1}{\pi\lambda_{\mathrm{K}}r_0}\,\sigma_a(1)\,I(x_{\mathrm{K}}) \end{aligned}$$

The magnitude of the discontinuity in the absorption cross-section at the edge, $\sigma_a(1)$, can be found from Eq. (6.15), which for *two* K electrons reads

$$\sigma_a(1) = 2\times\left(\frac{256\pi}{3\mathrm{e}^4}\right)\lambda_{\mathrm{K}}r_0$$

The real part of the dispersion correction close to the K edge is therefore given by

$$\boxed{f'(\omega) = -\left(\frac{512}{3\mathrm{e}^4}\right)\,I(x_{\mathrm{K}}) = -3.13\;I(x_{\mathrm{K}})} \tag{7.17}$$

In order to obtain a numerical value for f' it is necessary to perform the principal value integration given in Eq. (7.16). This is readily achieved using one of the modern mathematical software packages, such as Mathematica[3] or

[3] For example, the curve for f' in Fig. 7.4 was generated from three lines of code in Mathematica: g[x_,xk_]=-3.13/(x(x-xk)(x+xk)); f[xk_]:= Integrate[g[x,xk],{x,1,Infinity},PrincipalValue→True]; Plot[f[xk],{xk,0,2}].

Maple, and the results are shown in Fig. 7.4. It can be verified that this curve has the correct asymptotic behaviour. From Eq. (7.16), for $x_{\mathrm{K}} \rightarrow \infty$ the integral $I(x_{\mathrm{K}} \rightarrow \infty) \rightarrow 0$, in accordance with the vanishing dispersion corrections at high photon energies. In the limit $x_{\mathrm{K}} \rightarrow 0$ we have $I(x_{\mathrm{K}} \rightarrow 0) = \int_1^{\infty} x^{-3}\, dx = 1/2$, and at low energies $f'(\omega \ll \omega_{\mathrm{K}})$ tends to the value -1.565. In other words the contribution of the two K electrons to the Thomson scattering is partly quenched. Thus the curve of $f'(x_{\mathrm{K}})$ *vs* x_{K} is not expected to be symmetric around $x_{\mathrm{K}} = 1$, as indeed is evident in Fig. 7.4.

In summary, the total atomic scattering amplitude is

$$f(\mathbf{Q}, \hbar\omega) = f^0(\mathbf{Q}) + f'(\hbar\omega) + i\, f''(\hbar\omega) \tag{7.18}$$

where $f^0(\mathbf{Q})$ is the form factor for *all* Z electrons in the atom, K electrons included, and the two extra terms are the dispersion corrections. As $f^0(\mathbf{Q})$ is the Fourier transform of the electrons density in the atom, normalized to Z at $\mathbf{Q} = 0$, it is independent of the photon energy. On the other hand, in our model the dispersion corrections are due to the K electrons only. These are localized close to the nucleus so that the Fourier transform of their wavefunction is essentially constant. This explains why to a good approximation the dispersion corrections are independent of the scattering vector $\mathbf{Q}$. For photon energies below the K edge, the K electrons are so tightly bound that the electromagnetic field of the incident X-ray cannot set them into full vibrations. In other words the scattering amplitude of the entire atom is reduced compared to the Thomson value, and it follows that $f'(\hbar\omega)$ must be negative. Since $f(\mathbf{Q})$ decreases with increasing $\mathbf{Q}$, whereas $f'(\hbar\omega)$ remains constant, the relative contribution of $f'(\hbar\omega)$ to the total atomic scattering amplitude increases with increasing $\mathbf{Q}$. Thus the relative importance of the dispersion corrections increases at large scattering angles.

For heavier elements (La and beyond) the L edges fall in the X-ray region. Since the L shell contains six $2p$ electrons compared to the two electrons in the K shell, the dispersion corrections are larger by a factor of approximately three.

It is of course desirable to perform more accurate calculations of the dispersion corrections than described here. Accurate values of the dispersion corrections are needed in several branches of crystallography, including, for example, the derivation of electron density maps. In Fig. 7.5 examples of the energy dependence of the dispersion corrections for the noble gases Ar and Kr are shown. These have been calculated within the self-consistent Dirac-Hartree-Fock framework [Chantler, 1995]. (See also Henke et al. [1993].) For Ar the K edge occurs at 3.203 keV. By comparing the curves shown in the left panel of Fig. 7.5 with Fig. 7.4 it can be seen that our simpler model captures the essential features of the energy dependence of the dispersion corrections.

In fact, even the most sophisticated theoretical methods are not always adequate. The reason is that close to an absorption edge the dispersion corrections become sensitive to the details of the environment of the resonantly scattering atom. For example, it has already been described in Chapter 6 how the absorption cross-section is modified by EXAFS oscillations. Under these circumstances the best that can be done is to measure the absorption cross-section of the atom of interest in the particular crystal being investigated over a range of energies around the edge, and then use the general Kramers-Kronig relation to derive $f'(\hbar\omega)$ from $\sigma_a(\omega)$. In this way the effects of core-hole lifetime, as well as experimental resolution, are also included automatically.

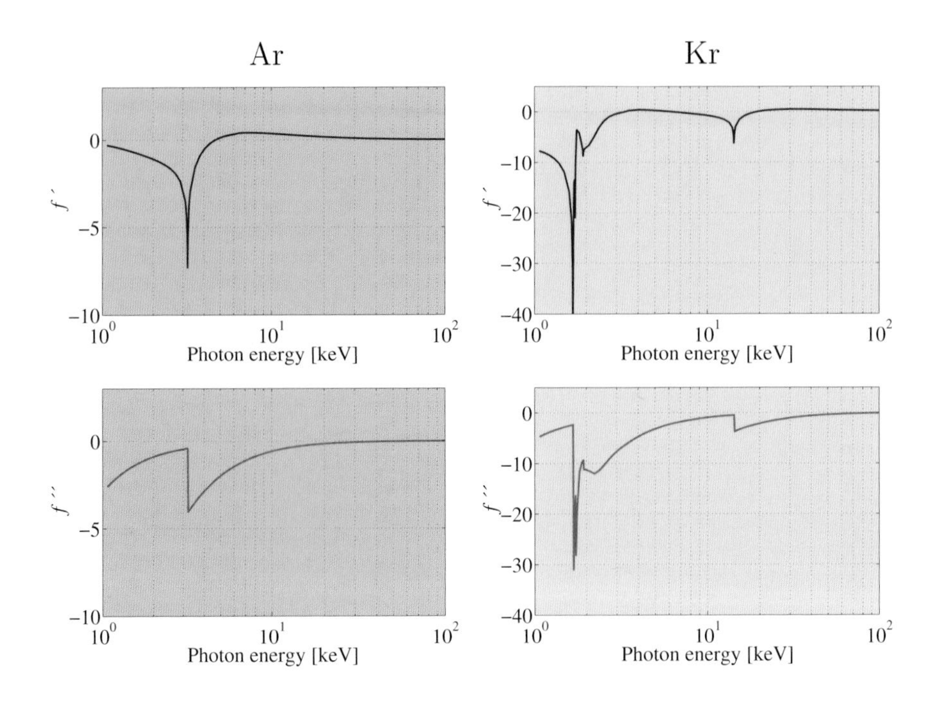

Figure 7.5: The calculated energy dependence of the real, f', and imaginary, f'', parts of the dispersion corrections (in units of r_0) for Ar and Kr. The calculations were performed within the self-consistent Dirac-Hartree-Fock framework, and are discussed in the text. The K edge of Ar is evident at 3.203 keV. For Kr the K edge occurs at 14.32 keV, while the L edges are centred around 1.8 keV.

7.5 Breakdown of Friedel's law and Bijvoet pairs

At the start of Chapter 4 it was shown that several important concepts in diffraction from materials could be understood by considering the interference of waves scattered by a simple two electron system. Here a similar approach is adopted to explain some of the important consequences that the existence of dispersion corrections have for diffraction experiments. Instead of two electrons, the scattering system is formed from two non-identical atoms, as indicated in Fig. 7.6.

The first issue to consider is whether it is possible in a diffraction experiment to determine the absolute configuration of a system. For the present discussion this boils down to the question of whether it is possible to deduce which atom sits to left, and which one to the right. One obvious way to attempt this is to perform two different scattering experiments: one with the wavevector transfer to the right (a), and one where it is to the left (b). Let the scattering amplitudes be f_1 and f_2, as indicated, in units of the Thomson scattering length $-r_0$. To start with the dispersion corrections are neglected, so that the scattering amplitudes are real, positive numbers. Further, let the distance between the two atoms be x, and the scattering vector component along this direction between the atoms

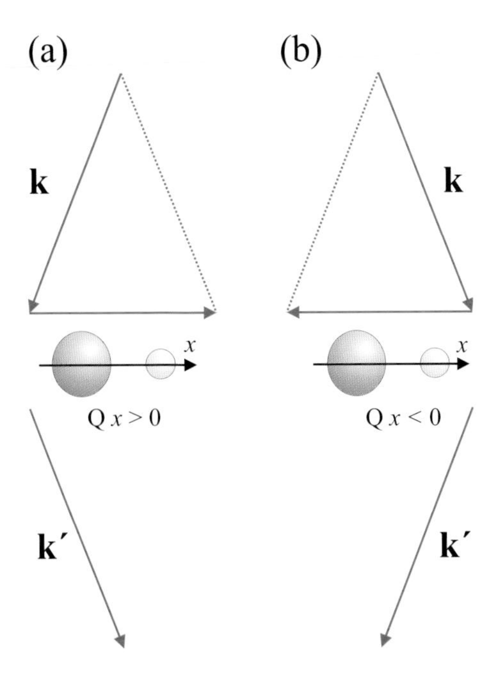

Figure 7.6: Diffraction by two non-identical atoms. Analysis of the scattered intensities when the scattering vector is parallel to the direction connecting the large and small atoms, or when it is in the opposite direction, shows that with the dispersion corrections taken into account it is possible to tell whether the large atom is to the right or left of the smaller one.

be +Q in case (a) and −Q in (b). For case (a) the total scattered amplitude is

$$A(\mathrm{Q}) = f_1 + f_2\, \mathrm{e}^{i\,\mathrm{Q}x} \tag{7.19}$$

and the intensity is

$$\begin{aligned} I(\mathrm{Q}) &= (f_1 + f_2\mathrm{e}^{i\,\mathrm{Q}x})(f_1 + f_2\mathrm{e}^{-i\,\mathrm{Q}x}) \\ &= f_1^2 + f_2^2 + 2f_1 f_2 \cos(\mathrm{Q}x) \end{aligned} \tag{7.20}$$

Under the stated assumptions it is obvious that for the wavevector −Q, case (b), the same scattered intensity is obtained, since $\cos(\mathrm{Q}x) = \cos(-\mathrm{Q}x)$. Therefore it is not possible from a diffraction experiment to determine the absolute position of the atoms. This argument can be generalized by letting the pair of atoms form the basis in a unit cell of a three dimensional crystal. In this case the result is written as

$$\boxed{I(\mathbf{Q}) = I(-\mathbf{Q})} \tag{7.21}$$

which is known as Friedel's law.

The assumption that the scattering amplitudes of the individual atoms are real positive numbers is now lifted; in other words we allow for the effect of the dispersion corrections. The scattering length of the two atoms is then written in the form

$$f_j = f_j^0 + f_j' + i\,f_j'' \qquad j = 1, 2 \tag{7.22}$$

This can be re-expressed more conveniently as

$$f_j = r_j \, \mathrm{e}^{i\,\phi_j} \tag{7.23}$$

where r_j=$|f_j|$. The amplitude $A(\mathrm{Q})$ in case (a) then becomes

$$A(\mathrm{Q}) = r_1 \, \mathrm{e}^{i\,\phi_1} + r_2 \, \mathrm{e}^{i\,\phi_2} \, \mathrm{e}^{i\,\mathrm{Q}x} \tag{7.24}$$

and Eq. (7.20) takes the form

$$I(\mathrm{Q}) = |f_1|^2 + |f_2|^2 + 2|f_1||f_2|\cos(\mathrm{Q}x + \phi_2 - \phi_1) \tag{7.25}$$

As in general $\phi_1 \neq \phi_2$ it follows that

$$I(\mathrm{Q}) \neq I(-\mathrm{Q}) \tag{7.26}$$

since $\cos(\mathrm{Q}x+\phi_2-\phi_1) \neq \cos(-\mathrm{Q}x+\phi_2-\phi_1)$. In other words Friedel's law breaks down when dispersion corrections are taken into account. Thus by measuring whether $I(\mathrm{Q})$ is larger or smaller than $I(-\mathrm{Q})$ it is possible to determine which atom is to the left and which atom to the right in Fig. 7.6.

It would, however, be wrong to conclude from this that Friedel's law is never fulfilled. If the unit cell is centrosymmetric, consisting for example of atoms of type 1 at $\pm x_1$, and atoms of type 2 at $\pm x_2$, then the unit cell structure factor is

$$\begin{aligned} F &= r_1 \, \mathrm{e}^{i\,(\phi_1+\mathrm{Q}x_1)} + r_1 \, \mathrm{e}^{i\,(\phi_1-\mathrm{Q}x_1)} + r_2 \, \mathrm{e}^{i\,(\phi_2+\mathrm{Q}x_2)} + r_2 \, \mathrm{e}^{i\,(\phi_2-\mathrm{Q}x_2)} \\ &= [r_1 \, 2\cos(\mathrm{Q}x_1)] \, e^{i\,\phi_1} + [r_2 \, 2\cos(\mathrm{Q}x_2)] \, e^{i\,\phi_2} \end{aligned} \tag{7.27}$$

and the intensity is

$$\begin{aligned} I(\mathrm{Q}) &= |F|^2 \\ &= 4|f_1|^2\cos^2(\mathrm{Q}x_1) \\ &+ 4|f_2|^2\cos^2(\mathrm{Q}x_2) \\ &+ 8|f_1|^2|f_2|^2\cos(\mathrm{Q}x_1)\cos(\mathrm{Q}x_2)\cos(\phi_2 - \phi_1) \end{aligned} \tag{7.28}$$

The intensity in this case is evidently an *even* function of Q, or in other words Friedel's law is reestablished for centrosymmetric structures.

The algebra presented here may appear to be a little complicated. If one is content with a qualitative discussion, then it often suffices to plot the scattering amplitudes in a so-called Argand diagram. An Argand diagram is a graphical representation of a complex number, where the real and imaginary parts are plotted on the x and y axes respectively. We start by assuming that there is one atom in the unit cell, and by neglecting dispersion corrections. The unit cell structure factor F_{hkl} is a complex number, $r\,\mathrm{e}^{i\,\phi_1}$, as shown in Fig. 7.7(a). The operation $\mathbf{Q} \to -\mathbf{Q}$ is equivalent to $\phi_1 \to -\phi_1$. It is obvious from the figure that the structure factor of the $(\overline{h},\overline{k},\overline{l})$ reflection is found by reflecting F_{hkl} about the real axis. The length of $|F_{hkl}\,|$ is unaffected by this operation, or in other words $|F_{hkl}\,|$=$|F_{\overline{hkl}}|$. Now imagine that an atom is added to the structure which has significant dispersion corrections due to the presence of resonant scattering terms. Let the scattering length of the atom be $f + if''$, where the real part of the scattering length is $f = f^0 + f'$ and the imaginary

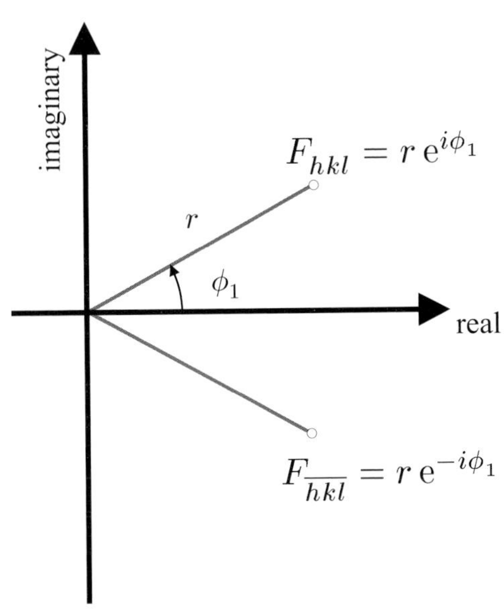

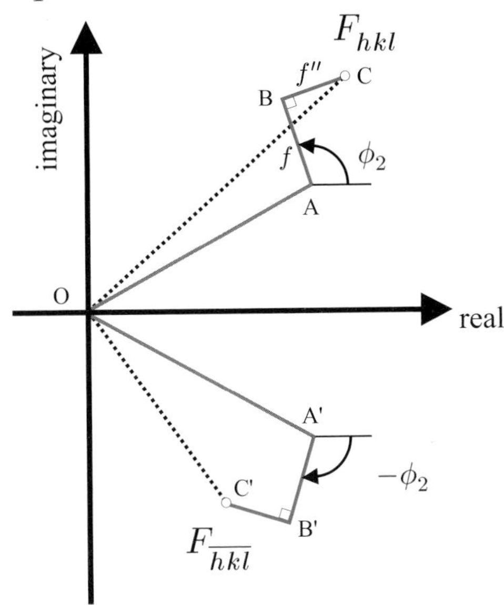

Figure 7.7: Argand diagram for a unit cell structure factor. (a) One atom per unit cell neglecting dispersion corrections. In this case $|F_{hkl}|=|F_{\overline{hkl}}|$. (b) Adding one more atom, with dispersion corrections, to the unit cell will change the structure factors to be F_{hkl}=OC and $F_{\overline{hkl}}$=OC$'$. In this case $|F_{hkl}| \neq |F_{\overline{hkl}}|$.

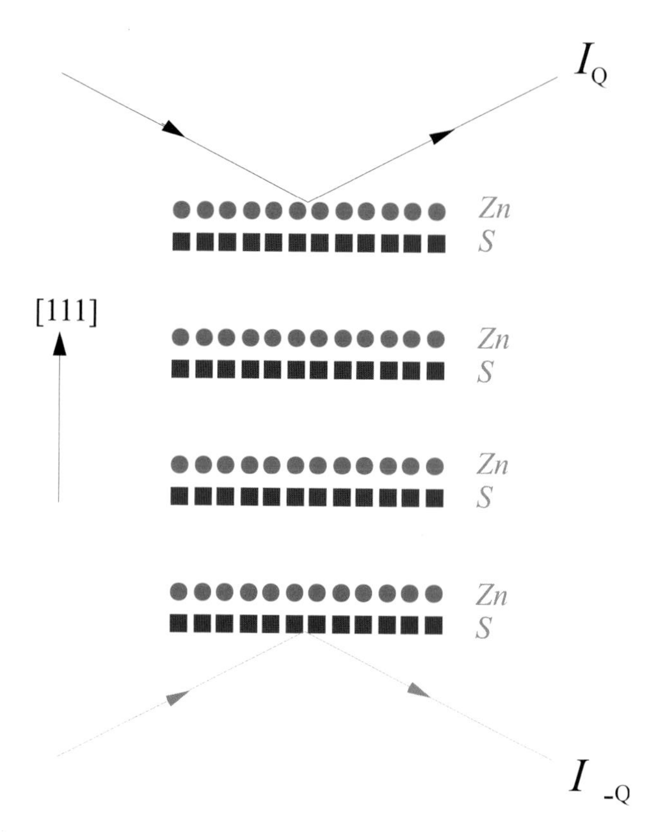

Figure 7.8: The magnitude of intensity in symmetric Bragg reflection from a {111} ZnS crystal depends on whether one reflects from the "Zn" (top) or the "S" side (bottom). In this way the absolute sense of the polar direction can be determined.

part is f''. When placed in the unit cell it acquires a phase factor $e^{i\,\mathbf{Q}\cdot\mathbf{r}_2}$, or for brevity $e^{i\,\phi_2}$. The construction of the total structure factor of the unit cell is shown in Fig. 7.7(b). Consider first the contribution from f. This is added as a line of length f originating from A at an angle ϕ_2 with respect to the real axis. In adding the contribution from if'', BC, it must be remembered that f'' is negative so that BC is turned clockwise with respect to AB. The total structure factor F_{hkl} is then OC. $F_{\overline{hkl}}$ is constructed in a similar manner, remembering that $\phi_2 \rightarrow -\phi_2$. From this geometrical structure it is clear that $|F_{hkl}| \neq |F_{\overline{hkl}}|$.

Example: the absolute polar direction in ZnS

A striking and simple experimental example of these considerations was obtained around 1930 independently by two groups [Nishikawa and Matsukawa, 1928, Coster et al., 1930]. In a crystal of ZnS with [111] faces they were able to determine which one of two opposite faces was terminated by Zn and which one by S. ZnS has the zinc-blende structure as shown in Fig. 4.13 – it consists of two inter-penetrating *fcc* lattices, with one being occupied by Zn atoms and the other by S atoms. The displacement of the two lattices is $\frac{1}{4}$ of a cube diagonal, so that along a diagonal the structure can be considered as double layers of

Zn and S as depicted in the Fig. 7.8. This structure obviously is not centrosymmetric, Friedel's law does not hold, and there will be a difference between the intensities $I_{(111)}$ and $I_{(\overline{111})}$. From this one can conclude which side of the crystal is the Zn side and which one is terminated by a plane of S atoms. The conclusions from such X-ray experiments have been confirmed by other means, such as ion scattering.

7.5.1 Bijvoet's experiment on chiral crystals

Having shown that it is possible to determine the absolute direction in a polar crystal, such as ZnS, by comparing the intensities of Friedel pairs of reflections, a question naturally arises: is it also possible to determine the absolute chirality of a molecule? It took approximately 20 years from the experiments on ZnS until around 1950 before Bijvoet and co-workers [Bijvoet et al., 1951] proved that this was indeed possible. Before describing this important experiment a few comments are appropriate on the concept chirality. The precise mathematical definition of a chiral structure reads: "A structure whose mirror image cannot be made to coincide with the original structure by rotation and translation is chiral." A familiar example from daily life is your hand: if you make a mirror image of a right-hand glove, say, it becomes a left-hand glove, and no matter how you try to twist and slide it, it will not fit on your right hand. Another comprehensible description is the comparison to a screw. A screw is characterized by a direction ("into or out of the wall") and a rotation direction. For example if a screw for going "in" has to be rotated clockwise, it is called a right-hand screw.

The screw picture can be applied to a very common kind of chiral molecule, which is based on a central carbon atom with four tetrahedrally bound moieties (see Fig. 4.4). The "direction" is then defined as the bond from the central carbon atom towards the lightest atom, often hydrogen. Figure 7.9(a) shows the tetrahedron looking along that direction, with the light atom and the central carbon in that particular projection coinciding at the point O. Imagine now that the carbon atom has bonds to three different atoms: Ol, Om, Os in Fig. 7.9(a)), or equivalently OA, OB and OC in Fig. 4.4. Here l stands for large, m for medium, and s for small. There are then two different possible rotation directions in going in the sequence $l-m-s$, namely clockwise or anti-clockwise. The normal convention is to refer to the first case as an R molecule and the second case as an S molecule.

The issue that Bijvoet considered was whether for a given chirality of molecule (R or S) there would be any difference in the diffracted intensity when scattering to the right (scattering vector +Q) or to the left (−Q), or equivalently whether for a given scattering sense (+Q, say) R and S molecules would produce different intensities. Instead of trying to answer this question by writing down the appropriate formulae, we shall use a geometrical construction, as it serves to further illustrate the usefulness of plotting structure factors in an Argand diagram.

The Argand diagrams for the S and R forms of the molecule are plotted in Fig. 7.9. To simplify things as much as possible, it is assumed that the scattering vector $\mathbf{Q}$ is parallel to the direction lm. There is then no phase factor $e^{i\,\mathbf{Q}\cdot\mathbf{r}}$ for atom s, as Os is perpendicular to $\mathbf{Q}$. With this assumption there are also symmetric phase factors $e^{i\,\mathbf{Q}\cdot\mathbf{r}} = e^{\pm i\,\phi}$ associated with atoms l and

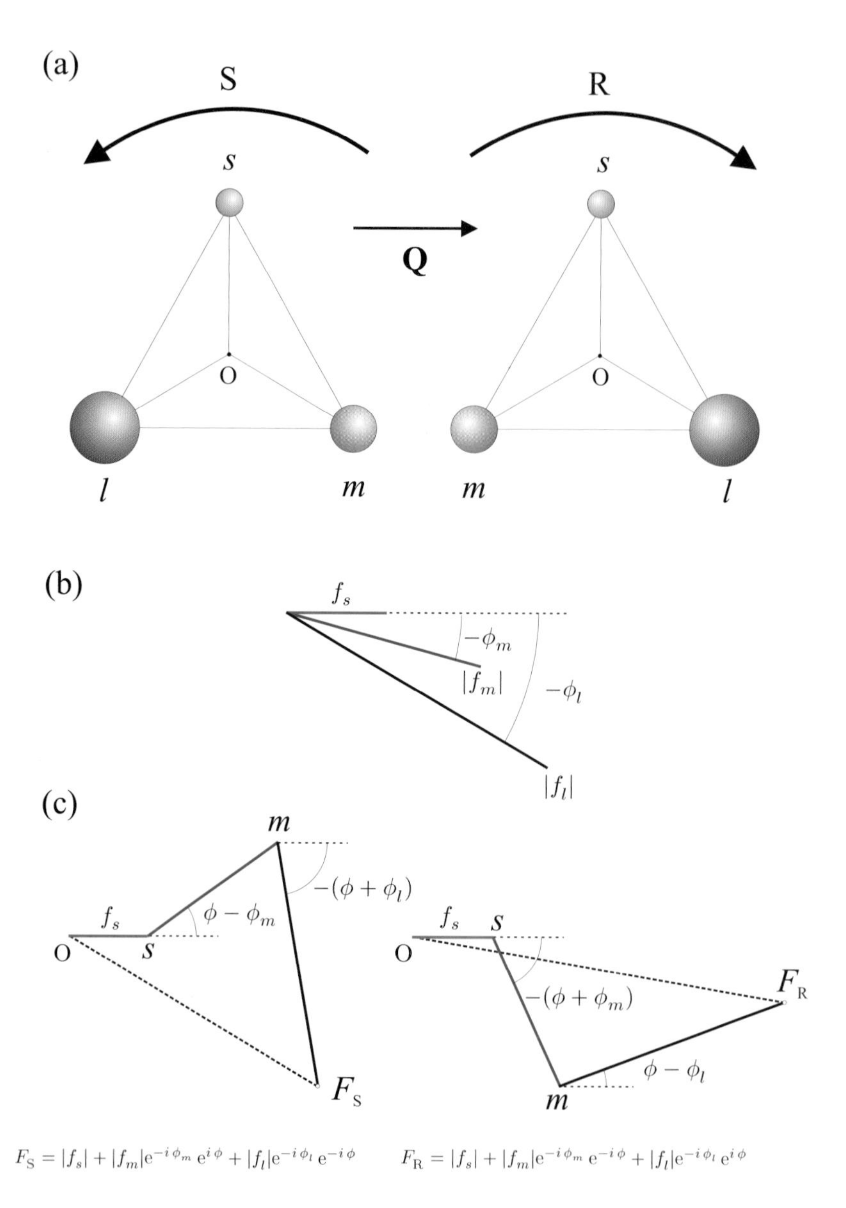

Figure 7.9: The effect of the dispersion corrections on the scattering from chiral molecules. (a) Definition of the chirality for a molecule of type S and its enantiomer, a R molecule. (b) Argand diagram for the individual atomic scattering lengths of the small (f_s), medium (f_m), and large atoms (f_l). (c) Construction of the total scattering lengths for the R and S molecules. It is evident that $|F_S| < |F_R|$, and hence by measuring Friedel pairs of reflections it is possible to determine the absolute chirality of a molecule.

m. The scattering lengths of the three atoms s, m and l are drawn in part (b). They are all complex, as all atoms are assumed to have a finite absorption cross-section. Furthermore, to go from the real part to the full complex value, one has to go clockwise in the complex plane, since $-f''$ is proportional to the absorption cross-section, which is itself positive (Eq. (3.15)). The angle one has to go clockwise increases with increasing atomic number Z, as, f^0, the dominant contribution to the real part varies in proportion to Z, but the imaginary part varies in proportion to Z^4. The three complex scattering lengths must therefore appear as shown in Fig. 7.9(b), i.e. turning clockwise with increasing magnitude in the sequence $s - m - l$. The absolute phase of this set of scattering lengths is irrelevant, and f_s has been chosen to be parallel to the real axis in the complex plane. With these pieces in place, we are now ready to construct the Argand diagrams for the S and R variants of the molecular structure, and these are shown in Fig. 7.9(c).

The scattering length of the "s" atom is represented by the line Os of length $|f_s|$. This line is along the real axis since the scattering vector $\mathbf{Q}$ is perpendicular to Os, and hence there is no phase factor associated with the scattering from atom "s" for either variant. Next consider the scattering from the m type atoms, which has a scattering length of $|f_m|e^{-i\phi_m}$. In case S the appropriate phase factor is $\mathrm{Q}x > 0{=}{+}\phi$, as the scalar product of Om and $\mathbf{Q}$ is positive. This means that f_m must be turned counter-clockwise by an amount $+\phi$. In the R case, it is the other way around, i.e. f_m must be turned clockwise by the same amount. The resulting scattering amplitude including the contributions from both "s" and "m" type atoms is Om in Fig. 7.9(c). To the point m it is now necessary to add the scattering length of atom "l" with the appropriate phase factor included. First consider case S. The phase is less than zero as the scalar product of Ol and $\mathbf{Q}$ is negative. This means that the line mF_S makes an angle $-(\phi + \phi_l)$ with the real axis. For case R the phase is greater than zero and the line mF_R makes an angle $(\phi - \phi_l)$ with the real axis. From this construction it is evident that $|F_\mathrm{S}| < |F_\mathrm{R}|$, or in other words the total scattering length from the three atoms is therefore different for the R and S molecules. It follows that by analysing the systematic difference of pairs of reflections it is possible to determine the absolute chirality of molecules. Pairs of reflections for which Friedel's law does not hold are known as Bijvoet pairs.

7.6 The phase problem in crystallography

In this section an outline is given of how the resonant scattering terms may be exploited to solve the phase problem in crystallography. Although the methods described here could be applied to solve the structure of any unit cell, they find their greatest utility in the crystallography of macromolecules, such as proteins, where there may be thousands of atoms in the unit cell.

The goal of macromolecular crystallography is to determine the structure of a large molecules on an atomic length scale. This is accomplished by diffraction techniques. The amplitude of the diffracted X-ray beam is proportional to the molecular structure factor:

$$\begin{aligned} F^{\mathrm{mol}}(\mathbf{Q}) &= \sum_j f_j(\mathbf{Q})\ \mathrm{e}^{-M_j}\ \mathrm{e}^{i\,\mathbf{Q}\cdot\mathbf{r}_j} \\ &= \sum_j (f_j^0 + f_j' + i\,f_j'')\,\mathrm{e}^{-M_j}\ \mathrm{e}^{i\,\mathbf{Q}\cdot\mathbf{r}_j} \\ &= \left|F^{\mathrm{mol}}(\mathbf{Q})\right|\,\mathrm{e}^{i\,\phi} \end{aligned} \tag{7.29}$$

Here as before $\mathbf{Q}$ is the wavevector transfer (also known as the scattering vector), $f_j(\mathbf{Q})$ is the atomic form factor, and e^{-M_j} is the Debye-Waller temperature factor for the jth atom at position $\mathbf{r}_j$ in the molecule (see Chapter 4). In the last equation above we have emphasized that the molecular structure factor is a complex number specified by a modulus, $|F^{\mathrm{mol}}(\mathbf{Q})|$, and a phase, ϕ. With a knowledge of the molecular structure factor one can determine the position vectors of the atoms in the molecule, or in other words the structure of the molecule.

Even in the strongest X-ray beams the diffracting power of a single molecule is insufficient to obtain a measurable diffraction pattern. However, when molecules are assembled into an array in a crystal, the diffracted waves from each molecule will interfere constructively whenever the scattering vector $\mathbf{Q}$ coincides with a reciprocal lattice vector $\mathbf{G}$; in other words the crystal acts as a diffraction amplifier for certain values of the scattering vector. For simplicity, let us assume a crystal structure with only one molecule per unit cell. The integrated intensity of a Bragg spot at $\mathbf{Q}$ around $\mathbf{G}$ is then

$$I(\mathbf{G}) \propto |F_{\mathrm{mol}}(\mathbf{G})|^2 \tag{7.30}$$

so the phase information of the molecular structure factor is lost by measuring the intensity rather than the amplitude of the diffracted ray. In the case of small molecules, a direct solution of the phase problem is possible from statistical relations among the intensities. However, these so-called direct methods do not work for the large number of atoms in a typical macromolecule.

7.6.1 The MAD method

It transpires that one way to solve the phase problem is to utilize the dispersion corrections by measuring the diffraction pattern at several wavelengths around the absorption edge of one of the types of atom in the molecule. The method that uses this technique is known as Multi-wavelength Anomalous Diffraction, or MAD for short [Karle, 1980, Hendrickson, 1985]. One obvious requirement for this technique is that the resonant scatterer should have its K edge or L edge in the X-ray region, and it thus needs to be a moderately heavy atom. For example, it could be a metal ion in a metallo-protein, or an isomorphous replacement atom, such as Selenium for Sulphur, in a derivative of the native molecule, or even a heavy rare earth metal replacing a Calcium atom. In any eventuality it is likely that the number of resonantly scattering atoms in the molecule is much smaller than the total number of atoms in the molecule. Nevertheless, as the scattering power of the resonant atom(s) can be varied in a controlled manner near an absorption edge, it will modulate the total scattering power of the molecule in such a way that the phase can be determined. Here it will be shown how this is possible by performing some relatively simple algebraic manipulations.

The summation in Eq. (7.29) is split into a sum over the resonantly (or anomalously) scattering atoms, A, and a sum over all the other atoms, B, which produce a non-resonant structure factor $F_B(\mathbf{G})$. The sum over the resonant scatterers, assumed to be identical, can be written as

$$\begin{aligned}\sum_{j'} (f_A^0 + f_A' + i\, f_A'')\, \mathrm{e}^{i\,\mathbf{G}\cdot\mathbf{r}_{j'}} &= (f_A^0 + f_A' + i\, f_A'') \sum_{j'} \mathrm{e}^{i\,\mathbf{G}\cdot\mathbf{r}_{j'}} \\ &= f_A^0 \sum_{j'} \mathrm{e}^{i\,\mathbf{G}\cdot\mathbf{r}_{j'}} + (f_A' + i\, f_A'') \sum_{j'} \mathrm{e}^{i\,\mathbf{G}\cdot\mathbf{r}_{j'}} \\ &= F_A(\mathbf{G}) + F_A(\mathbf{G}) \left[\frac{f_A'(\lambda)}{f_A^0} + \frac{f_A''(\lambda)}{f_A^0}\right] \end{aligned} \quad (7.31)$$

The second term on the right hand side is the resonant contribution from the atoms A, while the first term is the non-resonant contribution from the anomalous scatterers A. The latter can be added to $F_B(\mathbf{G})$ to give the total non-resonant structure factor

$$F_T(\mathbf{G}) = F_A(\mathbf{G}) + F_B(\mathbf{G}) \quad (7.32)$$

The molecular structure factor including both the resonant and non-resonant contributions becomes

$$F^{\mathrm{mol}}(\mathbf{G}) = |F_T|\, \mathrm{e}^{i\,\phi_T} + |F_A|\, \mathrm{e}^{i\,\phi_A} \left[\frac{f_A'(\lambda)}{f_A^0} + i\,\frac{f_A''(\lambda)}{f_A^0}\right] \quad (7.33)$$

The squared structure factor, determined by the measured intensity, is therefore

$$\boxed{\begin{aligned}|F_{\mathrm{mol}}(\mathbf{G})|^2 = &\quad |F_T|^2 \\ &+ \; a(\lambda)\; |F_A|^2 \\ &+ \; b(\lambda)\; |F_A|\,|F_T|\, \cos(\phi_T - \phi_A) \\ &+ \; c(\lambda)\; |F_A|\,|F_T|\, \sin(\phi_T - \phi_A)\end{aligned}} \quad (7.34)$$

with

$$a(\lambda) = \frac{(f_A')^2 + (f_A'')^2}{(f_A^0)^2}; \qquad b(\lambda) = \frac{2f_A'}{f_A^0}; \qquad c(\lambda) = \frac{2f_A''}{f_A^0} \quad (7.35)$$

The three coefficients $a(\lambda)$, $b(\lambda)$ and $c(\lambda)$ are determined in the following way. First, $f_A''(\lambda)$ is determined by assuming that it is proportional to the fluorescent yield. This allows $f_A'(\lambda)$ to be computed from $f_A''(\lambda)$ using the Kramers-Kronig relations. With this knowledge the three coefficients $a(\lambda)$, $b(\lambda)$ and $c(\lambda)$ may be evaluated, since the values of $f_A^0(\mathbf{G})$ are tabulated in many places. There are then three unknowns in the problem: $|F_T|$, $|F_A|$ and $(\phi_T - \phi_A)$. A complete data set of reflections is then recorded for at least three wavelengths, allowing the three unknowns to be determined. One can then proceed to solve the structure. From the values of $|F_A|$ one can find the positions of the few A atoms in the unit cell using the direct methods applicable to small molecules. This then allows the phases ϕ_A to be calculated. Since $|F_T|$ and $(\phi_T - \phi_A)$ have already been derived, the entire complex molecular structure factor can then be determined, thus facilitating the solving of the molecular structure.

The MAD method is illustrated in the Argand diagram shown in Fig. 7.10.

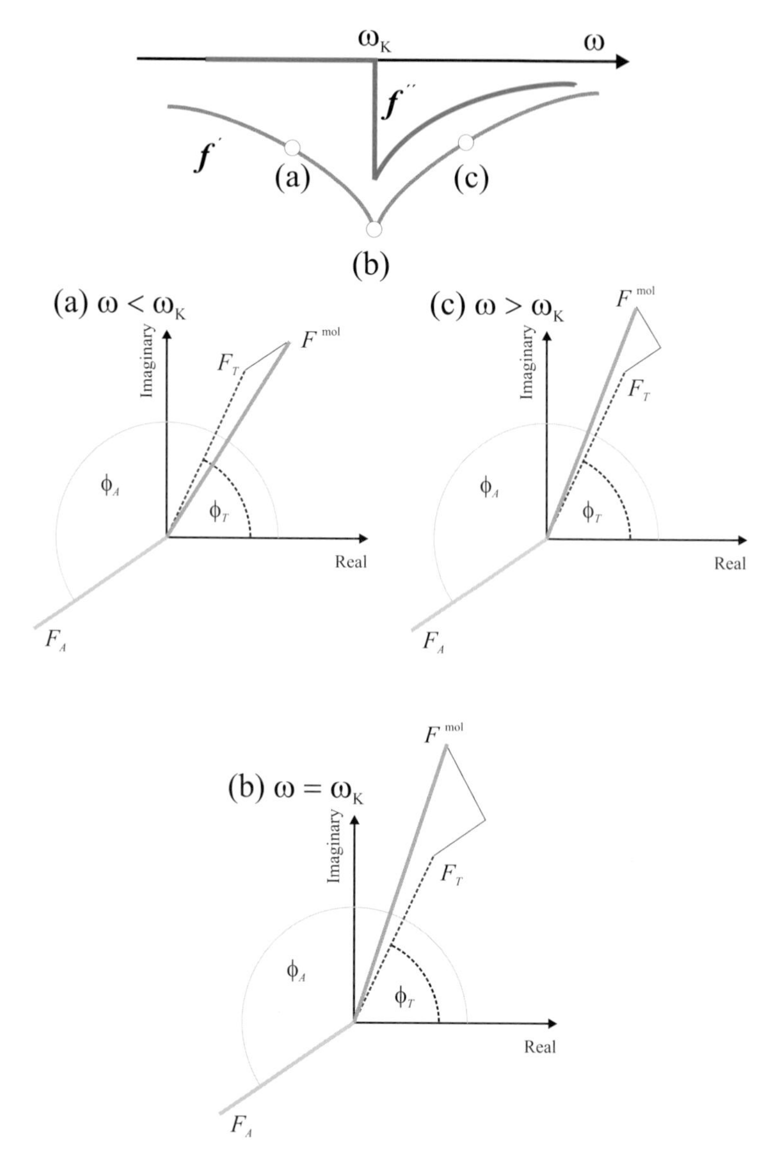

Figure 7.10: A summary of the MAD method where the structure factors are plotted in the complex plane. The atoms in the crystal are divided into two groups: the A atoms which produce the resonant, or anomalous, scattering, and all the other atoms. The non-anomalous contribution from the A atoms has a structure factor F_A (solid line) and phase ϕ_A. When the structure factor of all the other atoms are added to F_A the total non-resonant scattering structure factor F_T is obtained (dashed line) with phase ϕ_T. To obtain the molecular scattering factor the anomalous contribution from the A atoms must be added to F_T. The anomalous contribution has a component parallel to F_A of magnitude $|F_A|(f'/f^0)$, and a component perpendicular to F_A of magnitude $|F_A|(f''/f^0)$. (Note that both f' and f'' are negative.) Here it is illustrated what happens to the resulting molecular structure factor for three choices of the incident energy. (a) First, the energy of the photon is below the edge, so that f'' is zero, and F^{mol} is obtained from F_T by subtracting an amount $|F_A||f'/f^0|$ from F_T. (b) Second, the photon energy equals the edge energy, and the magnitude of both f' and f'' is a maximum, with the result that F^{mol} differs considerably from F_T. (c) Third, the photon energy is above the edge, and f' and f'' have a reduced, but still significant effect on F^{mol}.

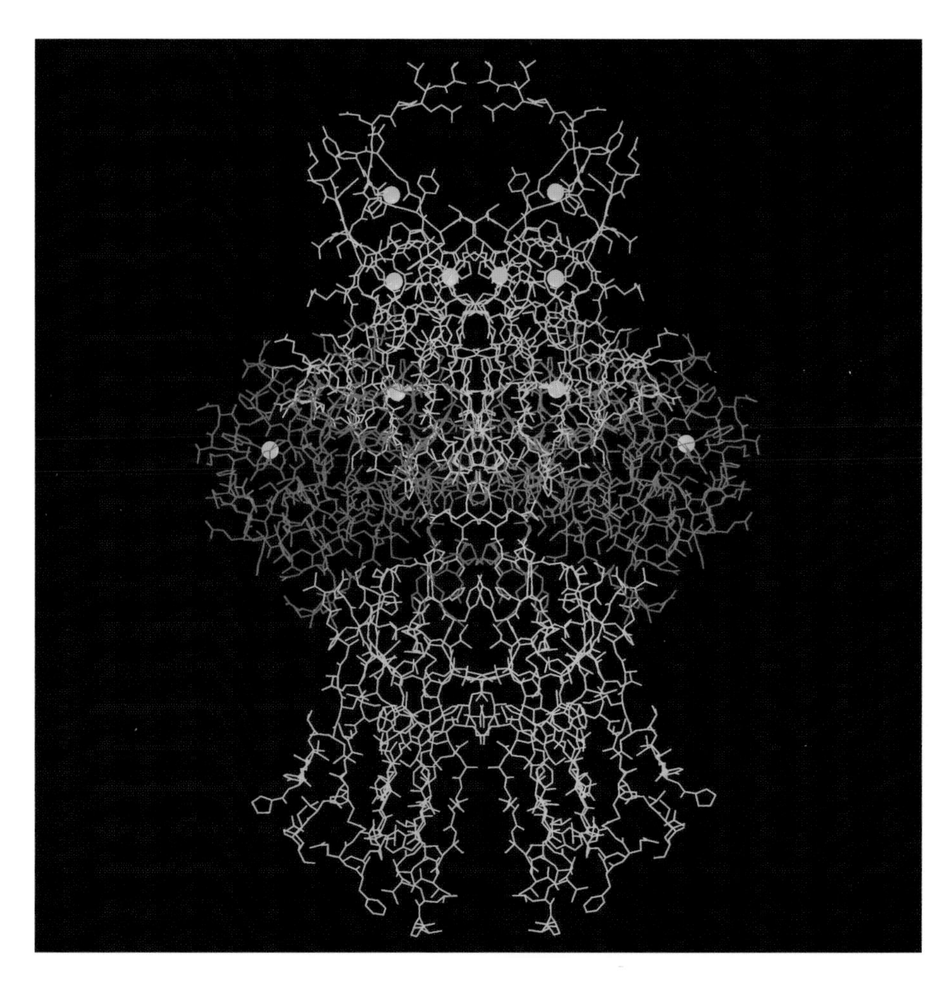

Figure 7.11: Atomic model of a protein complex determined by the MAD method. The structure is a dimeric complex between fibroblast growth factor (FGF1) and the ligand-binding portion of the receptor tyrosine kinase FGFR2. Crystals were grown from the complex between variant proteins in which the methionine residues were all replaced by selenomethionine, and diffraction data were measured at four wavelengths near the Se K absorption edge. These MAD data were first used to find the positions of the selenium atoms (five per asymmetric unit) and then to evaluate phases and produce an image used to locate all 6,162 non-hydrogen atoms of the dimeric complex (D.J. Stauber, A.D. DiGabriele, and W.A. Hendrickson, Proc. Natl. Acad. Sci. USA 97, 49, 2000). The selenium atoms are drawn as yellow spheres, the FGF ligand is represented by inter-atomic covalent bonds drawn in red, and the receptor is shown with bonds drawn in blue.

7.7 Quantum mechanical description

In this section a short introduction is given to the subject of the quantum mechanical description of resonant scattering. The intention here is to explain how resonant scattering fits into the general framework of the interaction of X-rays with matter that has been described in this book, and to give an idea of the additional possibilities that resonant scattering offers for studying ordering phenomena in condensed matter systems.

In a quantum mechanical derivation of the cross-section the quantity of interest is the transition rate probability, W, which in *first-order* perturbation theory is given by

$$W = \frac{2\pi}{\hbar} \left|\langle f|\,\mathcal{H}_I\,|i\rangle\right|^2 \rho(\mathcal{E}_f) \tag{7.36}$$

where $|i\rangle$ and $|f\rangle$ are the initial and final states of the combined system of X-ray photon plus target electron (Eq. (A.3)). The Hamiltonian, $\mathcal{H}_I$, describes the interaction between the photon and the electron. Neglecting the spin of the electron, the interaction Hamiltonian is given by

$$\mathcal{H}_I = \frac{e\mathbf{A}\cdot\mathbf{p}}{m} + \frac{e^2\mathrm{A}^2}{2m} \tag{7.37}$$

as described in Appendix C. The vector potential $\mathbf{A}$ of the photon field is *linear* in photon creation and annihilation operators (Eq. (C.6)). The first contribution to $\mathcal{H}_I$ is linear in $\mathbf{A}$, and it follows that it can either create or destroy a photon, but not both. It was shown in Chapter 6 that this term gives rise to photoelectric absorption. The second contribution to $\mathcal{H}_I$ is quadratic in $\mathbf{A}$, and as such can first destroy and then create a photon, while leaving the electron in the same state, $|a\rangle$ say. (Here it is important to note that $|a\rangle$ is the initial state of the electron, while $|i\rangle$ is the ground state of the combined system, photon plus electron.) This term therefore describes elastic Thomson scattering. These first-order processes are represented schematically in Fig. 7.12(a) and (b).

To obtain resonant scattering terms it is necessary to take the calculation to higher-order. In *second-order* perturbation theory the transition probability is given by

$$W = \frac{2\pi}{\hbar} \left| \langle f|\,\mathcal{H}_I\,|i\rangle + \sum_{n=1}^{\infty} \frac{\langle f|\,\mathcal{H}_I\,|n\rangle\langle n|\,\mathcal{H}_I\,|i\rangle}{E_i - E_n} \right|^2 \rho(\mathcal{E}_f) \tag{7.38}$$

where the sum is over all possible states with energy E_n. It can now be seen that the $\mathbf{A}\cdot\mathbf{p}$ term, which is linear in creation and annihilation operators, can produce scattering via an intermediate state. Reading the matrix element that appears in the numerator of the second term from right to left the scattering process can be described in the following way: the incident photon is first destroyed, and the electron makes a transition from the ground state, $|a\rangle$, to an intermediate state, $|n\rangle$. In an elastic scattering event the electron then makes a transition from $|n\rangle$ to $|a\rangle$ with the creation of the scattered photon. The resonant behaviour arises when the denominator tends to zero. This occurs when the total incident energy, $E_i = \hbar\omega + E_a$, is equal to the energy of the intermediate state E_n, or in other words when the energy of the incident photon is equal to the difference

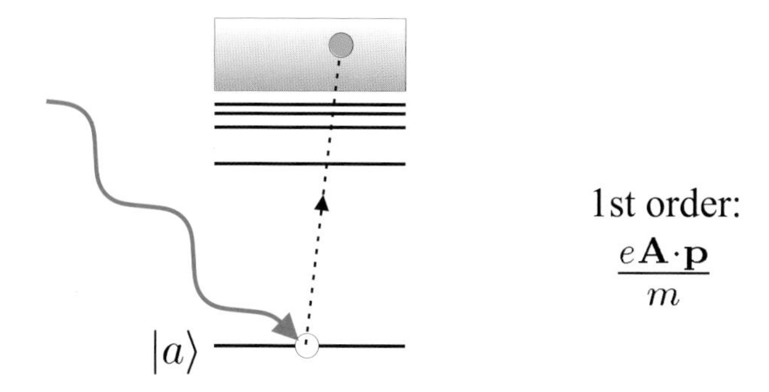

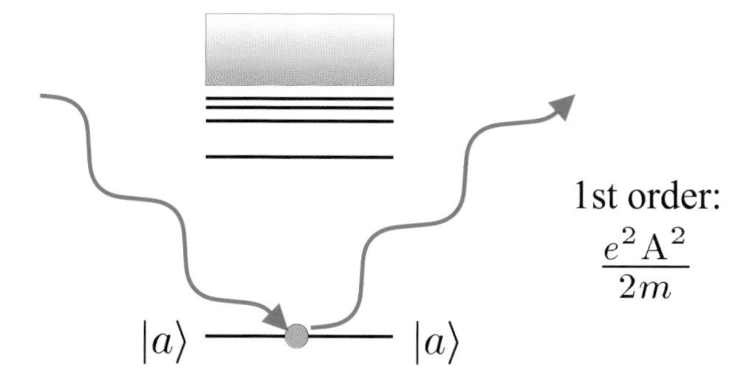

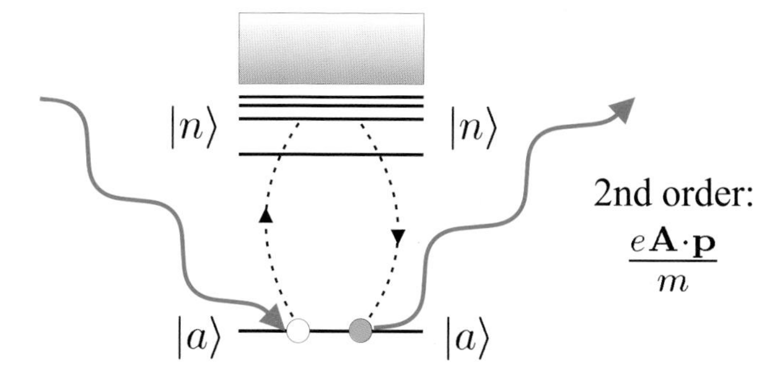

Figure 7.12: Summary of the quantum mechanical description of the interaction of a photon with an atomic electron. Photoelectric absorption (a) and Thomson scattering (b) can be explained by applying first-order perturbation theory to the terms in the interaction Hamiltonian which depend on $\mathbf{A}\cdot\mathbf{p}$ and A^2 respectively. Resonant scattering (c) is a second-order process and occurs via an intermediate electronic state. This picture should not be taken too literally, however, as resonant scattering is a virtual process, and does not occur in the two discrete steps suggested here.

in energy between the intermediate and ground states, $\hbar\omega = E_n - E_a$. The resonant scattering process is shown schematically in Fig. 7.12(c).

Resonant scattering may be thought of as a probe of the intermediate atomic states. Transitions to the intermediate states are controlled by two considerations. The Pauli exclusion principle requires that only unoccupied intermediate states can be accessed, while the usual quantum mechanical selection rules imply that electric dipole transitions dominate (as described on page 227). Interesting effects occur when the nature of the intermediate state is altered when atoms combine to form molecules, solids, etc. For example, the symmetry of the intermediate state may be lowered if it is involved in chemical bonding. Under this condition the dispersion corrections become dependent on the polarization geometry used in the experiment, and forbidden Bragg reflections may become observable which provide information on the phases of atoms in the unit cell [Templeton and Templeton, 1982]. Alternatively, the intermediate state may be split by magnetic interactions. Resonant scattering then becomes a probe of the magnetic order in the solid [Namikawa et al., 1985, Blume, 1985, Gibbs et al., 1988]. These and other aspects of resonant scattering, including inelastic processes, are presently at the forefront of experimental and theoretical X-ray science, and are still in a phase of rapid development [see, for example, Lovesey and Collins, 1996].

Further reading

Resonant anomalous X-ray scattering: Theory and Applications, Eds. G. Materlik, C.J. Sparks, and K. Fischer (Elsevier, 1994)

Determination of macromolecular structures from anomalous diffraction of synchrotron radiation, W.A. Hendrickson, Science **254**, 51 (1991)

A link between macroscopic phenomena and molecular chirality: Crystals as probes for the direct assignment of absolute configuration of chiral molecules, L. Addadi, Z. Berkovitch-Yellin, I. Weissbuch, M. Lahav, and L. Leiserowitz, Topics in Stereochemistry **16**, 1 (1986)

A. Scattering and absorption cross-sections

Basic definitions

In this section the basic definitions of the cross-section are recalled for processes that involve either the scattering or absorption of an X-ray photon. The cross-section is an important quantity, as it is the meeting point of experiment and theory. Although its definition is straightforward, confusion sometimes arises as there are several, but essentially equivalent, definitions. As illustrated in Fig. A.1 the definition depends on the situation considered, and in particular on whether or not the cross-sectional area of the beam is larger or smaller than that of the sample.

We start by considering the *scattering* event shown in Fig. A.1(a) in which an X-ray beam of intensity I_0 photons per second is incident on a sample, and where the sample is large enough that it intercepts the entire beam. Our objective is to calculate the number of X-ray photons, I_{sc}, scattered per second into a detector that subtends a solid angle $\Delta\Omega$. If there are N particles in the sample per unit area seen along the beam direction, then I_{sc} will be proportional to N and to I_0. It will of course also be proportional to $\Delta\Omega$. Most importantly it will depend on how efficiently the particles in the sample scatter the radiation. This is given by the *differential cross-section*, $(d\sigma/d\Omega)$, so that we may write

$$I_{\text{sc}} = I_0\, N\, \Delta\Omega \left(\frac{d\sigma}{d\Omega}\right)$$

Thus the differential cross-section per scattering particle is defined by

$$\boxed{\left(\frac{d\sigma}{d\Omega}\right) = \frac{\text{No. of X-ray photons scattered per second into } \Delta\Omega}{I_0\, N \Delta\Omega}} \tag{A.1}$$

Here no restriction has yet been placed on whether or not the scattering event is elastic or inelastic.

The corresponding absorption experiment is simpler to analyze, as the detector is placed directly in the incident beam, and the change in intensity recorded when a sample is introduced into the beam. The number of absorption events, $W_{4\pi}$, per second is proportional to I_0 and N as before. The subscript is used to remind us that the photoelectron liberated from the atom in the absorption process may be emitted into any direction in 4π steradians. The *absorption cross-section*, σ_a, is defined by

$$W_{4\pi} = I_0 N\, \sigma_a$$

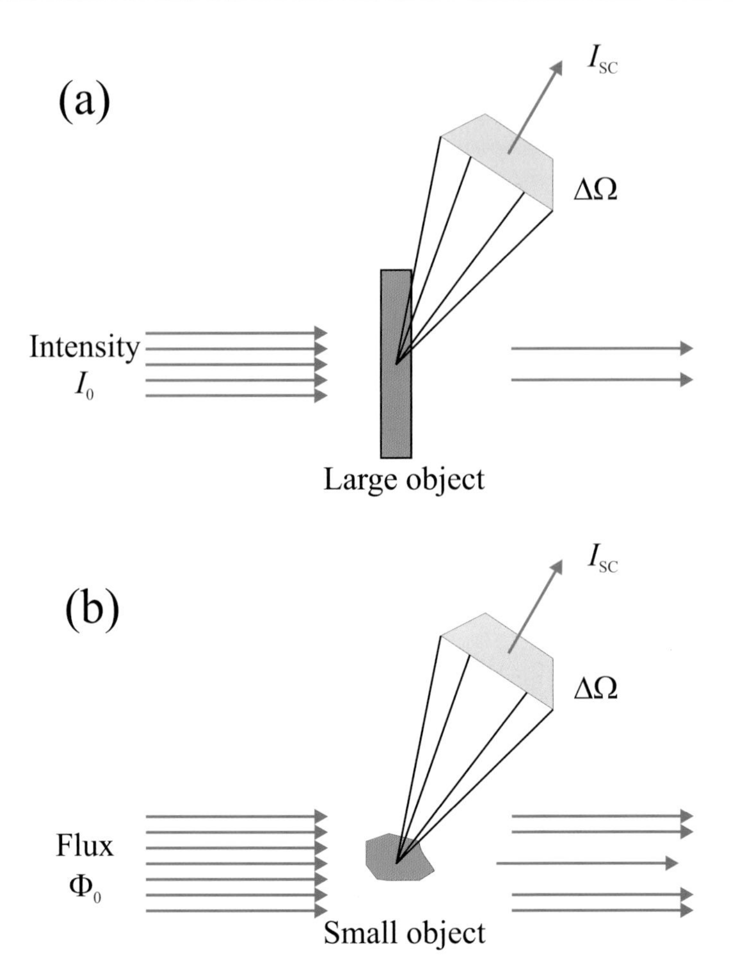

Figure A.1: (a) A beam is incident on a sample with a larger cross-sectional area. In this case the intensity I_{sc} scattered into a solid angle $\Delta\Omega$ is proportional to the incident *intensity* I_0 of the beam, i.e. the number of photons per second. (b) A beam is incident on a sample with a smaller cross-sectional area. Now the scattered intensity I_{sc} is proportional to the incident *flux* Φ_0 of the beam, i.e. the number of photons per second per unit area.

so that

$$\boxed{\sigma_a = \frac{W_{4\pi}}{I_0 N}}$$

Alternatively a different situation could have been imagined in which the incident beam is larger than the sample, as shown in Fig. A.1(b). It should be clear that in this case it is necessary to consider the *flux* of the incident beam, i.e. the number of photons per second per unit area, and not its intensity. The scattered intensity is now given by

$$I_{\mathrm{sc}} = \Phi_0 \, \Delta\Omega \left(\frac{d\sigma}{d\Omega} \right) \tag{A.2}$$

where Φ_0 is the flux of the incident beam, and the differential cross-section refers to the whole sample. The absorption cross-section for this geometry is

$$\sigma_a = \frac{W_{4\pi}}{\Phi_0}$$

These are the definitions of the cross-sections which are operational for analysis of an experiment. Having defined what exactly is meant by the cross-section the next question is how it may be calculated. The classical description of the scattering of an electromagnetic wave by a single electron is used in Chapter 1 and Appendix B to derive the Thomson scattering cross-section. This description is usually sufficient in many fields such as reflectivity, crystallography, etc. In contrast there is no classical model of the absorption process, and instead a quantum mechanical approach must be taken.

Quantum mechanical treatment

In a quantum mechanical treatment the scattering process is described by time-dependent perturbation theory. The interaction between the incident radiation and sample is specified by an Hamiltonian $\mathcal{H}_I$, which produces transitions between the initial $|i\rangle$ and final $|f\rangle$ states. Here $|i\rangle$ and $|f\rangle$ refer to the *combined* states of the X-ray field and sample. The number of transitions, W, per second between $|i\rangle$ and $|f\rangle$ is given in first-order perturbation theory by Fermi's Golden Rule as

$$W = \frac{2\pi}{\hbar} \left|M_{if}\right|^2 \rho(\mathcal{E}_f) \tag{A.3}$$

where the matrix element $M_{if}=\langle f|\mathcal{H}_I|i\rangle$, and $\rho(\mathcal{E}_f)$ is the density of states, defined such that $\rho(\mathcal{E}_f)d\mathcal{E}_f$ is the number of final states with energy in the interval $d\mathcal{E}_f$ centered around $\mathcal{E}_f$. (The correct dimensions of 1/[time] for W can be confirmed by inspection.)

Scattering

To evaluate the differential cross-section for scattering the number of transitions per second into the solid angle $\Delta\Omega$ needs to be found, and as we are mostly interested in elastic scattering, the restriction $\mathcal{E}_f = \mathcal{E}_i$ needs to be placed.

Here we follow the standard method to calculate the density of states $\rho(\mathcal{E}_f)$, where it is assumed that the total system (X-rays+sample) occupies a box of volume V. Periodic boundary conditions are applied to the X-ray wavefunctions, resulting in a uniform density of states in wavevector of $V/(2\pi)^3$. By definition $\rho(\mathcal{E}_f)d\mathcal{E}_f$ is the number of states with energy between $\mathcal{E}_f$ and $\mathcal{E}_f + d\mathcal{E}_f$, which is equal to the number of states with wavevectors between k_f and $\mathrm{k}_f + d\mathrm{k}_f$. We can therefore write

$$\rho(\mathcal{E}_f)d\mathcal{E}_f = \left(\frac{V}{8\pi^3}\right) d\mathbf{k}_f$$

or

$$\rho(\mathcal{E}_f) = \left(\frac{V}{8\pi^3}\right) \frac{d\mathbf{k}_f}{d\mathcal{E}_f} \tag{A.4}$$

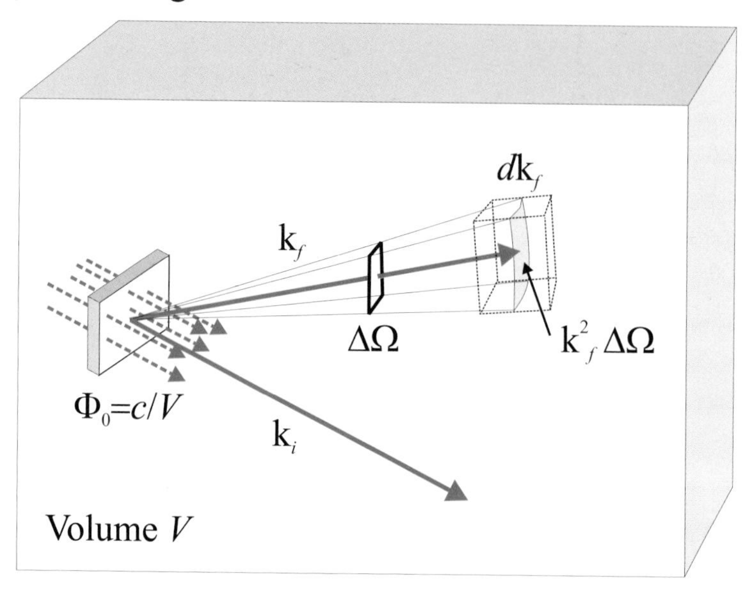

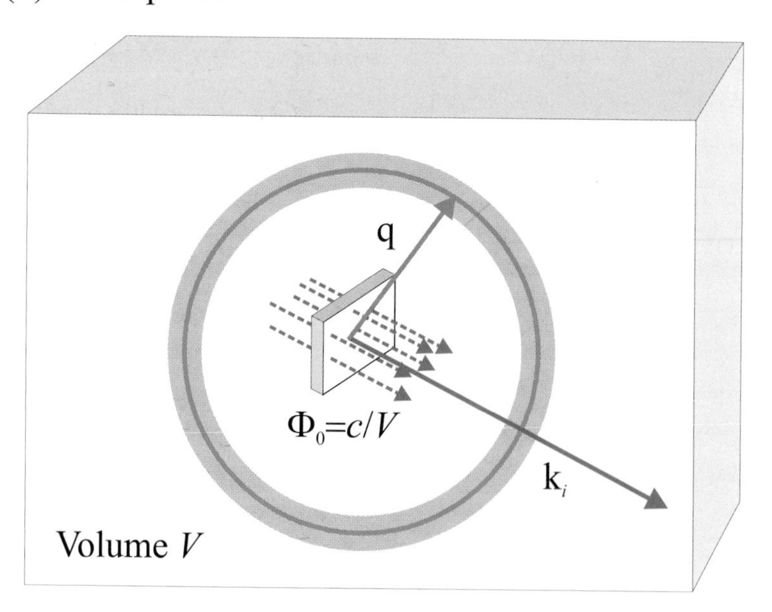

Figure A.2: Illustration of the quantum mechanical derivation of (a) the differential scattering and (b) the absorption cross-sections. For the differential scattering cross-section, one must integrate over the values of $\mathbf{k}_f$ accessible within the solid angle element $\Delta\Omega$. For the photo-electric absorption cross-section all directions of the expelled electron must be integrated over. In both cases, energy conservation is ensured by introducing a delta function in the integrand.

The differential cross-section can then be calculated from Eq. (A.2) and (A.3) as

$$\left(\frac{d\sigma}{d\Omega}\right) = \frac{W_{\Delta\Omega}}{\Phi_0 \Delta\Omega}$$

where $W_{\Delta\Omega} \equiv I_{\mathrm{sc}}$ is the number of transitions per second into $\Delta\Omega$. The restriction on elastic scattering events is introduced by including a delta function $\delta(\mathcal{E}_f - \mathcal{E}_i)$ in Eq. (A.3), and then integrating over all $\mathcal{E}_f$, which yields

$$W_{\Delta\Omega} = \frac{2\pi}{\hbar} \int \left|M_{if}\right|^2 \rho(\mathcal{E}_f)\, \delta(\mathcal{E}_f - \mathcal{E}_i)\, d\mathcal{E}_f \tag{A.5}$$

According to Eq. (A.4)

$$\rho(\mathcal{E}_f) = \left(\frac{V}{8\pi^3}\right) \mathrm{k}_f^2 \left(\frac{d\mathrm{k}_f}{d\mathcal{E}_f}\right) \Delta\Omega$$

where the differential volume element $d\mathbf{k}_f$ has been replaced by $\mathrm{k}_f^2 d\mathrm{k}_f \Delta\Omega$ as indicated in Fig. A.2. The expression for $W_{\Delta\Omega}$ may be simplified considerably. Since $\mathcal{E}_f = \hbar \mathrm{k}_f c$, it follows that

$$\mathrm{k}_f^2 \left(\frac{d\mathrm{k}_f}{d\mathcal{E}_f}\right) = \frac{1}{\hbar^3 c^3}\, \mathcal{E}_f^2 \tag{A.6}$$

With the flux given by $\Phi_0 = c/V$, we obtain

$$\boxed{\left(\frac{d\sigma}{d\Omega}\right) = \left(\frac{V}{2\pi}\right)^2 \frac{1}{\hbar^4 c^4} \int |M_{if}|^2 \mathcal{E}_f^2\, \delta(\mathcal{E}_f - \mathcal{E}_i)\, d\mathcal{E}_f} \tag{A.7}$$

Absorption

Calculation of the absorption cross-section proceeds along similar lines, with two main differences. The first is that the condition on the δ function that appears in Eq. (A.5) is altered, since the process is no longer elastic. In an absorption process the incident photon expels an electron from an atom with binding energy $\mathcal{E}_{\mathrm{b}}$. The difference between the energy of the incident photon, $\mathcal{E} = \hbar\omega$, and the binding energy of the electron, $\mathcal{E}_{\mathrm{b}}$, is the kinetic energy of the photoelectron, $\mathcal{E}_{\mathrm{pe}} = \hbar^2 \mathrm{q}^2/2m$. The second difference is that there is no restriction on the direction of $\mathbf{q}$, the wavevector of the photoelectron, so that instead of integrating over the states in $\Delta\Omega$, as was the case for the scattering cross-section, it is now necessary to integrate over the entire solid angle of 4π. Thus the absorption cross-section is

$$\sigma_a = \frac{W_{4\pi}}{\Phi_0}$$

with

$$W_{4\pi} = \int \frac{2\pi}{\hbar} \left|M_{if}\right|^2 \rho(\mathcal{E}_{\mathrm{pe}})\, \delta(\mathcal{E}_{\mathrm{pe}} - (\mathcal{E} - \mathcal{E}_{\mathrm{b}}))d\mathcal{E}_{\mathrm{pe}}$$

The density of states for the photoelectron is evaluated using the same box normalization introduced for the density of X-ray states in the case of scattering discussed above. The result for the photoelecton is

$$\rho(\mathcal{E}_{\mathrm{pe}}) = 2\left(\frac{V}{8\pi^3}\right)\left(\frac{d\mathbf{q}}{d\mathcal{E}_{\mathrm{pe}}}\right)$$

where the factor of 2 allows for the two possible spin states of the electron. The absorption cross-section is evaluated by replacing the volume element $d\mathbf{q}$ in the above by $\mathrm{q}^2 \sin\theta d\mathrm{q} d\theta d\varphi$, with the integral taken over the entire solid angle of 4π, as indicated in Fig. A.2(b). It follows that the absorption cross-section is

$$\sigma_a = \frac{2\pi}{\hbar c}\frac{V^2}{4\pi^3}\int \left|M_{if}\right|^2 \delta(\mathcal{E}_{\mathrm{pe}} - (\mathcal{E} - \mathcal{E}_{\mathrm{b}}))\, \mathrm{q}^2 \sin\theta d\mathrm{q} d\theta d\varphi \tag{A.8}$$

where once again the incident flux is given by $\Phi_0 = c/V$.

It remains of course to calculate the matrix elements M_{if} of the interaction Hamiltonian: this is described in Appendix C for the Thomson scattering cross-section, and in Chapter 6 for the absorption cross-section.

Further reading

Quantum Mechanics, A.I.M. Rae (Adam Hilger, 1986)

Quantum Mechanics, F. Mandl (John Wiley, 1992)

B. Classical electric dipole radiation

In Chapter 1 a classical model was used to describe the scattering of X-rays by electrons. The equation relating the strength of the radiated to incident X-ray electric fields (Eq. 1.4) was stated without proof. Here the derivation of this equation is outlined more fully.

We imagine that an electromagnetic plane wave with an electric field $\mathbf{E}_{\text{in}}$ is incident on a charge distribution, which oscillates in response to this driving field, and hence acts as a source of radiation. The problem then is to evaluate the radiated electric field at some observation point X, as shown in Fig. B.1(a). This is simplified considerably if it is assumed that r is much greater than the spatial extent of the charge distribution, and also if r is much greater than the wavelength of the radiation λ. The first of these is the *dipole approximation*, while the second assures that we can interpret the electromagnetic effects at X as radiation. Here it is further assumed that the electrons forming the charge distribution are free.

The electric and magnetic fields at X can be derived from the scalar potential Φ and the vector potential $\mathbf{A}$:

$$\mathbf{E} = -\nabla\Phi - \frac{\partial \mathbf{A}}{\partial t}$$

and

$$\mathbf{B} = \nabla \times \mathbf{A} \tag{B.1}$$

The task of evaluating the fields at X is further simplified if it is recalled that electromagnetic waves are transverse, with the fields being perpendicular to the propagation direction $\mathbf{n}$, as shown in Fig. B.1(b). We then have that $\mathbf{n}$ is colinear to $\mathbf{E} \times \mathbf{B}$, and by solving the wave equation it can be shown that $|\mathbf{E}| = c|\mathbf{B}|$. It is therefore sufficient to derive $\mathbf{B}$ from $\mathbf{A}$ (Eq. (B.1)), and then $\mathbf{E}$ follows immediately.

The vector potential is given by

$$\mathbf{A}(\mathbf{r}, t) = \frac{1}{4\pi\epsilon_0 c^2} \int_V \frac{\mathbf{J}(\mathbf{r}', t - |\mathbf{r} - \mathbf{r}'|/c)}{|\mathbf{r} - \mathbf{r}'|}\, d\mathbf{r}'$$

where $\mathbf{J}(\mathbf{r}', t)$ is the current density of the source. As the fields propagate at a finite velocity, the fields experienced at the observation point X at time t depend on the position of the electron at an earlier time $t - |\mathbf{r} - \mathbf{r}'|/c$. For this reason $\mathbf{A}$ given by the above is known as the *retarded* vector potential.

The *dipole approximation* allows us to ignore $\mathbf{r}'$ in comparison with $\mathbf{r}$, so that

$$\mathbf{A}(\mathbf{r}, t) \approx \frac{1}{4\pi\epsilon_0 c^2 \mathrm{r}} \int_V \mathbf{J}(\mathbf{r}', t - \mathrm{r}/c)\, d\mathbf{r}'$$

To proceed it is noted that the current density is equal to the product of the charge density ρ and the velocity $\mathbf{v}$,

$$\mathbf{J} = \rho\mathbf{v}.$$

For a distribution of discrete charges q_i the integral is replaced by a sum so that

$$\int_V \mathbf{J}\, d\mathbf{r}' = \int_V \rho\mathbf{v}\, d\mathbf{r}' = \sum_i q_i \mathbf{v}_i = \frac{d}{dt'}\sum_i q_i \mathbf{r}'_i$$

The last term is recognizable as the time derivative of the electric dipole moment which is written as $\dot{\mathbf{p}}$.

We now let the incident beam be linearly polarized along the z axis, so that the dipole moment and hence the vector potential will have a component along this direction only (Fig. B.1(b)). Thus for a single dipole we have

$$\mathrm{A}_z = \left(\frac{1}{4\pi\epsilon_0 c^2 \mathrm{r}}\right) \dot{\mathrm{p}}(t')$$

and $\mathrm{A}_x = \mathrm{A}_y = 0$. From Eq. (B.1) the components of the $\mathbf{B}$ field follow as

$$\mathrm{B}_x = \frac{\partial \mathrm{A}_z}{\partial y}; \qquad \mathrm{B}_y = -\frac{\partial \mathrm{A}_z}{\partial x}; \qquad \mathrm{B}_z = 0 \tag{B.2}$$

For the x component of the $\mathbf{B}$ field we evaluate the partial derivative of A_z with respect to y as

$$\begin{aligned}\frac{\partial A_z}{\partial y} &= \left(\frac{1}{4\pi\epsilon_0 c^2}\right)\frac{\partial}{\partial y}\left(\frac{\dot{\mathrm{p}}(t')}{\mathrm{r}}\right) \\ &= \left(\frac{1}{4\pi\epsilon_0 c^2}\right)\left[\frac{1}{\mathrm{r}}\frac{\partial \dot{\mathrm{p}}(t')}{\partial y} - \frac{\dot{\mathrm{p}}(t')}{\mathrm{r}^2}\frac{\partial \mathrm{r}}{\partial y}\right]\end{aligned}$$

Since we are interested in the far-field limit of $\mathbf{B}$, we can neglect the second term in the above, while the partial derivative of the first term with respect to y can be evaluated by noting that

$$\begin{aligned}\frac{\partial}{\partial y} &= \frac{\partial}{\partial t'}\frac{\partial t'}{\partial y} \\ &= \frac{\partial}{\partial t'}\frac{\partial}{\partial y}\left(t - \frac{1}{c}\sqrt{x^2+y^2+z^2}\right) \\ &= -\frac{1}{c}\left(\frac{y}{\mathrm{r}}\right)\frac{\partial}{\partial t'}\end{aligned}$$

Hence the x component of the $\mathbf{B}$ field in the far field limit is

$$\mathrm{B}_x \approx -\left(\frac{1}{4\pi\epsilon_0 c^2}\right)\frac{1}{c\mathrm{r}}\ddot{\mathrm{p}}(t')\left(\frac{y}{\mathrm{r}}\right)$$

and the y component follows by interchanging x and y, and allowing for the minus sign in Eq. (B.2). Recalling that $\ddot{\mathrm{p}}(t')$ is implicitly along the z axis we can generalize to any direction of $\ddot{\mathbf{p}}(t')$ by writing

$$\mathbf{B} \approx \left(\frac{1}{4\pi\epsilon_0 c^2}\right)\frac{1}{c\mathrm{r}}\ddot{\mathbf{p}}(t') \times \hat{\mathbf{r}}$$

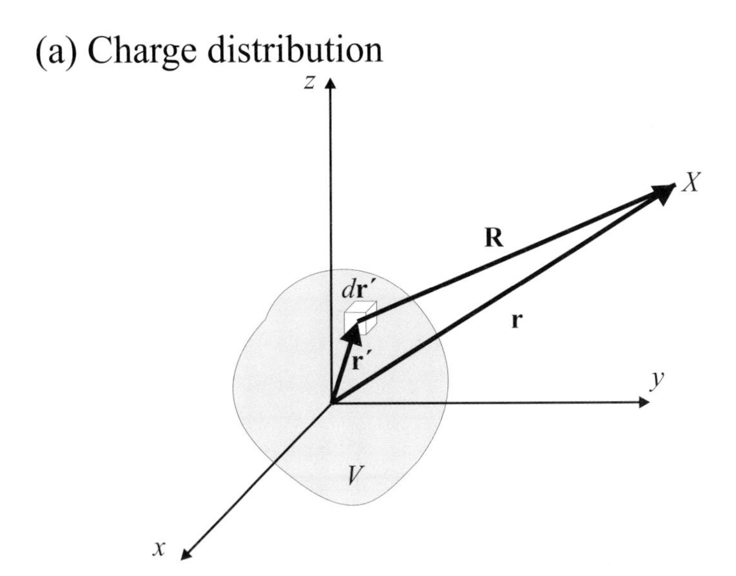

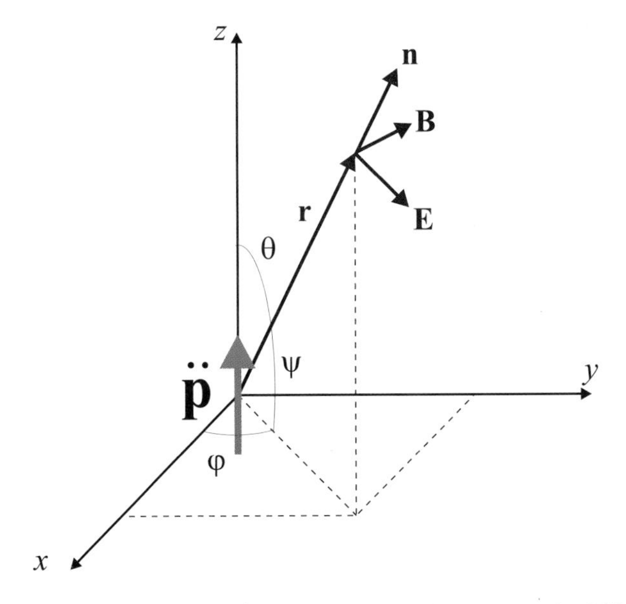

Figure B.1: (a) The coordinate system used to calculate the electromagnetic field radiated from a charge distribution when placed in an incident plane wave. (b) An electromagnetic plane wave polarized with its electric field along the z axis forces an electric dipole at the origin to oscillate. In the far-field limit the field radiated from the dipole is approximately a plane wave with the **E** and **B** fields perpendicular to the propagation direction as indicated in the figure.

where $\hat{\mathbf{r}}$ is the unit vector $(x/\mathrm{r}, y/\mathrm{r}, z/\mathrm{r})$. The numerical value of the vector cross product is $\ddot{\mathrm{p}}\cos\psi$ where ψ is defined in Fig. B.1(b). The direction of the electric field is perpendicular to both $\hat{\mathbf{r}}$ and $\mathbf{B}$ in such a way that the cross product of $\mathbf{E}\times\mathbf{B}$ is along $\hat{\mathbf{r}}$. In particular we note that for $\psi = 0$ the $\mathbf{E}$ field has the *opposite* direction of $\ddot{\mathrm{p}}$. Its magnitude is given by $|\mathbf{E}| = c|\mathbf{B}|$ so that

$$\mathrm{E}(t) = -\left(\frac{1}{4\pi\epsilon_0 c^2}\right)\frac{1}{\mathrm{r}}\,\ddot{\mathrm{p}}(t')\cos\psi \tag{B.3}$$

The next step is to calculate the magnitude of $\ddot{\mathrm{p}}$ in terms of the incident driving field $\mathrm{E_{in}} = \mathrm{E_0 e}^{-i\,\omega(t-\mathrm{r}/c)}$. By definition we have

$$\ddot{\mathrm{p}} = q\ddot{z} = q\,\frac{\text{Force}}{\text{mass}} = q\,\frac{q\,\mathrm{E_{in}}}{m} = \frac{q^2}{m}\mathrm{E_0 e}^{-i\,\omega(t-\mathrm{r}/c)}$$

which when inserted into Eq. (B.3) with $q = -e$, and remembering that $\omega/c = \mathrm{k}$, leads to

$$\mathrm{E}(t) = -\left(\frac{e^2}{4\pi\epsilon_0 mc^2}\right)\left(\frac{\mathrm{e}^{i\,\mathrm{kr}}}{\mathrm{r}}\right)\mathrm{E_{in}}(t)\cos\psi$$

The prefactor is the Thomson scattering length r_0, so that the ratio of the radiated to incident electric fields is given by

$$\frac{\mathrm{E}(t)}{\mathrm{E_{in}}(t)} = -r_0\left(\frac{\mathrm{e}^{i\,\mathrm{kr}}}{\mathrm{r}}\right)\cos\psi \tag{B.4}$$

The factor $\cos\psi$ in Eq. (B.3) is the origin of the polarization factor for X-ray scattering, as $\ddot{\mathrm{p}}(t')\cos\psi$ may be thought of as the apparent acceleration as seen by the observer. This is clear if we return to the case when $\mathbf{E}_{\mathrm{in}}$ is along the z axis. If ψ=0 the maximum acceleration is observed, whereas for $\psi = 90°$ the apparent acceleration is zero. The polarization factor is discussed further in Chapter 1.

We note that the minus sign means that there is a phase shift of π between the incident and scattered fields, and it follows that the index of refraction is necessarily less than unity (see Chapter 3). This result holds in the X-ray region, where most if not all of the atomic electrons may be treated as though they are essentially free. In the visible part of the spectrum, however, we have to allow for the fact that the electrons are bound. This produces resonances in the frequency dependence of the index of refraction, and on the low frequency side of the resonances, corresponding to the visible part of the spectrum, the index of refraction is greater than one.

One way to characterize the efficiency with which an electron scatters the incident radiation is to calculate the total scattering cross-section. The power per unit area is proportional to $|\mathrm{E}|^2$, and by definition the differential cross-section is the power scattered into the solid angle $d\Omega$, normalized by the incident flux (see Appendix A). From Eq. (B.4) it follows that the differential cross-section is

$$\boxed{\left(\frac{d\sigma}{d\Omega}\right) = r_0^2\cos^2\psi}$$

where the factor of r in the denominator of (B.4) cancels on taking the square with a factor of r^2 that arises in converting from surface area to solid angle. The total cross-section for Thomson scattering is found by integration over the polar angles φ and θ:

$$\begin{aligned}\sigma_{\mathrm{T}} &= r_0^2 \int \cos^2\psi \sin\theta d\theta d\varphi = r_0^2 \int \sin^2\theta \sin\theta d\theta d\varphi = \left(\frac{8\pi}{3}\right) r_0^2 \\ &= 0.665 \times 10^{-24}\mathrm{cm}^2 \\ &= 0.665 \text{ barn} \end{aligned} \tag{B.5}$$

The classical cross-section for the scattering of an electromagnetic wave by a free electron is therefore a constant, independent of energy.

Further reading

Foundations of Electromagnetic Theory, J.R. Reitz, F.J. Milford, and R.E. Christy (Addison-Wesley Publishing Company, 1992)

Classical Electromagnetic Radiation, M.A. Heald and J.B. Marion (Saunders College Publishing, 1995)

C. Quantization of the electromagnetic field

The cross-section for either scattering or absorption is evaluated from time-dependent perturbation theory, as outlined in Appendix A. In any perturbation problem it is of course first necessary to specify completely the non-interacting Hamiltonian, $\mathcal{H}_0$, of the system, before the effect of the perturbing Hamiltonian, $\mathcal{H}_I$, may be calculated. For the scattering or absorption of an X-ray this amounts to establishing a quantum mechanical description of both the electromagnetic field and the sample. The former may well be unfamiliar to many readers and here we explain briefly how this is achieved.

The starting point in quantizing the electromagnetic field is the classical expression for its energy in terms of the electric and magnetic fields, both of which may be derived from the vector potential **A** (Appendix B). When seeking a quantum mechanical description of the electromagnetic field it would therefore seem natural to focus on **A**. Indeed quantizing the electromagnetic field amounts to quantizing the vector potential. It also transpires that the Hamiltonian, $\mathcal{H}_I$, that describes the interaction of the X-ray and the sample, is a simple function of **A**. As a consequence the matrix elements of $\mathcal{H}_I$ that enter into the perturbation theory may be calculated readily, and in the last section we work through the example of the Thomson cross-section.

Classical energy density of the radiation field

The total energy of the electromagnetic field in *free space* is

$$\begin{aligned}\mathcal{E}_{\text{rad}} &= \frac{1}{2}\int_V \left[\epsilon_0\langle \mathrm{E}^2\rangle + \mu_0\langle \mathrm{H}^2\rangle\right] dV \\ &= \int_V \epsilon_0\langle \mathrm{E}^2\rangle\, dV\end{aligned}$$

Here it is assumed that the field is confined to some volume V, and $\epsilon_0\langle \mathrm{E}^2\rangle$ $=\mu_0\langle \mathrm{H}^2\rangle$, where the brackets $\langle\cdots\rangle$ indicate a temporal average. The **E** field is related to vector potential **A** through

$$\mathbf{E} = -\frac{\partial \mathbf{A}}{\partial t}$$

The most general approach for dealing with **A** would be to write it as a Fourier sum of plane waves. For clarity we shall consider just one term in this series and write the vector potential as

$$\mathbf{A}(\mathbf{r},t) = \hat{\boldsymbol{\varepsilon}}\mathrm{A}_0\left[a_{\mathbf{k}}\,\mathrm{e}^{i(\mathbf{k}\cdot\mathbf{r}-\omega t)} + a_{\mathbf{k}}^*\,\mathrm{e}^{-i(\mathbf{k}\cdot\mathbf{r}-\omega t)}\right] \tag{C.1}$$

The direction of $\mathbf{A}$ is specified by the polarization unit vector $\hat{\varepsilon}$, and in addition to the amplitude coefficients $a_{\mathbf{k}}$ we have introduced a normalization factor A_0. The electric field is

$$\mathbf{E} = \hat{\varepsilon}\mathrm{A}_0 \left[(i\omega)a_{\mathbf{k}}\, \mathrm{e}^{i(\mathbf{k}\cdot\mathbf{r}-\omega t)} - (i\omega)a_{\mathbf{k}}^*\, \mathrm{e}^{-i(\mathbf{k}\cdot\mathbf{r}-\omega t)} \right]$$

and its modulus squared is

$$\mathrm{E}^2 = \mathbf{E}\cdot\mathbf{E} = 4\omega^2 A_0^2\, a_{\mathbf{k}}^* a_{\mathbf{k}} \cos^2(\mathbf{k}\cdot\mathbf{r} - \omega t)$$

The temporal average of the modulus squared of the field is

$$\langle \mathrm{E}^2 \rangle = 2\omega^2 \mathrm{A}_0^2\, a_{\mathbf{k}}^* a_{\mathbf{k}}$$

since $\langle \cos^2(\mathbf{k}\cdot\mathbf{r} - \omega t)\rangle = \frac{1}{2}$. The total energy of the electromagnetic field is therefore equal to

$$\begin{aligned} \mathcal{E}_{\mathrm{rad}} &= \epsilon_0\, 2\omega^2 \mathrm{A}_0^2\, a_{\mathbf{k}}^* a_{\mathbf{k}}\, V \\ &= \hbar\omega a_{\mathbf{k}}^* a_{\mathbf{k}} \end{aligned} \tag{C.2}$$

where A_0 has been chosen to be equal to

$$\mathrm{A}_0 = \sqrt{\frac{\hbar}{2\epsilon_0 V\omega}}$$

It should be emphasized that so far we have only considered one particular $\mathbf{k}$ and polarization state, and in general we would need to sum over these quantities to obtain the total energy.

Quantization of the vector potential, A

The normalization constant A_0 in the last section was chosen to reveal the formal equivalence of the Hamiltonian of the electromagnetic field to that of the harmonic oscillator. The quantum mechanical Hamiltonian of the latter is usually written in the form

$$\mathcal{H}_{\mathrm{osc}} = \hbar\omega \left(a^\dagger a + \frac{1}{2} \right) \tag{C.3}$$

A direct comparison of this expression with Eq. (C.2) should at least make this equivalence plausible (apart from the additive term of $\frac{1}{2}$ which we shall return to later). The reason for this equivalence is that when our form for $\mathbf{A}$ (Eq. (C.1)) is substituted into the wave equation, the coefficients $a_{\mathbf{k}}$ obey the equation of motion of the harmonic oscillator.

The operators a and $a^\dagger$ appearing in Eq. (C.3) are known as the annihilation and creation operators, since they have the properties

$$a|n\rangle = \sqrt{n}|n-1\rangle \tag{C.4}$$

and

$$a^\dagger|n\rangle = \sqrt{n+1}|n+1\rangle \tag{C.5}$$

where $|n\rangle$ is an eigenfunction of $\mathcal{H}_{\rm osc}$ with an eigenvalue

$$\mathcal{E}_n = \hbar\omega(n + \frac{1}{2})$$

and n is an integer $0, 1, 2, \cdots$. We thus quantize the electromagnetic field by requiring that the coefficients $a_{\mathbf{k}}$ in Eq. (C.2) become operators that obey the same commutation relations as the annihilation and creation operators of the harmonic oscillator. Here we must extend our notation to allow for the different possible polarization states of the photon, so that the commutation relations read

$$\left[a_{u\mathbf{k}}, a^{\dagger}_{v\mathbf{k}'}\right] = \delta_{\mathbf{k}\mathbf{k}'}\delta_{uv}$$
$$\left[a_{u\mathbf{k}}, a_{v\mathbf{k}'}\right] = \left[a^{\dagger}_{u\mathbf{k}}, a^{\dagger}_{v\mathbf{k}'}\right] = 0$$

where the first subscript, u or v, refers to the polarization state.

The Hamiltonian of the radiation field is thus given by

$$\mathcal{H}_{\rm rad} = \sum_u \sum_{\mathbf{k}} \hbar\omega_{\mathbf{k}} a^{\dagger}_{u\mathbf{k}} a_{u\mathbf{k}}$$

For a given value of $\mathbf{k}$ and polarization u, the eigenfunctions of $\mathcal{H}_{\rm rad}$ are $|n_{u\mathbf{k}}\rangle$, where $n_{u\mathbf{k}}$ is the number of photons in that state. The $n_{u\mathbf{k}}$'s are sometimes referred to as the occupation numbers. It follows that a general state of the field, involving photons with different wavevectors and polarizations is a product of such states, since they are all independent. In writing down $\mathcal{H}_{\rm rad}$ we have followed convention and set the energy of the vacuum state (all $n_{u\mathbf{k}}$=0) equal to $\frac{1}{2}\sum_u \sum_{\mathbf{k}} \hbar\omega_{\mathbf{k}}$.

The operator form of the vector potential is

$$\mathbf{A}(\mathbf{r}, t) = \sum_u \sum_{\mathbf{k}} \hat{\varepsilon}_u \sqrt{\frac{\hbar}{2\epsilon_0 V \omega_{\mathbf{k}}}} \left[a_{u\mathbf{k}}\, \mathrm{e}^{i(\mathbf{k}\cdot\mathbf{r}-\omega t)} + a^{\dagger}_{u\mathbf{k}}\, \mathrm{e}^{-i(\mathbf{k}\cdot\mathbf{r}-\omega t)}\right] \tag{C.6}$$

The interaction Hamiltonian, $\mathcal{H}_I$

In the absence of any interaction between the photon field of the X-ray and the electrons in the sample the Hamiltonian is

$$\mathcal{H}_0 = \mathcal{H}_e + \mathcal{H}_{\rm rad}$$

where $\mathcal{H}_e$ refers to the electrons and $\mathcal{H}_{\rm rad}$ is given above. The eigenfunctions of $\mathcal{H}_0$ are a product of the eigenfunctions of $\mathcal{H}_e$ and $\mathcal{H}_{\rm rad}$.

Classically, it can be shown[1] that the interactions between an electromagnetic field and a charge q may be allowed for by replacing the momentum $\mathbf{p}$ by $\mathbf{p} - q\mathbf{A}$. For simplicity we shall consider the case of a free electron for which

[1] The substitution of $\mathbf{p}$ by $\mathbf{p} - q\mathbf{A}$ can be shown to produce the correct equation of motion of a charged particle in an electromagnetic field.

$\mathcal{H}_e = \mathrm{p}^2/2m$. This allows us to write down the Hamiltonian of the interacting system as

$$\begin{aligned}\mathcal{H} &= \frac{(\mathbf{p}+e\mathbf{A})^2}{2m} + \mathcal{H}_{\mathrm{rad}} \\ &= \frac{\mathrm{p}^2}{2m} + \frac{e\mathbf{A}\cdot\mathbf{p}}{m} + \frac{e^2\mathrm{A}^2}{2m} + \mathcal{H}_{\mathrm{rad}} \\ &= \mathcal{H}_e + \mathcal{H}_I + \mathcal{H}_{\mathrm{rad}}\end{aligned}$$

where $\mathcal{H}_I$ is the interacting Hamiltonian

$$\boxed{\mathcal{H}_I = \frac{e\mathbf{A}\cdot\mathbf{p}}{m} + \frac{e^2\mathrm{A}^2}{2m}} \tag{C.7}$$

The first term is linear in **A** and gives rise to absorption of the X-ray, whereas the second is quadratic in **A** and gives rise to scattering, as we shall now explain.

The operator for **A** is linear in the annihilation and creation operators. Hence when it acts on a state $|n_{u\mathbf{k}}\rangle$ it can either destroy or create a photon in that state. Absorption corresponds to the former, and it is clear that the first term in $\mathcal{H}_I$ results in absorption. Scattering on the other hand involves the destruction of a photon in one state (labelled by **k** say), and the creation of a new photon in a state (labelled by $\mathbf{k}'$). This process then requires a combination of operators of the form $a^\dagger_{\mathbf{k}'}a_{\mathbf{k}}$ to act on the product states $|n_{\mathbf{k}}\rangle|n_{\mathbf{k}'}\rangle$ that are the eigenfunctions of $\mathcal{H}_{\mathrm{rad}}$. Such combinations of operators can arise only from a term in the Hamiltonian that is quadratic in **A**. In the next section we explicitly calculate the cross-section arising from the second term in $\mathcal{H}_I$, and show that it is equivalent to the classical Thomson scattering of an X-ray by an electron. The absorption cross-section is derived from the first term in $\mathcal{H}_I$ in Chapter 6.

Thomson scattering cross-section

We imagine that an X-ray photon of wavevector **k** and polarization $\hat{\varepsilon}_u$ is scattered by an electron to $\mathbf{k}'$ and $\hat{\varepsilon}_v$. The scattering is restricted to be elastic, i.e. $\hbar\omega = \hbar\omega'$, which is equivalent to assuming that the energy of the X-ray is large compared to the binding energy of the electron. As the scattering is elastic the electron can be assumed to remain in its ground state $|\mathrm{p}\rangle$. The eigenfunction of the photon field on the other hand changes so that one photon is removed from the state $|n_{u\mathbf{k}}\rangle$, and one photon added to the state $|n_{v\mathbf{k}'}\rangle$. Before the scattering event we have $n_{u\mathbf{k}} = 1$ and $n_{v\mathbf{k}'} = 0$, and after $n_{u\mathbf{k}} = 0$ and $n_{v\mathbf{k}'} = 1$. We therefore write the initial and final eignestates of $\mathcal{H}_0$ as $|i\rangle = |\mathrm{p}\rangle|u1, v0\rangle$ and $|f\rangle = |\mathrm{p}\rangle|u0, v1\rangle$ respectively.

The scattering cross-section is calculated by taking the matrix element of the second term in Eq. (C.7), and using Eq. (A.7). The matrix element we need to evaluate is

$$M_{if} = \langle u0, v1|\langle \mathrm{p}|\frac{e^2}{2m}\mathrm{A}^2|\mathrm{p}\rangle|u1, v0\rangle$$

When we form the square of **A** from Eq. (C.6) there will be cross terms in the annihilation and creation operators such as $a^\dagger_{i\mathbf{k}_k}a_{j\mathbf{k}_l}$, which can destroy a photon in state $|n_{j\mathbf{k}_l}\rangle$, and create one in state $|n_{i\mathbf{k}_k}\rangle$. These terms give rise to scattering and, as there are two such terms, the matrix element becomes

$$M_{if} = \frac{e^2\hbar}{2m\epsilon_0 V}\frac{[\hat{\varepsilon}_u \cdot \hat{\varepsilon}_v]}{(\omega\omega')^{\frac{1}{2}}}\langle \mathrm{p}|\mathrm{e}^{i(\omega-\omega')t}\mathrm{e}^{i(\mathbf{k}-\mathbf{k}')\cdot\mathbf{r}}|\mathrm{p}\rangle$$

The cross-section given in Eq. (A.7) involves an integral with respect to the final X-ray energy, $\mathcal{E}_f \equiv \hbar\omega'$, whereas the matrix element is given in terms of ω. We therefore rewrite Eq. (A.7) as

$$\left(\frac{d\sigma}{d\Omega}\right) = \left(\frac{V}{2\pi}\right)^2 \frac{1}{\hbar^2 c^4}\int |M_{if}|^2 \omega'^2\, \delta(\omega-\omega')\, d\omega'$$

Taking the square of the matrix element and inserting it into the above gives

$$\boxed{\left(\frac{d\sigma}{d\Omega}\right) = \left(\frac{e^2}{4\pi\epsilon_0 mc^2}\right)^2 [\hat{\varepsilon}_u \cdot \hat{\varepsilon}_v]^2\, |f(\mathbf{Q})|^2}$$

This is the Thomson scattering cross-section, and should be compared with the classical result given in Chapter 1. The two descriptions are entirely equivalent, with the polarization factor given by $P = [\hat{\varepsilon}_u \cdot \hat{\varepsilon}_v]^2$, and the form factor by

$$f(\mathbf{Q}) = \langle \mathrm{p}|\mathrm{e}^{i\,\mathbf{Q}\cdot\mathbf{r}}|\mathrm{p}\rangle$$

Further reading

Quantum Field Theory, F. Mandl and G. Shaw (John Wiley, 1996)

D. Gaussian statistics

In scattering theory the problem of evaluating the double sum

$$I(\mathrm{q}) = \sum_{n,m} \mathrm{e}^{i\mathrm{q}\mathrm{r}_n} \mathrm{e}^{-i\mathrm{q}\mathrm{r}_m}$$

is often encountered, where the atomic positions r_n vary statistically around some average value. One example is the thermal vibrations of atoms, which leads to a reduction of the intensity of Bragg peaks with increasing wavevector Q as described by the Debye-Waller factor (see Section 4.4.7). By invoking translational invariance the double sum reduces to $N\langle \mathrm{e}^{i\,\mathrm{qR}} \rangle$, where $\mathrm{R} = \mathrm{r}_n - \mathrm{r}_m$, and the brackets indicate an average formed by moving the origin of R over all lattice sites. We shall now prove that if the statistical variation of R is Gaussian then

$$\boxed{\left\langle \mathrm{e}^{i\mathrm{qR}} \right\rangle = \mathrm{e}^{-\mathrm{q}^2 \langle \mathrm{R}^2 \rangle /2}} \tag{D.1}$$

This is known as the Baker-Hausdorff theorem.

The proof of this theorem follows from the fact that the Fourier transform of a Gaussian is a Gaussian, as shown in Appendix E. Here a normalised Gaussian in one dimension is used, which has a Fourier transform given by

$$\frac{1}{\sqrt{2\pi\sigma^2}} \int_{-\infty}^{\infty} \mathrm{e}^{-x^2/(2\sigma^2)}\, \mathrm{e}^{i\,\mathrm{q}x}\, dx = \mathrm{e}^{-\mathrm{q}^2\sigma^2/2}$$

The left hand side of this equation is by definition the average value of $\mathrm{e}^{i\,\mathrm{q}x}$:

$$\left\langle \mathrm{e}^{i\,\mathrm{q}x} \right\rangle = \frac{1}{\sqrt{2\pi\sigma^2}} \int_{-\infty}^{\infty} \mathrm{e}^{-x^2/(2\sigma^2)}\, \mathrm{e}^{i\,\mathrm{q}x}\, dx$$

Also $\langle x^2 \rangle = \sigma^2$ (see the section on Gaussian integrals below), and it therefore follows that

$$\left\langle \mathrm{e}^{i\mathrm{qx}} \right\rangle = \mathrm{e}^{-\mathrm{q}^2 \langle \mathrm{x}^2 \rangle /2}$$

This establishes the validity of the Baker-Hausdorff theorem.

Gaussian integrals

Here the recursion relation between Gaussian integrals defined by

$$I_m(a = 1) = \int_{-\infty}^{\infty} x^m \mathrm{e}^{-x^2}\, dx$$

is derived for $m = 0, 2, 4, \cdots$. For $m = 0$ the integral is most readily evaluated by considering its square:

$$\begin{aligned} I_0^2 &= \int_{-\infty}^{\infty} \mathrm{e}^{-x^2}\, dx \int_{-\infty}^{\infty} \mathrm{e}^{-y^2}\, dy = \int_{-\infty}^{\infty} \int_{-\infty}^{\infty} \mathrm{e}^{-(x^2+y^2)}\, dx dy \\ &\equiv \int_0^{2\pi} d\theta \int_0^{\infty} r\,\mathrm{e}^{-r^2}\, dr = 2\pi\,\frac{1}{2} = \pi \end{aligned}$$

The integral for $m = 0$ is therefore

$$I_0 = \int_{-\infty}^{\infty} \mathrm{e}^{-x^2}\, dx = \sqrt{\pi} \tag{D.2}$$

and

$$I_0(a) = f(a) = \int_{-\infty}^{\infty} \mathrm{e}^{-ax^2}\, dx = \sqrt{\pi}\, a^{-1/2} \tag{D.3}$$

By differentiating Eq. (D.3) with respect to a one finds

$$\begin{aligned} -I_2(a) = f'(a) &= -\int_{-\infty}^{\infty} x^2\, \mathrm{e}^{-ax^2}\, dx = \sqrt{\pi}\left(\frac{-1}{2}\right) a^{-3/2} \\ I_4(a) = f''(a) &= +\int_{-\infty}^{\infty} x^4\, \mathrm{e}^{-ax^2}\, dx = \sqrt{\pi}\left(\frac{-1}{2}\right)\left(\frac{-3}{2}\right) a^{-5/2} \\ -I_6(a) = f'''(a) &= -\int_{-\infty}^{\infty} x^6\, \mathrm{e}^{-ax^2}\, dx = \sqrt{\pi}\left(\frac{-1}{2}\right)\left(\frac{-3}{2}\right)\left(\frac{-5}{2}\right) a^{-7/2} \\ &\text{etc.} \end{aligned}$$

and in general the integrals obey the recursion relation

$$I_{2m} = I_{2m-2}\frac{2m-1}{2a}$$

E. Fourier transforms

The Fourier transform of the one dimensional function $f(x)$ is defined by

$$F(\mathrm{q}) = \int_{-\infty}^{\infty} f(x)\, \mathrm{e}^{i\,\mathrm{q}x}\, dx$$

and the inverse transform is

$$f(x) = \frac{1}{2\pi} \int_{-\infty}^{\infty} F(\mathrm{q})\, \mathrm{e}^{-i\,\mathrm{q}x}\, d\mathrm{q}.$$

Evaluation of the Fourier transform is simplified if the function is either symmetric or antisymmetric with respect to the line $x = 0$. For a symmetric function $f^{\mathrm{S}}(x)$ the Fourier transform is

$$\begin{aligned} F(\mathrm{q}) &= \int_{-\infty}^{\infty} f^{\mathrm{S}}(x)\, \mathrm{e}^{i\,\mathrm{q}x}\, dx \\ &= \int_{-\infty}^{\infty} f^{\mathrm{S}}(x)\, \cos(\mathrm{q}x)\, dx + i \int_{-\infty}^{\infty} f^{\mathrm{S}}(x)\, \sin(\mathrm{q}x)\, dx \end{aligned}$$

The second integral on the right hand side is identically zero, since the product of $f^{\mathrm{S}}(x)$ and sine is itself antisymmetric, and the integral of an antisymmetric function over a symmetric domain is zero. Therefore the Fourier transform of a symmetric function is real and is given by the cosine transform

$$F(\mathrm{q}) = 2 \int_{0}^{\infty} f^{\mathrm{S}}(x)\, \cos(\mathrm{q}x)\, dx$$

Similar arguments can be used to show that the Fourier transform of an antisymmetric function $f^{\mathrm{A}}(x)$ is purely imaginary and is the sine transform

$$F(\mathrm{q}) = i\,2 \int_{0}^{\infty} f^{\mathrm{A}}(x)\, \sin(\mathrm{q}x)\, dx$$

Gaussian

Here the Gaussian function is written as

$$f(x) = A\, \mathrm{e}^{-a^2 x^2} \tag{E.1}$$

and is plotted in Fig. E.1(a). As it is a symmetric function its Fourier transform is

$$F(\mathrm{q}) = 2 \int_{0}^{\infty} A\, \mathrm{e}^{-a^2 x^2}\, \cos(\mathrm{q}x)\, dx$$

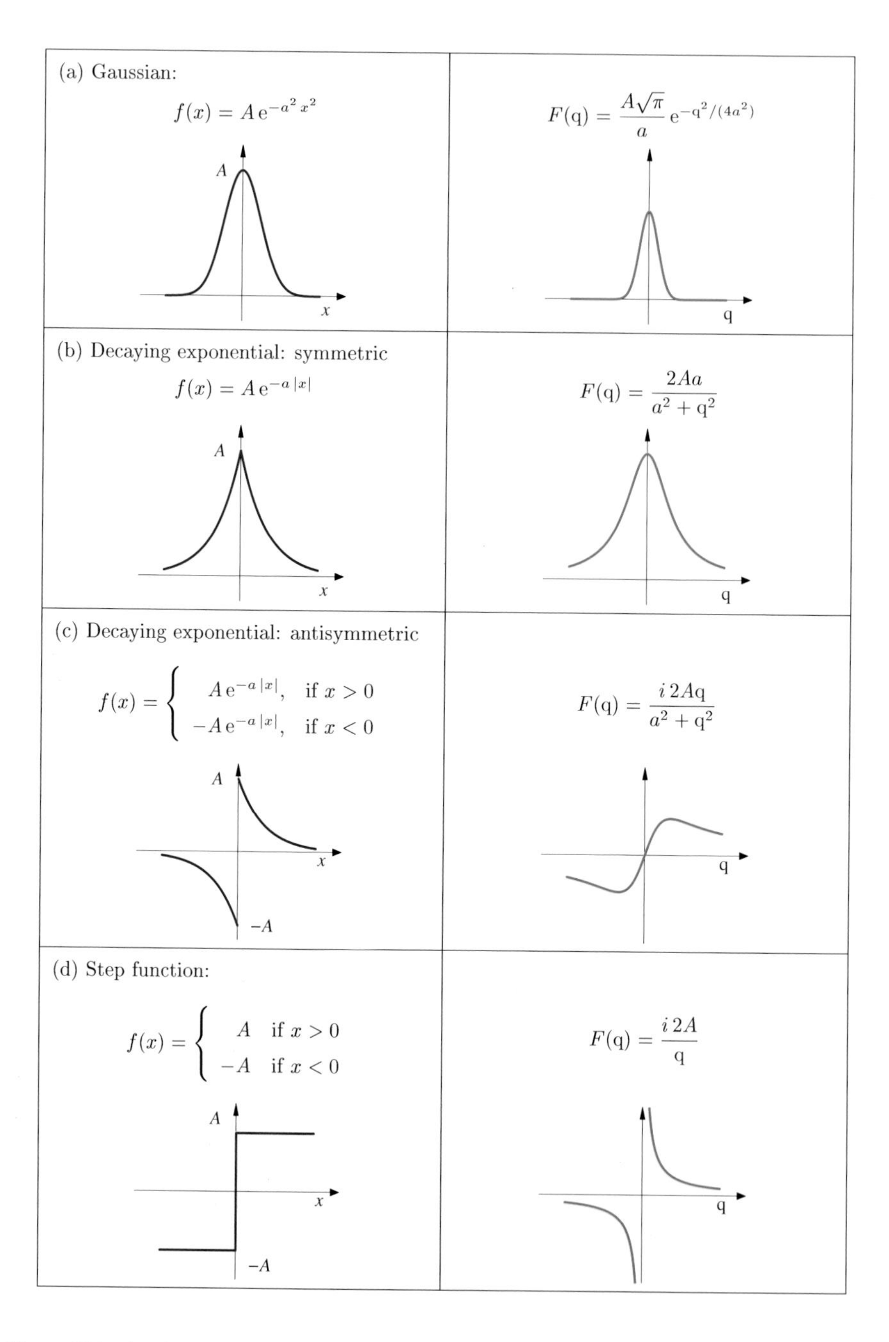

Figure E.1: A selection of functions (left panel) and their Fourier transform (right panel).

This may be evaluated by writing the cosine as the real part of a complex exponential, $\cos(\mathrm{q}x) = Re\{\mathrm{e}^{i\,\mathrm{q}x}\}$. The Fourier integral then becomes

$$F(\mathrm{q}) = 2A\,Re\left\{\int_0^\infty \mathrm{e}^{-a^2x^2}\,\mathrm{e}^{i\,\mathrm{q}x}\,dx\right\} = 2A\,Re\left\{\int_0^\infty \mathrm{e}^{-a^2x^2+i\,\mathrm{q}x}\,dx\right\}$$

$$= 2A\,\mathrm{e}^{-\mathrm{q}^2/(4a^2)}\,Re\left\{\int_0^\infty \mathrm{e}^{-(ax-i\,\mathrm{q}/(2a))^2}\,dx\right\}$$

$$= 2A\,\mathrm{e}^{-\mathrm{q}^2/(4a^2)}\,Re\left\{\frac{1}{a}\int_0^\infty \mathrm{e}^{-\kappa^2}\,d\kappa\right\}$$

where κ is a complex variable defined by $\kappa = (ax - i\,\mathrm{q}/(2a))^2)$. The real part of the last integral is equal to the standard integral

$$\int_0^\infty \mathrm{e}^{-y^2}\,dy = \frac{\sqrt{\pi}}{2}$$

(see Eq. D.2). The Fourier transform of a Gaussian is thus

$$\boxed{F(\mathrm{q}) = \frac{A\sqrt{\pi}}{a}\,\mathrm{e}^{-\mathrm{q}^2/(4a^2)}} \tag{E.2}$$

which is itself a Gaussian.

It is instructive to consider the width Δx (full width at half maximum) of the Gaussian function and the width $\Delta\mathrm{q}$ of its Fourier transform. From Eq. (E.1), $\Delta x{=}2\sqrt{\log_e(2)}/a$, and from Eq. (E.2), $\Delta\mathrm{q}{=}4a\sqrt{\log_e(2)}$. The product of the widths is a constant equal to

$$\Delta x\,\Delta\mathrm{q} = 8\log_e(2) \tag{E.3}$$

This illustrates the reciprocal nature of the description of an object in real or direct space, and the description of its Fourier transform in q space, also known as reciprocal space. If an object is extended in real space, Δx is large, and its Fourier transform is well localized in reciprocal space, i.e. $\Delta\mathrm{q}$ is small. Correspondingly, if an object is well localized in real space, then its Fourier transform is extended in reciprocal space. One extreme limit of this is a 2D object. This is infinitely thin in one direction, and hence its Fourier transform in this direction is perfectly delocalized, or in other words it has a constant value. This explains why the scattering from a two dimensional sheet of atoms forms rods perpendicular to the sheet.

Decaying exponential: symmetric

The symmetric decaying exponential is defined by

$$f(x) = A\,\mathrm{e}^{-a|x|} \tag{E.4}$$

and is plotted in Fig. E.1(b). Its Fourier transform is

$$F(\mathrm{q}) = 2A\int_0^\infty \mathrm{e}^{-ax}\cos(\mathrm{q}x)\,dx$$

The integral may be integrating by parts once to yield

$$\int_0^\infty \mathrm{e}^{-ax}\cos(\mathrm{q}x)\,dx = \frac{a}{\mathrm{q}}\int_0^\infty \mathrm{e}^{-ax}\sin(\mathrm{q}x)\,dx$$

The right hand side may also be integrated by parts again with the result that the cosine transform of e^{-ax} is

$$\int_0^\infty \mathrm{e}^{-ax}\cos(\mathrm{q}x)\,dx = \frac{a}{\mathrm{q}}\left[\frac{1}{\mathrm{q}} - \frac{a}{\mathrm{q}}\int_0^\infty \mathrm{e}^{-ax}\cos(\mathrm{q}x)\,dx\right]$$

This can be rearranged to give

$$\int_0^\infty \mathrm{e}^{-ax}\cos(\mathrm{q}x)\,dx = \frac{a}{a^2+\mathrm{q}^2} \tag{E.5}$$

It follows that the Fourier transform of a symmetric decaying exponential is a Lorentzian:

$$\boxed{F(\mathrm{q}) = \frac{2Aa}{a^2+\mathrm{q}^2}} \tag{E.6}$$

The product of the widths in real and reciprocal space for the symmetric decaying exponential function is

$$\Delta x\,\Delta\mathrm{q} = 4\log_e(2) \tag{E.7}$$

Decaying exponential: antisymmetric

The antisymmetric decaying exponential function is defined by

$$f(x) = \begin{cases} A\,\mathrm{e}^{-a|x|}, & \text{for } x > 0 \\ -A\,\mathrm{e}^{-a|x|}, & \text{for } x < 0 \end{cases}$$

and is plotted in Fig. E.1(c). Its Fourier transform is purely imaginary and is

$$F(\mathrm{q}) = i\,2\int_0^\infty A\,\mathrm{e}^{-ax}\sin(\mathrm{q}x)\,dx$$

The sine transform of e^{-ax} is evaluated by integrating by parts twice, which yields

$$\int_0^\infty \mathrm{e}^{-ax}\sin(\mathrm{q}x)\,dx = \frac{\mathrm{q}}{a^2+\mathrm{q}^2} \tag{E.8}$$

The Fourier transform of the antisymmetric decaying exponential is plotted in the right hand panel of Fig. E.1 and is given by

$$\boxed{F(\mathrm{q}) = \frac{i\,2A\mathrm{q}}{a^2+\mathrm{q}^2}} \tag{E.9}$$

Step function

The step function

$$f(x) = \begin{cases} A, & \text{for } x > 0 \\ -A, & \text{for } x < 0 \end{cases}$$

is plotted in Fig. E.1(d). Its Fourier transform is equal to the Fourier transform of the antisymmetric decaying exponential in the limit that $a \to 0$. From Eq. (E.9) the Fourier transform of the step function is

$$\boxed{F(\mathrm{q}) = \frac{i\,2A}{\mathrm{q}}} \tag{E.10}$$

Further reading

A Handbook of Fourier Transforms, D.C. Champeney (Cambridge University Press, 1987)

A Student's Guide to Fourier Transforms, J.F. James (Cambridge University Press, 1995)

F. Comparison of X-rays with neutrons

The relationship given in Eq. (3.1) between the refractive index, n, and the scattering length density ρr_0 can be derived in a different way when considering neutrons entering a material. Here the material is viewed as a continuum of nuclei, each having the scattering length b, and with a density of ρ.

An incident particle that experiences a potential $\mathcal{V}(\mathbf{r})$ from the material will be scattered. The scattering length b of a particle is related to the scattering potential $\mathcal{V}(\mathbf{r})$ in first-order perturbation theory by

$$\mathcal{V}(\mathbf{Q}) = 4\pi \left(\frac{\hbar^2}{2m_n} \right) b$$

Here $\mathcal{V}(\mathbf{Q})$ is the Fourier transform of the scattering potential:

$$\mathcal{V}(\mathbf{Q}) = \int \mathcal{V}(\mathbf{r})\, e^{i\,\mathbf{Q}\cdot\mathbf{r}}\, d\mathbf{r}$$

and as usual $\mathbf{Q} = \mathbf{k} - \mathbf{k}'$ is the wavevector transfer in the scattering process. This relation appears plausible when it is recalled that

1. there is a phase difference of $\mathbf{Q} \cdot \mathbf{r}$ between the scattering from volume elements around the origin and around $\mathbf{r}$
2. the scattering can be thought of as a weighted superposition of the scattering from such volume elements, the weight being $\mathcal{V}(\mathbf{r})$ times the phase factor
3. the dimensionality in the equation relating b and $\mathcal{V}(\mathbf{Q})$ is correct, i.e. the term $(\hbar^2/2m_n)$ occurs naturally.

Of course, the factor of 4π must rely on a more rigorous treatment.

Fermi suggested that one could define a pseudo-potential between thermal neutrons and nuclei in such a way that this general first-order perturbation result would reproduce the correct scattering length. The nuclear scattering of neutrons is due to a short range potential between the neutron and the nucleus. The range of the potential is extremely short (of order 10^{-15} m) in comparison with the wavelength of thermal neutrons (of order 10^{-10} m) so the shape of the potential is well approximated by a Dirac delta function allowing us to write

$$\mathcal{V}_{\mathrm{F}}(\mathbf{r}) = C\,\delta(\mathbf{r})$$

so that

$$\mathcal{V}_{\mathrm{F}}(\mathbf{Q}) = C \times 1$$

since the Fourier transform of a Dirac delta function is unity. (The properties of the Dirac delta function are reviewed in Chapter 4 on page 147.) Equating this expression for $\mathcal{V}_{\mathrm{F}}(\mathbf{Q})$ with the general one given above for $\mathcal{V}(\mathbf{Q})$ allows us to identify the constant C and to write

$$\mathcal{V}_{\mathrm{F}}(\mathbf{r}) = 4\pi \left(\frac{\hbar^2}{2m_n} \right) b\,\delta(\mathbf{r})$$

In refraction, the medium is considered to be a homogeneous continuum, and the change in wavenumber from k outside to nk inside the medium is due to a corresponding change in kinetic energy from $(\hbar^2/2m_n)\mathrm{k}^2$ to $(\hbar^2/2m_n)(n\mathrm{k})^2$. Energy conservation then immediately leads to

$$\left(\frac{\hbar^2}{2m_n} \right) \mathrm{k}^2 = \left(\frac{\hbar^2}{2m_n} \right) (n\mathrm{k})^2 + \langle \mathcal{V} \rangle$$

or

$$\mathrm{k}^2 - (n\mathrm{k})^2 = \left(\frac{2m_n}{\hbar^2} \right) \langle \mathcal{V} \rangle$$

Inserting the average potential

$$\langle \mathcal{V} \rangle = \frac{\int_V \mathcal{V}_F(r)\, d\mathbf{r}}{\int_V \, d\mathbf{r}} = 4\pi \left(\frac{\hbar^2}{2m_n} \right) b\rho$$

yields

$$\mathrm{k}^2(1 - n^2) = 4\pi\, b\rho$$

and since $(1 - n^2) = (1 + n)(1 - n) \approx 2\delta$ we obtain Eq. (3.2) with the neutron scattering length b substituting for the X-ray scattering length r_0.

Further reading

Introduction to the Theory of Thermal Neutron Scattering, G.L. Squires (Dover Publications, 1996)

Neutron Optics, V.F. Sears (Oxford University Press, 1989)

G. MATLAB computer programs

Listings are given here of MATLAB files which have been used to generate some of the figures in this book.

The files may be downloaded from the World Wide Web by following the link given on the home page of this book at the official John Wiley and Son site (http://www.wiley.co.uk).

MATLAB is a registered trademark of The MathWorks, Inc. Further information can be found at http://www.mathworks.com.

Chapter 2: Sources of X-rays

Undulator characteristics, Fig. 2.11 on page 49

```
function wout=undulator
%
% MATLAB function from:
% "Elements of Modern X-ray Physics" by Jens Als-Nielsen and Des McMorrow
%
% Calculates: Undulator characteristics
% Calls to: w1tcalc (observer phase from emitter phase)

set(gcf,'papertype','a4','paperunits','centimeters','units','centimeters',...
        'position',[0.1 -8 21 26],'paperposition',[0.1 0.1 21 26]);

%%%%%%%%%%%%%%%%%%%%%%%%%%%%%%%%%%%%%%%%%%%%%%%%%%%%%%%%%%%%%%%%%%%%%%%%%%%%%%%%
% Bottom left, on-axis harmonic content for K=2
%%%%%%%%%%%%%%%%%%%%%%%%%%%%%%%%%%%%%%%%%%%%%%%%%%%%%%%%%%%%%%%%%%%%%%%%%%%%%%%%

axes('position',[0.1 0.1 0.35 0.35]);

wutp=0:0.01:2*pi;                              % Emitter  phase
K=2; w1t=w1tcalc(wutp,K,0);                    % Observer phase

xn=0:0.01:2*pi;
yn=spline(w1t,2*sin(wutp),xn);
f=fft(yn);                                     % Fourier transform displacement
line(xn,-2*imag(f(2))/length(xn)*sin(xn),'linestyle','--','linewidth',1.0)
line(xn,-2*imag(f(4))/length(xn)*sin(3*xn),'linestyle','--','linewidth',1.0)
line(xn,-2*imag(f(6))/length(xn)*sin(5*xn),'linestyle','--','linewidth',1.0)
line(xn,-2*imag(f(2))/length(xn)*sin(xn)...
        -2*imag(f(4))/length(xn)*sin(3*xn)-2*imag(f(6))/length(xn)*sin(5*xn),...
        'color','m','linewidth',1.0)

axis([0 2*pi -2 2]); axis square
set(gca,'Xtick',[0 pi/2 pi 3*pi/2 2*pi],'Xticklabel',[],'Ytick',[-2 0 2])
set(gca,'FontName','Times','Fontsize',16,'xgrid','on','ygrid','on','box','on')
ylabel('Transverse displacement')
xlabel('\omega t'' (or \omega_1t)','position',[pi -2.5 0]);
title('On axis')
text(0,-2.3,'0','horizontalalignment','center','Fontname','Times','Fontsize',16)
text(pi/2,-2.3,'\pi/2','horizontalalignment','center','Fontname','Times','Fontsize',16)
text(pi,-2.3,'\pi','horizontalalignment','center','Fontname','Times','Fontsize',16)
text(1.5*pi,-2.3,'3\pi/2','horizontalalignment','center','Fontname','Times','Fontsize',16)
text(2*pi,-2.3,'2\pi','horizontalalignment','center','Fontname','Times','Fontsize',16)
text(3*pi/2,1.55,'{\it K=2}','horizontalalignment',...
      'center','Fontname','Times','Fontsize',16)

```

```
45 %%%%%%%%%%%%%%%%%%%%%%%%%%%%%%%%%%%%%%%%%%%%%%%%%%%%%%%%%%%%%%%%%%%%%%%%%%%%%
46 % Bottom right, Harmonic content for K=2
47 %%%%%%%%%%%%%%%%%%%%%%%%%%%%%%%%%%%%%%%%%%%%%%%%%%%%%%%%%%%%%%%%%%%%%%%%%%%%%
48
49 axes('position',[0.6 0.1 0.35 0.35]);
50
51 hc=([-2*imag(f(2))  -2*imag(f(4))  -2*imag(f(6)) -2*imag(f(8))]/length(xn))';
52 hc=abs(hc);
53 ac=[hc(1) hc(2)*9 hc(3)*25 hc(4)*49]';
54 h=bar([hc ac],0.85,'grouped');
55 axis([0.5 4.5 0 2.5]); axis square
56
57 set(gca,'Box','on','xgrid','on','ygrid','on','FontName','Times','Fontsize',16,...
58        'xticklabel',['1';'3';'5';'7'])
59 set(h(1),'facecolor','w')
60 set(h(2),'facecolor','k')
61 xlabel('Harmonic')
62 ylabel('Displacement/acceleration')
63 title('On axis')
64 text(1,2.2,'{\it K=2}','horizontalalignment','center','Fontname','Times','Fontsize',16)
65
66 %%%%%%%%%%%%%%%%%%%%%%%%%%%%%%%%%%%%%%%%%%%%%%%%%%%%%%%%%%%%%%%%%%%%%%%%%%%%%
67 % Top left, on-axis  for K=1, 2 and 5
68 %%%%%%%%%%%%%%%%%%%%%%%%%%%%%%%%%%%%%%%%%%%%%%%%%%%%%%%%%%%%%%%%%%%%%%%%%%%%%
69
70 axes('position',[0.1 0.5 0.35 0.35]);
71
72 wutp=0:0.01:2*pi;
73
74 w1t=w1tcalc(wutp,1,0); % K=1
75 hl1=line(wutp,sin(wutp),'linestyle','--','color','r','linewidth',1.0);
76 hl2=line(w1t,sin(wutp),'linestyle','-','color','r','linewidth',1.0);
77
78 w1t=w1tcalc(wutp,2,0); % K=2
79 hl1=line(wutp,2*sin(wutp),'linestyle','--','color','b','linewidth',1.0);
80 hl2=line(w1t,2*sin(wutp),'linestyle','-','color','b','linewidth',1.0);
81
82 w1t=w1tcalc(wutp,5,0); % K=5
83 hl1=line(wutp,5*sin(wutp),'linestyle','--','color','m','linewidth',1.0);
84 hl2=line(w1t,5*sin(wutp),'linestyle','-','color','m','linewidth',1.0);
85
86 axis([0 2*pi -6 6]); axis square
87 set(gca,'Xtick',[0 pi/2 pi 3*pi/2 2*pi],'Xticklabel',[],'box','on','ygrid','on')
88 set(gca,'Ytick',[-6 -4 -2 0 2 4 6],'FontName','Times','Fontsize',16,'Xgrid','on')
89 ylabel('Transverse displacement')
90 xlabel('\omega t'' (or \omega_1t)','position',[pi -8.0 0 ]);
91 title('On axis')
92 arrow([3.3 3.7],[2.1 3.7],10)
93 arrow([pi/2 -2],[pi/2 0.5],10)
94 text(0,-7.0,'0','horizontalalignment','center','Fontname','Times','Fontsize',16)
95 text(pi/2,-7.0,'\pi/2','horizontalalignment','center','Fontname','Times','Fontsize',16)
96 text(pi,-7.0,'\pi','horizontalalignment','center','Fontname','Times','Fontsize',16)
97 text(1.5*pi,-7.0,'3\pi/2','horizontalalignment','center','Fontname','Times','Fontsize',16)
98 text(2*pi,-7.0,'2\pi','horizontalalignment','center','Fontname','Times','Fontsize',16)
99 text(pi/2,-3,'{\it K}=1','FontName','times','fontsize',16,'horizontalalignment','center')
100 text(5*pi/4,3.7,'{\it K}=5','FontName','times','fontsize',16,'horizontalalignment','center')
101
102 %%%%%%%%%%%%%%%%%%%%%%%%%%%%%%%%%%%%%%%%%%%%%%%%%%%%%%%%%%%%%%%%%%%%%%%%%%%%%
103 % Top right, off-axis  for K=2
104 %%%%%%%%%%%%%%%%%%%%%%%%%%%%%%%%%%%%%%%%%%%%%%%%%%%%%%%%%%%%%%%%%%%%%%%%%%%%%
105
106 axes('position',[0.6 0.5 0.35 0.35]);
107
108 wutp=0:0.01:2*pi;
109
110 w1t=w1tcalc(wutp,2,0);
111 line(w1t,2*sin(wutp),'linestyle','-','color','k','linewidth',1.0);
112
113 w1t=w1tcalc(wutp,2,1);
114 line(w1t,2*sin(wutp),'linestyle','--','color','m','linewidth',1.0);
115
116 axis([0 2*pi -6 6]); axis square
117 set(gca,'Xtick',[0 pi/2 pi 3*pi/2 2*pi],'Xticklabel',[])
118 set(gca,'Ytick',[-6 -4 -2 0 2 4 6],'FontName','Times','Fontsize',16);
```

```
119 set(gca,'xgrid','on','ygrid','on','box','on')
120 ylabel('Transverse displacement')
121 xlabel('\omega t'' (or \omega_1t)','position',[pi -8 0 ]);
122 title('On and off axis')
123 text(0,-7.0,'0','horizontalalignment','center','Fontname','Times','Fontsize',16)
124 text(pi/2,-7.0,'\pi/2','horizontalalignment','center','Fontname','Times','Fontsize',16)
125 text(pi,-7.0,'\pi','horizontalalignment','center','Fontname','Times','Fontsize',16)
126 text(1.5*pi,-7.0,'3\pi/2','horizontalalignment','center','Fontname','Times','Fontsize',16)
127 text(2*pi,-7.0,'2\pi','horizontalalignment','center','Fontname','Times','Fontsize',16)
128 text(3*pi/2,4.0,'{\it K=2}','horizontalalignment','center','Fontname','Times','Fontsize',16)
129
130 %%%%%%%%%%%%%%%%%%%%%%%%%%%%%%%%%%%%%%%%
131 function [w1t]=w1tcalc(wutp,K,ratio)
132 %
133 % MATLAB function from:
134 % "Elements of Modern X-ray Physics" by Jens Als-Nielsen and Des McMorrow
135 %
136 % Calculates: observer phase (w1t) from emitter phase (wutp)
137
138 w1t= wutp + 0.25*K^2/(1+ratio^2+K^2/2)*sin(2*wutp)-2*K/(1+(ratio^2)+K^2/2)*ratio*sin(wutp);
```

Chapter 3: Reflection and refraction

Fresnel reflectivity characteristics, Fig. 3.6 on page 74

```
1  function FresnelR
2  %
3  % MATLAB function from:
4  % "Elements of Modern X-ray Physics" by Jens Als-Nielsen and Des McMorrow
5  %
6  % Calculates: Fresnel reflectivity characteristics
7
8  figure;
9
10 %%%%%%%%%%%%%%%%%%%%%%%%%%%%%%%%%%%%%%%%%%%%%%%%%%%%%%%%%%%%%%%%%%%%%%%%%%%%%%%%%%%%%%%%%%%%%%%%%
11 %----- Intensity reflectivity for different values of b_mu
12 %%%%%%%%%%%%%%%%%%%%%%%%%%%%%%%%%%%%%%%%%%%%%%%%%%%%%%%%%%%%%%%%%%%%%%%%%%%%%%%%%%%%%%%%%%%%%%%%%
13
14 axes('position',[0.35 0.7 0.3 0.2])
15
16 q=0.01:0.001:2.5;
17
18 b=0.1; qp=sqrt(q.^2-1+2*sqrt(-1)*b); rq=(q-qp)./(q+qp);
19 Rq=rq.*conj(rq); Rn=Rq.*(q.^4)*(2^4);
20 iq1=find(q<1.5); iq2=find(q>1.5);
21 [ax,h1,h2]=plotyy(q(iq1),Rq(iq1),q(iq2),Rn(iq2));
22 set(h1,'color','r'); set(h2,'color','r')
23
24 axis([0 2.6 0 1.1])
25 yl=str2mat('    R(q)     ','  Fresnel   ','reflectivity');
26 text(-1.15,0.5,yl,'FontName','Times','rotation',90,'horizontalalignment','center');
27
28 b=0.05; qp=sqrt(q.^2-1+2*sqrt(-1)*b); rq=(q-qp)./(q+qp);
29 Rq=rq.*conj(rq); Rn=Rq.*(q.^4)*(2^4);
30 line(q(iq1),Rq(iq1),'color','g')
31 axes(ax(2)); line(q(iq2),Rn(iq2),'color','g')
32
33 b=0.01;  qp=sqrt(q.^2-1+2*sqrt(-1)*b); rq=(q-qp)./(q+qp);
34 Rq=rq.*conj(rq); Rn=Rq.*(q.^4)*(2^4);
35 axes(ax(1)); line(q(iq1),Rq(iq1),'color','b')
36 axes(ax(2)); line(q(iq2),Rn(iq2),'color','b')
37
38 b=0.001; axes(ax(1)); qp=sqrt(q.^2-1+2*sqrt(-1)*b); rq=(q-qp)./(q+qp);
39 Rq=rq.*conj(rq); Rn=Rq.*(q.^4)*(2^4);
40 line(q(iq1),Rq(iq1),'color','m')
41 axes(ax(2))
42 text(3.2,1.6,'R(q).(2Q/Q_c)^4','FontName','Times','rotation',90,...
43      'horizontalalignment','center')
44 axis([0 2.6 1 2.1])
45 set(ax,'Ycolor',[0 0 0],'Ytick',[0.5 1.0 1.5 2.0 2.5],'Xticklabels',[])
```

```
set(ax,'FontName','Times','Fontsize',12,'box','on');

%%%%%%%%%%%%%%%%%%%%%%%%%%%%%%%%%%%%%%%%%%%%%%%%%%%%%%%%%%%%%%%%%%%%%%%%%%%%
%----- Penetration length
%%%%%%%%%%%%%%%%%%%%%%%%%%%%%%%%%%%%%%%%%%%%%%%%%%%%%%%%%%%%%%%%%%%%%%%%%%%%

axes('position',[0.35 0.5 0.3 0.2])

q=0.01:0.01:1.4;

b=0.1; qp=sqrt(q.^2-1+2*sqrt(-1)*b);
line(q,1./imag(qp),'color','r')

b=0.05; qp=sqrt(q.^2-1+2*sqrt(-1)*b);
line(q,1./imag(qp),'color','g')

b=0.01; qp=sqrt(q.^2-1+2*sqrt(-1)*b);
line(q,1./imag(qp),'color','b')

b=0.001; qp=sqrt(q.^2-1+2*sqrt(-1)*b);
line(q,1./imag(qp),'color','m')

axis([0 2.6 0 1000]);
yl=str2mat('  \Lambda Q_c',' Penetration ','   length   ');
text(-1.1,30,yl,'FontName','Times','rotation',90,'horizontalalignment','center');
set(gca,'Xticklabels',[],'Yscale','log','Ytick',[1 10 100])
set(gca,'FontName','Times','FontSize',12,'box','on')

%%%%%%%%%%%%%%%%%%%%%%%%%%%%%%%%%%%%%%%%%%%%%%%%%%%%%%%%%%%%%%%%%%%%%%%%%%%%
%----- Evanescent intensity
%%%%%%%%%%%%%%%%%%%%%%%%%%%%%%%%%%%%%%%%%%%%%%%%%%%%%%%%%%%%%%%%%%%%%%%%%%%%

axes('position',[0.35 0.3 0.3 0.2])

q=0.01:0.01:2.5;

b=0.1; qp=sqrt(q.^2-1+2*sqrt(-1)*b); ttq=2*q./(q+qp);
h1=line(q,ttq.*conj(ttq),'color','r');

b=0.05; qp=sqrt(q.^2-1+2*sqrt(-1)*b); ttq=2*q./(q+qp);
line(q,ttq.*conj(ttq),'color','g')

b=0.01; qp=sqrt(q.^2-1+2*sqrt(-1)*b); ttq=2*q./(q+qp);
line(q,ttq.*conj(ttq),'color','b')

b=0.001; qp=sqrt(q.^2-1+2*sqrt(-1)*b); ttq=2*q./(q+qp);
line(q,ttq.*conj(ttq),'color','m')

axis([0 2.6 0 4]);
yl=str2mat('   T(q)    ','Evanescent','intensity ');
hl=text(-1.15,2,yl);
set(hl,'FontName','Times','rotation',90,'horizontalalignment','center');
set(gca,'Xticklabels',[],'Ytick',[1 2 3],'Yticklabel',['1';'2';'3'])
set(gca,'FontName','Times','FontSize',12,'box','on')

%%%%%%%%%%%%%%%%%%%%%%%%%%%%%%%%%%%%%%%%%%%%%%%%%%%%%%%%%%%%%%%%%%%%%%%%%%%%
%----- Phase shift of reflected wave
%%%%%%%%%%%%%%%%%%%%%%%%%%%%%%%%%%%%%%%%%%%%%%%%%%%%%%%%%%%%%%%%%%%%%%%%%%%%

axes('position',[0.35 0.1 0.3 0.2])

q=0.01:0.01:2.5;
b=0.1; qp=sqrt(q.^2-1+2*sqrt(-1)*b); rq=(q-qp)./(q+qp);
line(q,angle(rq),'color','r')

b=0.05; qp=sqrt(q.^2-1+2*sqrt(-1)*b); rq=(q-qp)./(q+qp);
line(q,angle(rq),'color','g')

b=0.01; qp=sqrt(q.^2-1+2*sqrt(-1)*b); rq=(q-qp)./(q+qp);
line(q,angle(rq),'color','b')

b=0.001; qp=sqrt(q.^2-1+2*sqrt(-1)*b); rq=(q-qp)./(q+qp);
line(q,angle(rq),'color','m')

```

```
120 axis([0 2.6 -pi pi/4]);
121 yl=str2mat('Phase shift of','   reflected  ','    wave      ');
122 text(-1.,-pi/2,yl,'FontName','Times','rotation',90,'horizontalalignment','center');
123 text(-0.25,0,'0','horizontalalignment','center','FontName','Times','FontSize',12);
124 text(-0.25,-pi,'-\pi','horizontalalignment','center','FontSize',14);
125 xlabel('Q/Q_c or  \alpha/\alpha_c')
126 set(gca,'Yticklabels',[],'Ytick',[-pi -pi/2 0 pi/2],'Xtick',[0.5 1.0 1.5 2.0 2.5])
127 set(gca,'FontName','Times','FontSize',12,'box','on')
```

Kiessig fringes from a thin film, Fig. 3.8 on page 77

```
1  function kiessig
2  %
3  % MATLAB function from:
4  % "Elements of Modern X-ray Physics" by Jens Als-Nielsen and Des McMorrow
5  %
6  % Calculates: Reflectivity from a thin film of tungsten
7
8  axes('position',[0.2 0.2 0.6 0.6]);
9
10 r0=2.82e-5;          % Thompson scattering length in Angs
11 rho=4.678;           % electron density in electrons/Angs^3
12 b=0.0409;            % parameter b_mu
13 Delta=10*2*pi;       % thickness of film in Angs
14 sigma=0.0;           % surface roughness in Angs
15
16 Qc=4*sqrt(pi*rho*r0);
17
18 Q=0:0.001:1;
19 q=Q/Qc;
20 Qp=Qc*sqrt(q.^2-1+2*sqrt(-1)*b);
21
22 rQ=(Q-Qp)./(Q+Qp);
23 r_slab=rQ.*(1-exp(i*Qp*Delta))./(1-rQ.^2.*exp(i*Qp*Delta));
24 r_slab=r_slab.*exp(-Q.^2*sigma^2/2);
25 line(Q,r_slab.*conj(r_slab),'LineWidth',1.0,'Color','b');
26
27 axis([0.0 1.0 1e-10 1.5]); grid on
28 set(gca,'FontName','Times','FontSize',16,'box','on')
29 set(gca,'Ytick',[1e-10 1e-8 1e-6 1e-4 1e-2 1e0],'yscale','log')
30 xlabel('Wavevector transfer Q (^{-1})')
31 ylabel('|{\it r}_{slab}|^2','position',[-0.175 1e-5 0])
```

Parratt and kinematical reflectivity, Fig. 3.10 on page 81

```
1  function par_kin
2  %
3  % MATLAB function from:
4  % "Elements of Modern X-ray Physics" by Jens Als-Nielsen and Des McMorrow
5  %
6  % Calculates: Parratt and kinematical reflectivities from a multilayer
7  %             Specific case of W/Si, 10 bilayers of [10 Angs W, 40 Angs Si]
8
9  r0=2.82e-5;                             % Thompson scattering length in Angs
10 Q=0.01:0.001:0.3;                       % Wavevector transfer in 1/Angs
11 lambda=1.54;                             % wavelength in Angs
12 rhoA=4.678; muA=33.235e-6;              % density and absorption coefficient of W
13 rhoB=0.699; muB=1.399e-6;               % density and absorption coefficient of Si
14
15 bl=[rhoA*r0+i*muA rhoB*r0+i*muB];                 % bilayer scattering factor
16 dbl=[10 40];                                      % bilayer d-spacings
17 ml=[bl bl bl bl bl bl bl bl bl bl 0.1e-20];       % multilayer scattering factor
18 dml=[dbl dbl dbl dbl dbl dbl dbl dbl dbl dbl];    % multilayer d-spacings
19 sml=[0 0 0 0 0 0 0 0 0 0 0 0 0 0 0 0 0 0 0 0 0]; % roughness at each interface
20
21 %------ Parratt reflectivity
22
23 R=parratt(Q,lambda,ml,dml,sml);
24
```

```
25  axes('position',[0.2 0.15 0.7 0.4]); line(Q,R)
26  axis([0 0.3 8e-6 2]);
27  set(gca,'FontName','Times','FontSize',18,'box','on')
28  set(gca,'Ytick',[1e-4 1e-3 1e-2 1e-1 1],'Yscale','log')
29  text(0.15,5e-7,'Wavevector transfer Q (^{-1})',...
30          'FontName','Times','FontSize',18,'horizontalalignment','center')
31  text(-0.05,1,'Reflectivity','FontName','Times',...
32     'FontSize',18,'horizontalalignment','center','rotation',90)
33  text(0.20,0.7,'(b) Parratt','FontName','Times','FontSize',16)
34
35  %----- kinematical reflectivity
36
37  sld=bl;
38  sigma=0; N=10; Lambda=50; Gamma=0.2;
39
40  R=kinematicalR(Q,lambda,sld,sigma,N,Lambda,Gamma);
41
42  axes('position',[0.2 0.55 0.7 0.4]); line(Q,R)
43  set(gca,'FontName','Times','FontSize',18,'box','on')
44  axis([0 0.3 8e-6 2]);
45  set(gca,'Xticklabel','')
46  set(gca,'Ytick',[1e-4 1e-3 1e-2 1e-1 1],'Yscale','log')
47  text(0.20,0.7,'(a) Kinematical ','FontName','Times','FontSize',16)
48
49  %%%%%%%%%%%%%%%%%%%%%%%%%%%%%%%%%%%%%%%%%%%%%%%%%%%%%%%%%%%%%%%%%%%%%%%%
50  function [R]=kinematicalR(Q,lambda,sld,sigma,N,Lambda,Gamma)
51  %
52  % MATLAB function from:
53  % "Elements of Modern X-ray Physics" by Jens Als-Nielsen and Des McMorrow
54  %
55  % Calculates: kinematical reflectivity of a multilayer
56  % Inputs:  Q       wavevector transfer           1/Angs
57  %          lambda  wavelength of radiation       Angs
58  %          sld     scattering length density     1/Angs^2
59  %                  sld=[sldA+i*muA sldB+i*muB]
60  %          sigma   rouhgness                     Angs
61  %          N       number of bilayers
62  %          Lambda  length of bilayer             Angs
63  %          Gamma   fraction of bilayer that is A
64  % Outputs: R       Intensity reflectivity
65
66  muA=imag(sld(1));
67  muB=imag(sld(2));
68
69  Dsld=real(sld(1))-real(sld(2));
70
71  zeta=Q/2/pi*Lambda;
72
73  beta=2*Lambda*Lambda*(muA*Gamma+muB*(1-Gamma))/lambda./zeta;
74
75  r_1=-2*i*Dsld*Lambda*Lambda*Gamma./zeta;
76  r_1=r_1.*sin(pi*Gamma*zeta)./(pi*Gamma*zeta);
77
78  r_N=r_1.*(1-exp(i*2*pi*zeta*N).*exp(-beta*N))./(1-exp(i*2*pi*zeta).*exp(-beta));
79
80  r_N=r_N.*exp(-((Q*sigma).^2/2));
81  R=r_N.*conj(r_N);
82
83  %%%%%%%%%%%%%%%%%%%%%%%%%%%%%%%%%%%%%%%%%%%%%%%%%%%%%%%%%%%%%%%%%%%%%%%%
84  function [RR]=parratt(Q,lambda,sld,d,sigma)
85  %
86  % MATLAB function from:
87  % "Elements of Modern X-ray Physics" by Jens Als-Nielsen and Des McMorrow
88  %
89  % Calculates: Parratt reflectivity of a multilayer
90  % Inputs: Q      wavevector transfer                1/Angs
91  %         lambda wavelength of radiation            Angs
92  %         sld    scattering length density          1/Angs^2
93  %                sld=[sld1+i*mu1 sld2+i*mu2 ....]
94  %         d      thickness of layer                 Angs
95  %                d=[d1 d2 .....];
96  %         sigma  rouhgness                          Angs
97  % Outputs:R      Intensity reflectivity
98
```

```
99  k=2*pi/lambda;
100
101 %----- Calculate refractive index n of each layer
102
103 delta=lambda^2*real(sld)/(2*pi); beta=lambda/(4*pi)*imag(sld);
104 n=size(sld,2);
105 nu=1-delta+i*beta;
106
107 %----- Wavevector transfer in each layer
108
109 Q=reshape(Q,1,length(Q));
110 x=asin(Q/2/k);
111 for j=1:n
112    Qp(j,:)=sqrt(Q.^2-8*k^2*delta(j)+i*8*k^2*beta(j));
113 end
114 Qp=[Q;Qp];
115
116 %----- Reflection coefficients (no multiple scattering)
117
118 for j=1:n
119    r(j,:)=((Qp(j,:)-Qp(j+1,:))./(Qp(j,:)+Qp(j+1,:))).*...
120       exp(-0.5*(Qp(j,:).*Qp(j+1,:))*sigma(j)^2);
121 end
122
123 %----- Reflectivity from first layer
124
125 RR=r(1,:);
126 if n>1
127    R(1,:)=(r(n-1,:)+r(n,:).*...
128       exp(i*Qp(n,:)*d(n-1)))./(1+r(n-1,:).*r(n,:).*exp(i*Qp(n,:)*d(n-1)));
129 end
130
131 %----- Reflectivity from more layers
132
133 if n>2
134   for j=2:n-1
135      R(j,:)=(r(n-j,:)+R(j-1,:).*...
136         exp(i*Qp(n-j+1,:)*d(n-j)))./(1+r(n-j,:).*R(j-1,:).*exp(i*Qp(n-j+1,:)*d(n-j)));
137   end
138 end
139
140 %------ Intensity reflectivity
141
142 if n==1
143   RR=r(1,:);
144 else
145   RR=R(n-1,:);
146 end
147
148 RR=(abs(RR).^2)';
149
150
151
152
153
```

Reflectivity from a Langmuir film, Fig. 3.14 on page 93

```
1  function lang_ref
2  %
3  % MATLAB function from:
4  % "Elements of Modern X-ray Physics" by Jens Als-Nielsen and Des McMorrow
5  %
6  % Calculates: Reflectivity from a lanngmuir layer
7  % Data: Langmuir Vol. 10 (1994) 826
8
9  axes('position',[0.15 0.20 0.60 0.75])
10
11 Q=0:0.01:1;              % wavevector transfer in 1/Angs
12
13 %----- (a) pH
14
```

```
15 rho_hw=2.28;              % density of head group, rho_head/rho_water
16 rho_tw=1.08;              % density of tail group, rho_tail/rho_water
17 l_h=6.2;                  % length of head  in Angs
18 l_t=22.0;                 % length of tail  in Angs
19 sigma=1.36;               % roughness in Angs
20 Qc=0.0217;                % critical Q for water in 1/Angs
21 mc=0.75;                  % monolayer coverage
22
23 phi1=Q*(l_h/2+l_t); phi2=Q*(l_h/2);
24 phi=exp(-Q.^2*sigma^2/2).*(rho_tw*exp(-i*phi1)+...
25     (rho_hw-rho_tw)*exp(-i*phi2)-(rho_hw-1)*exp(i*phi2));
26 R=mc*abs(phi).^2;
27 line(Q/Qc,R,'linewidth',1.5)
28
29 data_a=[0.90 1.11;1.20 1.19;1.50 1.35;1.58 1.41;1.88 1.51;2.09 1.60;2.35 1.73;...
30 2.48 1.82;2.65 1.97;2.86 2.09;3.25 2.17;3.89 2.40;4.27 2.43;4.61 2.33;4.91 2.19;...
31 5.25 2.03;5.51 1.87;5.94 1.52;6.28 1.25;6.45 1.04;6.79 0.82;7.18 0.58;7.35 0.46;...
32 7.65 0.31;8.12 0.17;8.54 0.12;9.06 0.19;9.44 0.34;9.70 0.52;10.12 0.92;10.51 1.36;...
33 10.72 1.81;11.15 2.43;11.53 3.03;11.83 3.38;12.22 3.92;12.52 4.49;12.77 4.97;...
34 13.20 5.23;13.58 5.66;13.88 5.61;14.18 5.73;14.48 6.02;14.82 5.96;15.04 5.89;...
35 15.55 5.80;15.89 5.43;16.23 5.00;16.62 4.53;16.96 4.00;17.26 3.65;17.60 3.14;...
36 17.86 2.78;18.28 2.48;18.54 1.90;18.92 1.86;19.18 1.68;19.57 1.39];
37 line(data_a(:,1),data_a(:,2),'Marker','square','MarkerSize',8,'linestyle','none')
38
39 %----- (b) pH
40
41 rho_hw=3.35;              % density of head group, rho_head/rho_water
42 rho_tw=1.01;              % density of tail group, rho_tail/rho_water
43 l_h=2.7;                  % length of head  in Angs
44 l_t=23.4;                 % length of tail  in Angs
45 sigma=2.74;               % roughness in Angs
46 Qc=0.0217;                % critical Q for water in 1/Angs
47 mc=0.75;                  % monolayer coverage
48
49 phi1=Q*(l_h/2+l_t); phi2=Q*(l_h/2);
50 phi=exp(-Q.^2*sigma^2/2).*(rho_tw*exp(-i*phi1)+...
51     (rho_hw-rho_tw)*exp(-i*phi2)-(rho_hw-1)*exp(i*phi2));
52 R=mc*abs(phi).^2;
53 line(Q/Qc,R,'linewidth',1.5)
54
55 data_b=[1.03 1.03;1.24 1.17;1.46 1.27;1.88 1.41;2.40 1.54;2.65 1.63;2.95 1.70;...
56 3.34 1.71;3.64 1.76;4.15 1.76;4.41 1.66;4.67 1.62;5.18 1.47;5.52 1.31;5.69 1.12;...
57 6.12 1.00;6.51 0.81;6.89 0.60;7.28 0.46;7.49 0.34;7.79 0.23;8.26 0.11;8.56 0.06;...
58 8.82 0.12;9.25 0.19;9.54 0.34;9.76 0.50;10.10 0.69;10.40 0.90;10.87 1.19;...
59 11.13 1.44;11.56 1.70;11.94 1.98;12.20 2.13;12.71 2.51;12.93 2.70;13.40 2.62;...
60 13.57 2.73;14.12 2.79;14.34 2.75;14.81 2.79;15.19 2.86;15.54 2.71;15.88 2.62;...
61 16.22 2.46;16.69 2.32;17.08 1.78;17.38 1.57;17.76 1.16;18.15 1.36;18.49 0.87;...
62 18.96 0.79;19.18 1.09;19.56 0.81];
63 line(data_b(:,1),data_b(:,2),'Marker','diamond',...
64    'MarkerSize',8,'MarkerFaceColor','b','linestyle','none')
65
66 axis([0 20 0 7])
67 set(gca,'FontName','Times','FontSize',36,'Xtick',[5 10 15 20 ])
68 xlabel('Q/Q_c'); ylabel('{\it R}/{\it R_F}','position',[-2.5 3.5 0])
69 box on; grid on
70 text(13.5,2.0,'(b)','FontName','Times','FontSize',24)
71 text(13.5,5.,'(a)','FontName','Times','FontSize',24)
```

Chapter 4: Kinematical diffraction

The Fibonacci chain, Fig. 4.15 on page 134

```
1 function [xn]=quasi
2 %
3 % MATLAB function from:
4 % "Elements of Modern X-ray Physics" by Jens Als-Nielsen and Des McMorrow
5 %
6 % Calculates: Positionso of atoms xn in a Fibonacci chain from the strip projection method,
7 %             and calculates the scattered intensity
8 % Calls to: pline, isinpoly, arrow
```

```
 9
10  figure; axes('position',[0.15 0.15 0.8 0.8],'visible','off'); axis equal
11
12  tau=(1+sqrt(5))/2;           % golden mean
13  latp=sqrt(1+tau^2);          % lattice parameter of 2D lattice
14  Nx=10; Ny=10;                % number of lattice points
15  angle=atan(1/tau)*180/pi;    % angle of strip
16  Delta=1+tau;                 % width of strip
17
18  %----- Draw strip and rotate
19  h=patch([0 Nx*latp/cos(angle*pi/180) Nx*latp/cos(angle*pi/180) 0],...
20          [0 0 Delta Delta],[0.7 0.7 1]);
21  rotate(h,[0 0 1],angle,[0 0 0])
22  vp=get(h,'Vertices');
23
24  %----- Draw lattice
25  x=[]; for i=0:Nx; for j=0:Ny; x=[x;i j]; end; end
26  x=x*latp;
27  line(x(:,1),x(:,2),'linestyle','none','marker','o','markerfacecolor','g','markersize',6);
28
29  %----- Find lattice points that lie in strip
30  isp=isinpoly(x(:,1),x(:,2),vp(:,1),vp(:,2)); x(find(isp~=1),:)=[];
31
32  b=vp(2,1:2);                  % end point of line xn
33  %----- Draw perpendicular lines from points in strip to xn
34  xn=[0]; yn=[0];
35  for ix=1:length(x)
36      [intx,inty]=pline(b,x(ix,:));
37      xn=[xn; intx]; yn=[yn; inty];
38  end
39  line(x(:,1),x(:,2),'linestyle','none','marker','o','markerfacecolor','w','markersize',6);
40
41  %------ Label the graph
42  xnd=diff(xn);L=max(xnd); S=min(xnd);
43  for id=1:length(xnd)
44  if abs(xnd(id)-L)< 0.02 ,col=[0.6 0.6 0.6]; lab='L'; else col=[1 0 0]; lab='S'; end
45  line([xn(id) xn(id+1)],[yn(id) yn(id+1)],'color',col,'linewidth',2.0)
46  text(0.5*(xn(id)+xn(id+1)),-2,lab,'color',col,'horizontalalignment','center',...
47       'Fontsize',18,'FontName','Times')
48  end
49  arrow([-2 5*latp],[-2 6*latp],8,'ends','both')
50  text(-7,5.5*latp,'\surd(1+\tau^2)','Fontsize',24,'FontName','Times')
51  arrow(b,b*1.10,10)
52  text(b(1)*1.10,b(2)*1.0-0.2,'{\it x}_n','Fontsize',24,'FontName','Times',...
53       'horizontalalignment','center')
54  h=arrow(-0.1*b,-0.1*b+[0 1+tau],10,'ends','both'); rotate(h,[0 0 1],angle,[-0.1*b 0])
55  text(-0.1*b(1),-0.1*b(1)+2.5,'\Delta','Fontsize',24,'FontName','Times')
56  circ=4.5*latp; ax=circ*cos(angle*pi/180):0.005:circ; ay=sqrt(circ^2-ax.^2); line(ax,ay);
57
58  xn=xn./cos(angle*pi/180);
59
60  %----- Calculate scttareing from chain
61
62  figure; axes('position',[0.15 0.15 0.8 0.8])
63  Q=[0:0.01:20];
64  F=sum(exp(sqrt(-1)*xn*Q));
65  plot(Q,F.*conj(F))
66  set(gca,'FontName','Times','FontSize',24,'Xtick',[0 5 10 15 20])
67  xlabel('Wavevector transfer (^{-1})'); ylabel('Intensity')
68
69  function [intx,inty]=pline(b,c)
70  %
71  % MATLAB function from:
72  % "Elements of Modern X-ray Physics" by Jens Als-Nielsen and Des McMorrow
73  %
74  % Calculates: Draws a perpendicular line from point c(x,y)
75  %             to line that starts at origin and ends at point b(x,y)
76
77  if norm(c)==0
78     tc=0;
79  elseif c(1)==0 & c(2)~=0
80     tc=pi/2;
81  else
82     tc=atan(c(2)./c(1));
```

```
83  end
84  tb=atan(b(2)./b(1)); dt=tc-tb;
85
86  intx=norm(c)*cos(dt)*cos(tb); inty=norm(c)*cos(dt)*sin(tb);
87  line([c(1) intx],[c(2) inty],'color','w','linewidth',1.5)
88
89  function  isin = isinpoly(x,y,xp,yp)
90  % ISIN = ISINPOLY(X,Y,XP,YP)   Finds whether points with coordinates X and Y are inside
91  %        or outside of a polygon with vertices XP, YP. Returns matrix ISIN of the same
92  %        size as X and Y with 0 for points outside a polygon, 1 for inside points and
93  %        0.5 for points belonging to a polygon XP, YP itself.
94  %  Copyright (c) 1995  by Kirill K. Pankratov
95  %       kirill@plume.mit.edu, 4/10/94, 8/26/94.
96
97  %----- Handle input
98  if nargin<4
99    fprintf('\n  Error: not enough input arguments.\n\n')
100   return
101 end
102 %----- Make the contour closed and get the sizes
103 xp = [xp(:); xp(1)]; yp = [yp(:); yp(1)];
104 sz = size(x); x = x(:); y = y(:);
105 lp = length(xp); l = length(x);
106 ep = ones(1,lp); e = ones(1,l);
107 %----- Calculate cumulative change in azimuth from points x,y to all vertices
108 A = diff(atan2(yp(:,e)-y(:,ep)',xp(:,e)-x(:,ep)'))/pi;
109 A = A+2*((A<-1)-(A>1));
110 isin = any(A==1)-any(A==-1);
111 isin = (abs(sum(A))-isin)/2;
112 %----- Check for boundary points
113 A = (yp(:,e)==y(:,ep)')&(xp(:,e)==x(:,ep)');
114 fnd = find(any(A));
115 isin(fnd) = .5*ones(size(fnd));
116 isin = round(isin*2)/2;
117 %----- Reshape output to the input size
118 isin = reshape(isin,sz(1),sz(2));
```

Crystal Truncation Rod properties, Fig. 4.17 on page 137

```
1  function ctr
2  %
3  % MATLAB function from:
4  % "Elements of Modern X-ray Physics" by Jens Als-Nielsen and Des McMorrow
5  %
6  % Calculates: Properties of the Crystal Truncation Rod
7
8  figure
9  set(gcf,'papertype','a4','paperunits','centimeters','units','centimeters',...
10         'position',[0.1 -8 21 26],'paperposition',[0.1 0.1 21 26]);
11
12 %%%%%%%%%%%%%%%%%%%%%%%%%%%%%%%%%%%%%%%%%%%%%%%%%%%%%%%%%%%%%%%%%%%%%%%%%%%%%%%%%%%%
13 %Plot rod from flat surface without (beta=0) and with (beta=0.2) absorption
14 %%%%%%%%%%%%%%%%%%%%%%%%%%%%%%%%%%%%%%%%%%%%%%%%%%%%%%%%%%%%%%%%%%%%%%%%%%%%%%%%%%%%
15
16 axes('position',[0.2 0.55 0.6 0.35])
17
18 ell1=[0.01:0.001:0.99];       % beta=0, l range chosen to avoid Bragg peak at l=1
19 F_CTR=1./(1-exp(i*2*pi*ell1));
20 h1=line(ell1,F_CTR.*conj(F_CTR),'color','b','linewidth',1,'linestyle','-')
21 ell2=[1.01:0.001:1.99];
22 F_CTR=1./(1-exp(i*2*pi*ell2));
23 line(ell2,F_CTR.*conj(F_CTR),'color','b','linewidth',1,'linestyle','-')
24
25 ell=[0.01:0.001:1.99];        % beta=0.2, l range now includes Bragg peak at l=1
26 beta=0.2;
27 F_CTR=1./(1-exp(i*2*pi*ell)*exp(-beta));
28 h2=line(ell,F_CTR.*conj(F_CTR),'color','b','linewidth',1,'linestyle','-.')
29
30 set(gca,'Fontsize',16,'FontName','Times')
31 [h,obj]=legend([h1 h2],'\beta=0','\beta=0.2')
32 set(gca,'Fontsize',18,'FontName','Times')
33 set(gca,'FontName','Times','FontSize',18)
```

```
34 xlabel('{\it l} (r.l.u.)'); ylabel('|{\it F }^{CTR}|^2'); box on
35 axis([0.0 2.0 0.1 1000])
36 set(gca,'Ytick',[0.1 1 10 100 1000],'Yscale','Log')
37
38 %%%%%%%%%%%%%%%%%%%%%%%%%%%%%%%%%%%%%%%%%%%%%%%%%%%%%%%%%%%%%%%%%%%%%%%%%%%%%%%%
39 %Plot rod from flat surface + overlayer at different relative diplacements, z0
40 %%%%%%%%%%%%%%%%%%%%%%%%%%%%%%%%%%%%%%%%%%%%%%%%%%%%%%%%%%%%%%%%%%%%%%%%%%%%%%%%
41
42 axes('position',[0.2 0.12 0.6 0.35])
43
44 ell1=[0.01:0.001:0.99];          %l range chosen to avoid Bragg peak at l=1
45 F_CTR=1./(1-exp(i*2*pi*ell1));
46 line(ell1,F_CTR.*conj(F_CTR),'color','b','linewidth',1,'linestyle','-')
47 ell2=[1.01:0.001:1.99];
48 F_CTR=1./(1-exp(i*2*pi*ell2));
49 h1=line(ell2,F_CTR.*conj(F_CTR),'color','b','linewidth',1,'linestyle','-')
50
51 z0=0.05;                         % relative displacement of overlayer, z0=0.05
52 F_CTR=1./(1-exp(i*2*pi*ell1));
53 F_T=F_CTR+exp(-i*2*pi*(1+z0)*ell1);
54 line(ell1,F_T.*conj(F_T),'color','b','linewidth',1,'linestyle','--')
55 F_CTR=1./(1-exp(i*2*pi*ell2));
56 F_T=F_CTR+exp(-i*2*pi*(1+z0)*ell2);
57 h2=line(ell2,F_T.*conj(F_T),'color','b','linewidth',1,'linestyle','--')
58
59 z0=-0.05;                        % relative displacement of overlayer, z0=-0.05
60 F_CTR=1./(1-exp(i*2*pi*ell1));
61 F_T=F_CTR+exp(-i*2*pi*(1+z0)*ell1);
62 line(ell1,F_T.*conj(F_T),'color','b','linewidth',1,'linestyle','-.')
63 F_CTR=1./(1-exp(i*2*pi*ell2));
64 F_T=F_CTR+exp(-i*2*pi*(1+z0)*ell2);
65 h3=line(ell2,F_T.*conj(F_T),'color','b','linewidth',1,'linestyle','-.')
66
67 set(gca,'FontName','Times','FontSize',18)
68 xlabel('{\it l} (r.l.u.)'); ylabel('|{\it F }^{CTR}|^2'); box on
69 set(gca,'Fontsize',14,'FontName','Times')
70 [h,obj]=legend([h1 h2 h3],'{\it z}_0=0','{\it z}_0=0.05','{\it z}_0=-0.05');
71 set(gca,'Fontsize',18,'FontName','Times')
72 axis([0. 2.0 0.1 1000])
73 set(gca,'Ytick',[0.1 1 10 100 1000],'Yscale','log')
```

Debye-Waller factor of Aluminium, Fig. 4.19 on page 143

```
1  function DebyeWaller
2  %
3  % MATLAB function from:
4  % "Elements of Modern X-ray Physics" by Jens Als-Nielsen and Des McMorrow
5  %
6  % Calculates:  The Debye-Waller factor for Aluminium
7  % Calls to: phiDebye
8
9  figure
10 set(gcf,'papertype','a4','paperunits','centimeters','units','centimeters',...
11         'position',[0.1 -8 21 26],'paperposition',[0.1 0.1 21 26]);
12
13 %%%%%%%%%%%%%%%%%%%%%%%%%%%%%%%%%%%%%%%%%%%%%%%%%%%%%%%%%%%%%%%%%%%%%%%%%%%%%%%%
14 % Plot of phi(x) vs x.
15 %%%%%%%%%%%%%%%%%%%%%%%%%%%%%%%%%%%%%%%%%%%%%%%%%%%%%%%%%%%%%%%%%%%%%%%%%%%%%%%%
16
17 axes('position',[0.30 0.70 0.45 0.225]);
18
19 x=0.01:0.02:8;
20 for il=1:length(x), phi(il)=phiDebye(x(il));   end
21
22 line(x,phi,'color','b','linewidth',1.5)
23
24 axis([0 8 0 1.1]); grid on
25 set(gca,'FontName','Times','FontSize',16,'box','on')
26 ylabel('phi(x)','position',[-1.5 0.55 0]); xlabel('Theta/T')
27
28 %%%%%%%%%%%%%%%%%%%%%%%%%%%%%%%%%%%%%%%%%%%%%%%%%%%%%%%%%%%%%%%%%%%%%%%%%%%%%%%%
29 % Plot of sqrt(u^2) vs Temperature for Al
```

```
%%%%%%%%%%%%%%%%%%%%%%%%%%%%%%%%%%%%%%%%%%%%%%%%%%%%%%%%%%%%%%%%%%%%%%%%%%%%%%%%%

axes('position',[0.30 0.40 0.45 0.225]);

Theta_Al=394;            % Debye temperature of Al
A=27;                    % Atomic mass
nnd=4.04/sqrt(2);        % Nearest neighbour distance
T=x*394;
B_Al=11492.*T.*phi/A/Theta_Al/Theta_Al+2873/A/Theta_Al;
rms=sqrt(3/8/pi/pi.*B_Al);
iT=find(T<933); iTg=find(T>=933);
line(T(iT),sqrt(2)*rms(iT)/4.04,'color','g','linewidth',1.5)
line(T(iTg),sqrt(2)*rms(iTg)/4.04,'color','g','linewidth',1.5,'linestyle',':')

axis([0 1050 0 0.1]); grid on
set(gca,'FontName','Times','FontSize',16,'box','on')
xlabel('Temperature (K)','position',[500 -0.017 0])
ylabel('rms','position',[-200 0.05 0])

%%%%%%%%%%%%%%%%%%%%%%%%%%%%%%%%%%%%%%%%%%%%%%%%%%%%%%%%%%%%%%%%%%%%%%%%%%%%%%%%%
% Temperature dependence at different Q's
%%%%%%%%%%%%%%%%%%%%%%%%%%%%%%%%%%%%%%%%%%%%%%%%%%%%%%%%%%%%%%%%%%%%%%%%%%%%%%%%%

axes('position',[0.30 0.10 0.45 0.225]);

I400=exp(-(8/4.04/4.04.*B_Al))./exp(-(8/4.04/4.04.*B_Al(1)));
I800=exp(-(32/4.04/4.04.*B_Al))./exp(-(32/4.04/4.04.*B_Al(1)));

line(T(iT),I400(iT),'color','r','LineWidth',1.5)
line(T(iT),I800(iT),'color','r','Linestyle','--','LineWidth',1.5)

axis([0 1050 0 1.1]); grid on
set(gca,'FontName','Times','FontSize',16,'box','on')
xlabel('Temperature (K)','position',[500 -0.2 0]);
ylabel('Relative Intensity','position',[-200 0.55 0])
text(650,0.30,'(8,0,0)','FontName','Times','FontSize',16)
text(650,0.70,'(4,0,0)','FontName','Times','FontSize',16)

function phi=phiDebye(x)
%
% MATLAB function from:
% "Elements of Modern X-ray Physics" by Jens Als-Nielsen and Des McMorrow
%
% Calculates: Evaluates the integral to calculate phi(x)
% Calls to: phiDebyeInt

phi=quad8('phiDebyeInt',0.000000001,x)./x;

function y=phiDebyeInt(xi)
%
% MATLAB function from:
% "Elements of Modern X-ray Physics" by Jens Als-Nielsen and Des McMorrow
%
% Calculates: Defines the integrabd used to evaluate phi(x)
% Note: Must be placed in a separate file called phiDebyeInt.m

y=xi./(exp(xi)-1);
```

Fibre diffraction from DNA, Fig. 4.31 on page 164

```
function dna
%
% MATLAB function from:
% "Elements of Modern X-ray Physics" by Jens Als-Nielsen and Des McMorrow
%
% Calculates: Fibre diffraction pattern from DNA

[x,y]=meshgrid(-11:0.2:11,-11:0.2:11);

iw=20;    %iw is the inverse width of a Bragg peak, here modelled as a Gaussian
z=zeros(size(x));
for il=-11:11
```

```
13   z=z+ abs((1+exp(i*il*2*pi*0.125)).*besselj(abs(il),x)).^2.*exp(-iw*(y+il).^2);
14 end
15
16 pcolor(x,y,z)
17 shading interp
18
19 axis equal; axis([-11 11 -11 11]); caxis([-0.1 2]); box on
20 set(gca,'FontName','Times','FontSize',18,'Position',[0.15 0.15 0.7 0.7])
21
22 colormap(1-gray)
23 caxis([-0.05 0.3])
24 set(gca,'dataaspectratio',[1*34/20 1 1])
```

Crystal truncation rods from O on Cu(110), Fig. 4.34 on page 170

```
1  function cuoctr
2  %
3  % MATLAB function from:
4  % "Elements of Modern X-ray Physics" by Jens Als-Nielsen and Des McMorrow
5  %
6  % Calculates: CTR of O on Cu (110) and compares with data
7  % Calls to: ff
8  % Data: Feidenhans'l et al., Phys. Rev. B., vol. 41, page 5420 (1990)
9
10 %----- Cu real and reciprocal lattice parameters
11
12 ac=3.615; ar=2*pi/(ac/sqrt(2)); br=2*pi/ac; cr=2*pi/(ac/sqrt(2));
13
14 %%%%%%%%%%%%%%%%%%%%%%%%%%%%%%%%%%%%%%%%%%%%%%%%%%%%%%%%%%%%%%%%%%%%%%%%%%%%%%%%
15 %(1,1) rod
16 %%%%%%%%%%%%%%%%%%%%%%%%%%%%%%%%%%%%%%%%%%%%%%%%%%%%%%%%%%%%%%%%%%%%%%%%%%%%%%%%
17
18 figure; axes('position',[0.55 0.15 0.35 0.8]);
19
20 h=1; k=1; l=0.05:0.001:1.0; Q=sqrt(h^2*ar^2+k^2*br^2+l.^2*cr^2);
21
22 %----- Cu form factor
23
24 a=[13.338 7.1676 5.6158 1.6735]; b=[3.5828 0.2470 11.3966 64.82]; c=[1.1910];
25 f_Cu=ff(a,b,c,Q);
26
27 %------ Cu Debye-Waller factor for bulk (B) and surface (S)
28
29 DW_Cu_B=exp(-0.55*(Q/4/pi).^2); DW_Cu_S=exp(-1.70*(Q/4/pi).^2);
30
31 %----- O form factor
32
33 a=[3.0485 2.2868 1.5463 0.8670]; b=[13.2771 5.7011 0.3239 32.9089]; c=[0.2508];
34 f_O=ff(a,b,c,Q);
35
36 %----- Bulk CTR
37
38 Phi=pi*(h+k+l); F_CTR=f_Cu.*DW_Cu_B./(1-exp(i*Phi));
39 semilogy(l,4*abs(F_CTR).^2,'-.');
40 axis([0 1 6 100000 ])
41 set(gca,'FontName','Times','FontSize',16,'Ytick',[1e1 1e2 1e3 1e4 1e5])
42 text(0.1,50000,'(b) (1,1) rod','FontName','Times','FontSize',16)
43
44 %----- Add 1/2 a monolayer of Cu (no relaxation)
45
46 F_S=0.5*f_Cu.*DW_Cu_S*exp(i*pi*(h+k)).*exp(-i*2*pi*0.5*l);
47 F_T=F_CTR+F_S;
48 line(l,4*abs(F_T).^2,'linestyle','--');
49
50 %----- Add 1/2 a monolayer of Cu (relaxed to z0) plus O layer (relaxed to -z1)
51
52 z0=0.1445; z1=z0-0.133;
53 F_S=0.5*exp(i*pi*(h+k))*(f_Cu.*exp(-i*2*pi*(0.5+z0)*l).*DW_Cu_S...
54                         +f_O.*exp(i*pi*k).*exp(-i*2*pi*(0.5+z1)*l));
55 F_T=F_CTR+F_S;
```

```
56  line(l,4*abs(F_T).^2,'linestyle','-');
57
58  %----- Add 1/2 a monolayer of Cu (relaxed to z0) plus O layer (relaxed to +z1)
59
60  z0=0.1445; z1=z0+0.133;
61  F_S=0.5*exp(i*pi*(h+k))*(f_Cu.*exp(-i*2*pi*(0.5+z0)*l).*DW_Cu_S...
62                           +f_O.*exp(i*pi*k).*exp(-i*2*pi*(0.5+z1)*l));
63  F_T=F_CTR+F_S;
64  line(l,4*abs(F_T).^2,'linestyle',':');
65
66  data=[0.0787 4.5555;0.1517 3.8388;0.2247 3.5032;0.2978 3.1584;0.3539 3.0586;...
67        0.3708 2.8954;0.4382 2.6051;0.5169 2.3692;0.5787 2.1878;0.6517 1.9700;...
68        0.7360 1.9156];
69  line(data(:,1),10.^data(:,2),'marker','o','linestyle','none','markerfacecolor','w')
70
71  %%%%%%%%%%%%%%%%%%%%%%%%%%%%%%%%%%%%%%%%%%%%%%%%%%%%%%%%%%%%%%%%%%%%%%%%%%%%%%%%
72  %(1,0) rod
73  %%%%%%%%%%%%%%%%%%%%%%%%%%%%%%%%%%%%%%%%%%%%%%%%%%%%%%%%%%%%%%%%%%%%%%%%%%%%%%%%
74
75  axes('position',[0.10 0.15 0.35 0.8]);
76
77  h=1; k=0; l=0.0:0.001:0.95; Q=sqrt(h^2*ar^2+k^2*br^2+l.^2*cr^2);
78
79  %----- Cu form factor
80
81  a=[13.338 7.1676 5.6158 1.6735]; b=[3.5828 0.2470 11.3966 64.82]; c=[1.1910];
82  f_Cu=ff(a,b,c,Q);
83
84  %----- Cu Debye-Waller factor for bulk (B) and surface (S)
85
86  DW_Cu_B=exp(-0.55*(Q/4/pi).^2); DW_Cu_S=exp(-1.70*(Q/4/pi).^2);
87
88  %----- O form factor
89
90  a=[3.0485 2.2868 1.5463 0.8670]; b=[13.2771 5.7011 0.3239 32.9089]; c=[0.2508];
91  f_O=ff(a,b,c,Q);
92
93  %----- Bulk unit cell SF
94
95  Phi=pi*(h+k+l); F_CTR=f_Cu.*DW_Cu_B./(1-exp(i*Phi));
96
97  semilogy(l,4*F_CTR.*conj(F_CTR),'-.');
98  axis([0 1 6 100000 ])
99  set(gca,'FontName','Times','FontSize',16,'Ytick',[1e1 1e2 1e3 1e4 1e5])
100 text(0.1,50000,'(a) (1,0) rod','FontName','Times','FontSize',16)
101 ylabel('Intensity (electron units)','Fontsize',18)
102 text(1.15,2,'l (r.l.u.)','FontName','Times','FontSize',18,...
103          'HorizontalAlignment','Center')
104
105 %----- Add 1/2 a monolayer of Cu (not relaxed)
106
107 F_S=0.5*exp(i*pi*(h+k))*f_Cu.*DW_Cu_S.*exp(-i*2*pi*(0.5+z0)*l);
108 F_T=F_CTR+F_S.*DW_Cu_B;
109 line(l,4*abs(F_T).^2,'linestyle','--');
110
111 %----- Add 1/2 a monolayer of Cu (relaxed to z0) plus O layer (relaxed to -z1)
112
113 z0=0.1145; z1=z0-0.133;
114 F_S=0.5*exp(i*pi*(h+k))*(f_Cu.*exp(-i*2*pi*(0.5+z0)*l).*DW_Cu_S...
115                          +f_O.*exp(i*pi*k).*exp(-i*2*pi*(0.5+z1)*l));
116 F_T=F_CTR+F_S;
117 line(l,4*abs(F_T).^2,'linestyle','-');
118
119 %----- Add 1/2 a monolayer of Cu (relaxed to z0) plus O layer (relaxed to +z1)
120
121 z0=0.1145; z1=z0+0.133;
122 F_S=0.5*exp(i*pi*(h+k))*(f_Cu.*exp(-i*2*pi*(0.5+z0)*l).*DW_Cu_S...
123                          +f_O.*exp(i*pi*k).*exp(-i*2*pi*(0.5+z1)*l));
124 F_T=F_CTR+F_S;
125 line(l,4*abs(F_T).^2,'linestyle',':');
126
127 %----- Plot data
128
129 data=[0.0226 1.8793;0.0960 1.9973;0.1695 2.2785;0.2429 2.4781;0.3220 2.7139;...
```

```
130        0.3955 2.8228;0.4689 2.9861;0.5480 3.1222;0.6158 3.2310;0.6949 3.4124;...
131        0.7740    3.6574];
132  line(data(:,1),10.^data(:,2),'marker','o','linestyle','none','markerfacecolor','w')
```

In-plane Bragg reflections of O on Cu(110), Table 4.3 on page 169

```
1   function Iout=cuFS(hp,k)
2   %
3   % MATLAB function from:
4   % "Elements of Modern X-ray Physics" by Jens Als-Nielsen and Des McMorrow
5   %
6   % Calculates: In-plane Bragg peak intensities for O on Cu(110)
7   % Inputs: (hp,k), Miller indices of Bragg peak
8   % Outputs: Iout, Intensity
9   % Calls to: ff
10
11  %----- Cu real and reciprocal lattice parameters
12
13  ac=3.615; ar=2*pi/(2*ac/sqrt(2)); br=2*pi/ac; cr=2*pi/(ac/sqrt(2));
14
15  l=0; Q=sqrt(hp^2*ar^2+k^2*br^2+l.^2*cr^2);
16
17  %----- Cu form factor
18
19  a=[13.338 7.1676 5.6158 1.6735]; b=[3.5828 0.2470 11.3966 64.82]; c=[1.1910];
20  f_Cu=ff(a,b,c,Q);
21
22  %----- Cu Debye-Waller factor for bulk and surface
23
24  DW_Cu_B=exp(-0.55*(Q/4/pi).^2);
25  DW_Cu_S=exp(-1.70*(Q/4/pi).^2);
26
27  %----- O form factor
28
29  a=[3.0485 2.2868 1.5463 0.8670]; b=[13.2771 5.7011 0.3239 32.9089]; c=[0.2508];
30  f_O=ff(a,b,c,Q);
31
32  delta=0.031/(2*ac/sqrt(2));
33
34  F_1=f_Cu.*DW_Cu_S+f_O*exp(i*pi*k);
35  F_2=(-1)^(hp/2+k+0.5)*2.*f_Cu.*DW_Cu_B*sin(2*pi*hp*delta);
36  F_S=F_1+F_2;
37  Iout=1.047539547173480e-002*abs(F_S).^2;
```

Form factor

```
1   function fofQ=ff(a,b,c,Q)
2   %
3   % MATLAB function from:
4   % "Elements of Modern X-ray Physics" by Jens Als-Nielsen and Des McMorrow
5   %
6   % Calculates: X-ray form factor as a function of Q
7   % Inputs: (a,b,c), coeffics. from ITC, Q
8   % Outputs: fofQ, form factor
9   % Note: Q is given by 4*pi*sin(theta)/lambda.
10
11  %----- Convert Q to be compatible with the definition
12  %      in the International Tables of Crystallography
13
14  Q=Q/(4*pi);
15  fofQ=a(1)*exp(-b(1)*Q.^2)+...
16       a(2)*exp(-b(2)*Q.^2)+a(3)*exp(-b(3)*Q.^2)+a(4)*exp(-b(4)*Q.^2)+c;
```

Chapter 5: Diffraction by perfect crystals

Darwin curve including absorption, Fig. 5.8 on page 193

```
function darabs
%
% MATLAB function from:
% "Elements of Modern X-ray Physics" by Jens Als-Nielsen and Des McMorrow
%
% Calculates: Darwin reflectivity curve of Si (111), including absorption

set(gcf,'papertype','a4','paperunits','centimeters','units','centimeters')
set(gcf,'position',[0.1 -8 21 26],'paperposition',[0.1 0.1 21 26])

axes('Position',[0.2 0.60 0.6 0.40])

% Case 1: lambda=1.5405 Angs
r0=2.82E-5;                                 % Thompson scattering length in Angs
V=160.1966;                                 % unitcell volume in Ang^3 e.g. 160.1966 for Si
d=3.13562;                                  % d spacing for Si (111)
m=1;                                        % order of reflection, ie 1 for (111), 3 for (333)
F_hkl=abs(4-4*i)*(10.54+0.25-i*0.33);       % Complex structure factor for 111
F_0=8*(14+0.25-i*0.33);                     % Complex structure factor for 000
g=(2*d*d/m)*(r0/V)*F_hkl;
g0=g*(F_0/F_hkl);
[x,R]=darwin(g,g0,m); line(x,R,'color','b','linestyle','--')

axis([-2 2 0 1.1])
set(gca,'Ytick',[0 0.5 1],'FontSize',20,'FontName','Times')
xlabel('x','position',[0 -0.1 0]); ylabel('Intensity reflectivity','position',[-2.75 0.5 0])
box on; grid on

% Case 2: lambda=0.70926 Angs
F_0=8*(14+0.082-i*0.071);                   % Complex structure factor for 000
F_hkl=abs(4-4*i)*(10.54+0.082-i*0.071);     % Complex structure factor for 111
g=(2*d*d/m)*(r0/V)*F_hkl;
g0=g*(F_0/F_hkl);
[x,R]=darwin(g,g0,m);
line(x,R,'color','b','linestyle','-');
text(-1.80,0.95,'(a)','FontName','Times','Fontsize',24)

%%%%%%%%%%%%%%%%%%%%%%%%%%%%%%%%%%%%%%%%%%%%%%%%%%%%%%%%%%%%%%%%%%%%%%%%%%%%%%%%%%%%%%%%%%%%
% Plot as a function of energy and angular variable in milli degrees

axes('Position',[0.2 0.12 0.6 0.40])

lambda=12.398/5.000;                        % 5 keV
theta=asin(m*lambda/2/d);
F_hkl=abs(4-4*i)*(10.54+0.38-i*0.8029);     % Complex structure factor for 111
F_0=8*(14+0.3807-i*0.8029);                 % Complex structure factor for 000
g=(2*d*d/m)*(r0/V)*F_hkl;
g0=g*(F_0/F_hkl);
[x,R]=darwin(g,g0,m);
line(x*real(g/m/pi)*tan(theta)*180/pi*1e3,R,'color','b','linestyle','--')

lambda=12.398/10.000;                        % 10 keV
theta=asin(m*lambda/2/d);
F_hkl=abs(4-4*i)*(10.54+0.1943-i*0.2169);    % Complex structure factor for 111
F_0=8*(14+0.1943-i*0.2169);                  % Complex structure factor for 000
g=(2*d*d/m)*(r0/V)*F_hkl;
g0=g*(F_0/F_hkl);
[x,R]=darwin(g,g0,m);
line(x*real(g/m/pi)*tan(theta)*180/pi*1e3,R,'color','b','linestyle','--')

lambda=12.398/50.000;                          % 50 keV
theta=asin(m*lambda/2/d);
F_hkl=abs(4-4*i)*(10.54+0.0027-i*0.0076);      % Complex structure factor for 111
F_0=8*(14+0.0027-i*0.0076);                    % Complex structure factor for 000
g=(2*d*d/m)*(r0/V)*F_hkl;
g0=g*(F_0/F_hkl);
[x,R]=darwin(g,g0,m);
line(x*real(g/m/pi)*tan(theta)*180/pi*1e3,R,'color','b','linestyle','-')

```

```
70  axis([-3 3 0 1.1])
71  set(gca,'Xtick',[-2 -1 0 1 2],'Ytick',[0 0.5 1],'FontSize',20,'FontName','Times')
72  xlabel('\omega (milli degrees)');
73  ylabel('Intensity reflectivity','position',[-4.15 0.55 0])
74  box on; grid on
75  text(-2.85,0.25,'5 keV','FontName','Times','Fontsize',18)
76  text(-2.10,0.15,'10 keV','FontName','Times','Fontsize',18)
77  text(-1.30,0.05,'50 keV','FontName','Times','Fontsize',18)
78  text(-2.70,0.95,'(b)','FontName','Times','Fontsize',24)
79
80  function [x,R]=darwin(g,g0,m);
81  %
82  % MATLAB function from:
83  % "Elements of Modern X-ray Physics" by Jens Als-Nielsen and Des McMorrow
84  %
85  % Calculates: Darwin reflectivity R vs x  (absorption effects included)
86
87  x_m=[-5:0.01:-1]; zeta=real((g*x_m+g0)/m/pi); xc_m=m*pi*zeta/g-g0/g;
88  rc_m=xc_m+sqrt(xc_m.^2-1);
89  x_t=[-1:0.01:1]; zeta=real((g*x_t+g0)/m/pi); xc_t=m*pi*zeta/g-g0/g;
90  rc_t=xc_t-i*sqrt(1-xc_t.^2);
91  x_p=[1:0.01:5]; zeta=real((g*x_p+g0)/m/pi); xc_p=m*pi*zeta/g-g0/g;
92  rc_p=xc_p-sqrt(xc_p.^2-1);
93  x=[x_m x_t x_p]; rc=[rc_m rc_t rc_p];
94  R=abs(rc).^2;
```

Chapter 6: Photoelectric Absorption

K absorption edge of Kr, Fig. 6.5 on page 214

```
1   function kedge
2   %
3   % MATLAB function from:
4   % "Elements of Modern X-ray Physics" by Jens Als-Nielsen and Des McMorrow
5   %
6   % Calculates: Photoelectric absorption cross-section of Kr - comparison
7   %             between hydrogen-like model of K shell contribution and the
8   %             self-consistent Dirac-Hartree-Fock theory
9   %             (C.T. Chantler, J. Phys. Chem. Ref. Data vol. 24, 71 (1995))
10  % Calls to: loaddata
11
12  [en,sigma,thom_comp]=loaddata;                  % load theoretical values
13
14  %----- Plot theoretical photoelectron and (Thomson+Compton) cross-sections
15
16  line(en,sigma,'linestyle','-.'); line(en,thom_comp,'linestyle','--')
17
18  %----- Find L shell photoelectric contribution below 14.3 keV
19
20  xl=find(en<14.3); xg=find(en>14.32);
21  xlog=log10(en(xl));
22  ylog=log10(sigma(xl));
23  [P,S]=polyfit(xlog,ylog,1);
24  ylogfit=polyval(P,log10(en));
25  y=10.^ylogfit;
26  line(en(xg),y(xg),'linestyle',':')             % Plot L contribution for E>14.32
27
28  %----- Plot theorectical absorption for K shell from Stoppe theory
29
30  en=en(xg); y=y(xg);
31  r0=2.82e-5;                                     % Thomson scattering length
32  ek=14.32;                                       % K edge of Kr energy in keV
33  lambda=12.398./en;
34  xi=sqrt(ek./(en-ek));
35  f=2*pi*sqrt(ek./en).*exp(-4*xi.*acot(xi))./(1-exp(-2*pi*xi));
36  sigmaa=256/3.*lambda.*(ek./en).^2.5.*f*r0*1e8;
37  line(en,sigmaa+y,'color','r','linewidth',2)
38  axis([5 40 50 100000])
39  set(gca,'FontName','Times','FontSize',18,'Xtick',[5 10 20 50],'Xscale','log','Yscale','log')
40  grid on; box on
```

```
41 xlabel('Photon energy [keV]');ylabel('Absorption cross-section [barn]')
42
43 text(6,150,'Thomson+Compton','FontName','Times','Fontsize',18)
44 text(10,20000,'L edges','FontName','Times','Fontsize',18,'horizontalalignment','center')
45 text(25,20000,'K + L edges','FontName','Times','Fontsize',18,'horizontalalignment','center')
46
47 function [en,sigma,thom_comp]=loaddata
48 %
49 % MATLAB function from:
50 % "Elements of Modern X-ray Physics" by Jens Als-Nielsen and Des McMorrow
51 %
52 % Photoelectric and Thomson+Compton cross-sections for Kr
53
54 en=[5.3 5.66 6.05 6.47 6.92 7.39 7.9 8.45 9.03 9.65 10.3 11 11.8 12.6 13.5 14 14.3...
55    14.3 14.4 14.6 15.4 16.5 17.6 18.8 20.1 21.5 23 24.6 26.3 28.1 30 32.1 34.3 36.7 39.2];
56 sigma=[3.83e+004 3.20e+004 2.67e+004 2.23e+004 1.87e+004 1.57e+004 1.31e+004 1.09e+004...
57        9.04e+003 7.43e+003 6.11e+003 5.03e+003 4.14e+003 3.42e+003 2.82e+003 2.51e+003...
58        2.40e+003 2.37e+003 1.81e+004 1.74e+004 1.49e+004 1.25e+004 1.05e+004 8.87e+003...
59        7.42e+003 6.20e+003 5.17e+003 4.32e+003 3.60e+003 3.00e+003 2.48e+003 2.05e+003...
60        1.69e+003 1.40e+003 1.15e+003];
61  thom_comp=[...
62        4.72e+002 4.49e+002 4.27e+002 4.04e+002 3.82e+002 3.61e+002 3.40e+002 3.20e+002...
63        3.00e+002 2.81e+002 2.63e+002 2.45e+002 2.29e+002 2.13e+002 1.98e+002 1.89e+002...
64        1.86e+002 1.85e+002 1.84e+002 1.81e+002 1.71e+002 1.58e+002 1.47e+002 1.36e+002...
65        1.26e+002 1.16e+002 1.07e+002 9.92e+001 9.18e+001 8.49e+001 7.85e+001 7.27e+001...
66        6.73e+001 6.24e+001 5.80e+001];
```

List of Tables

References

S.K. Andersen, J.A. Golovchenko, and G. Mair. *Phys. Rev. Lett.*, 37:1141, 1976.

S.R. Andrews and R.A. Cowley. *J. Phys. C: Solid State Phys.*, 18:6247, 1985.

B.W. Batterman. *Phys. Rev.*, 133:A759, 1964.

J.M. Bijvoet, A.F. Peerdeman, and J.A. van Bommel. *Nature*, 168:271, 1951.

M. Blume. *J. Appl. Phys.*, 57:3615, 1985.

P. Carra, B.T. Thole, M. Altarelli, and X. Wang. *Phys. Rev. Lett.*, 70:694, 1993.

C.T. Chantler. *J. Phys. Chem. Ref. Data*, 24:71, 1995.

C.T. Chen, Y.U. Idzerda, H.-J. Lin, N.V. Smith, G. Meigs, E. Chaban, G.H. Ho, E. Pellegrin, and F. Sette. *Phys. Rev. Lett.*, 75:152, 1995.

W. Cochran, F.H.C. Crick, and V. Vand. *Acta Cryst.*, 5:581, 1952.

D. Coster, K.S. Knol, and J.A. Prins. *Z. Phys.*, 63:345, 1930.

R.A. Cowley. *Nato ASI series C Mathematical and Physical Sciences*, 432:67, 1994.

F. de Bergevin and M Brunel. *Phys. Lett.*, 39A:141, 1972.

P. Debye. *Ann. Physik*, 46:809, 1915.

Ya. S. Derbenev, A.M. Kondratenko, and E.L. Saldin. *Nuc. Instr. and Meth.*, A193:415, 1982.

J.L. Erskine and E.A. Stern. *Phys. Rev. B*, 12:5016, 1975.

R. Feidenhans'l, F. Grey, R.L. Johnson, S.G.J. Mochrie, J. Bohr, and M. Nielsen. *Phys. Rev. B*, 41:5420, 1990.

R.E. Franklin and R.G. Gosling. *Nature*, 171:740, 1953.

D. Gibbs, D.R. Harshman, E.D. Isaacs, D.B. McWhan, D. Mills, and C. Vettier. *Phys. Rev. Lett.*, 61:1241, 1988.

W.A. Hendrickson. *Trans. Am. Crystallogr. Assoc.*, 21:11, 1985.

B.L. Henke, E.M. Gullikson, and J.C. Davis. *Atomic Data and Nuclear Data Tables*, 54:181, 1993.

M. Holt, Z. Wu, H. Hong, P. Zschack, P. Jemian, J. Tischler, H. Chen, and T.-C. Chiang. *Phys. Rev. Lett.*, 83:3317, 1999.

C. Janot. *Quasicrystals: a primer.* Oxford University Press, 1992.

J. Karle. *International Journal of Quantum Chemisry, Quantum Biology Symposium*, 7:357, 1980.

H. Kiessig. *Ann. Phys.*, 10:769, 1931.

F. Leveiller, C. Böhm, D. Jacquemain, H. Möhwald, L. Leiserowitz, K. Kjaer, and J. Als-Nielsen. *Langmuir*, 10:819, 1994.

S.W. Lovesey and S.P. Collins. *X-ray Scattering and Absorption by Magnetic Materials.* Oxford University Press, 1996.

J.B. Murphy and C. Pellegrini. *J. Opt. Soc. of America*, B2:259, 1985.

K. Namikawa, M. Ando, T. Nakajima, and H. Kawata. *J. Phys. Soc. Japan*, 54:4099, 1985.

S. Nishikawa and R. Matsukawa. *Proc. Imp. Acad. Japan*, 4:96, 1928.

B. Ocko, A. Braslau, P. S. Pershan, J. Als-Nielsen, and M. Deutsch. *Phys. Rev. Lett.*, 57:94, 1986.

L.G. Parratt. *Phys. Rev.*, 95:359, 1954.

L. Pauling, R.B. Corey, and H.R. Branson. *Proc. Nat. Acad. Sci.*, 37:205, 1951.

P. Pershan, A. Braslau, A. H. Weiss, and J. Als-Nielsen. *Phys. Rev. A*, 35: 4800, 1987.

H. M. Rietveld. *J. Appl. Cryst.*, 2:65, 1969.

I.K. Robinson. *Phys. Rev. B*, 33:3830, 1986.

J. Rockengerger, L. Tröger, A.L Rogach, M. Tischer, M. Grundmann, A. Eychmüller, and H. Weller. *J. Chem. Phys.*, 108:7807, 1998.

A.L. Rogach, L. Katsikas, A. Kornowski, Dansheng Su, A. Eychmüller, and H. Weller. *Ber. Bunsenges. Phys. Chem.*, 100:1772, 1996.

G. Schütz, W. Wagner, W. Wilhelm, P. Kienle, R. Zeller, R. Frahm, and G. Materlik. *Phys. Rev. Lett.*, 58:737, 1987.

D. Shechtman, I. Blech, D. Gratias, and J.W. Cahn. *Phys. Rev. Lett.*, 53:1951, 1984.

S.K. Sinha, E.B. Sirota, S. Garoff, and H.B. Stanley. *Phys. Rev. B*, 38:2297, 1988.

A. Snigirev, V. Kohn, I. Snigireva, and B. Lengeler. *Nature*, 384:49, 1996.

E.A. Stern. *Scientific American*, 234(4):96, 1976.

E.A. Stern and S.M. Heald. *Handbook of Synchrotron radiation*, chapter EXAFS, pages 1–10. A, 1900.

D.H. Templeton and L.K. Templeton. *Acta Cryst*, A38:62, 1982.

B.T. Thole, P. Carra, F. Sette, and G. van der Laan. *Phys. Rev. Lett.*, 68: 1943, 1992.

B.T. Thole, G. van der Laan, and G.A. Sawatzky. *Phys. Rev. Lett.*, 55:2086, 1985.

G. van der Laan, B.T. Thole, , G.A. Sawatzky, J.B. Goedkoop, J.C. Fuggle, J.-M. Esteva, R. Karnatak, J.P. Remeika, and H.A. Dabkowska. *Phys. Rev. B*, 34:6529, 1986.

J.D. Watson and F.H.C. Crick. *Nature*, 171:737, 1953.

M.H.F. Wilkins, A.R. Stokes, and H.R. Wilson. *Nature*, 171:738, 1953.

P. Wong. *Phys. Rev. B*, 32:7417, 1985.

W. Yun. *Rev. Sci. Instrum.*, 70:2238, 1999.

e	electron charge	1.602×10^{-19}	C
$\hbar = \frac{h}{2\pi}$	Planck's constant	$1.055\ \times10^{-34}$	J s
m	electron mass	9.109×10^{-31}	kg
m_n	neutron mass	1.675×10^{-27}	kg
m_p	proton mass	1.673×10^{-27}	kg
c	speed of light	2.998×10^{8}	m s^{-1}
ϵ_0	permittivity of vacuum	8.854×10^{-12}	A s V^{-1} m^{-1}
$\mu_0=\frac{1}{\epsilon_0 c^2}$	permeability of vacuum	$4\pi\times\ 10^{-7}$	V s A^{-1} m^{-1}
N_{A}	Avogadro's number	6.022×10^{26}	mols. kmole^{-1}
k_{B}	Boltzmann's constant	1.381×10^{-23}	J K^{-1}
$a_0 = \frac{4\pi\epsilon_0\hbar^2}{me^2}$	Bohr radius	5.292×10^{-11}	m
$r_0=\frac{e^2}{4\pi\epsilon_0 mc^2}$	Thomson scattering length	2.818×10^{-15}	m
$\lambda_{\mathrm{C}}=\frac{\hbar}{mc}$	Compton scattering length	3.860×10^{-13}	m
1 Å	Ångström	1×10^{-10}	m
1 barn		1×10^{-28}	m^2

Table G.1: Table of fundamental constants in SI, and other useful constants.

Index